Graduate Texts in Mathematics 232

Graduate Texts in Mathematics

(continued after index)

Graham Everest
Thomas Ward

An Introduction to Number Theory

With 16 Figures

 Springer

Graham Everest, BSc, PhD
School of Mathematics
University of East Anglia
Norwich
NR4 7TJ
UK

Thomas Ward, BSc, MSc, PhD
School of Mathematics
University of East Anglia
Norwich
NR4 7TJ
UK

Mathematics Subject Classification (2000): 11Y05/11/16/55

British Library Cataloguing in Publication Data
Everest, Graham, 1957–
  An introduction to number theory. — (Graduate texts in
  mathematics ; 232)
  1. Number theory
  I. Title II. Ward, Thomas, 1963–
  512.7

Graduate Texts in Mathematics series ISSN 0072-5285
ISBN-13: 978-1-84996-959-8
e-ISBN-13: 978-1-84628-044-3
Springer Science+Business Media
springeronline.com

Printed in the United States of America
12/3830-543210 Printed on acid-free paper

And he brought him forth abroad, and said, Look now toward heaven, and tell the stars, if thou be able to number them: and he said unto him, So shall thy seed be.

Genesis 15, verse 5

Contents

Introduction

This book is written from the perspective of several passionately held beliefs about mathematical education. The first is that mathematics is a good story. Theorems are not discovered in isolation, but happen as part of a culture, and they are generally motivated by paradigms. In this book we are going to show how one result from antiquity can be used to illuminate the study of much that forms the undergraduate curriculum in number theory at a typical U.K. university. The result is the Fundamental Theorem of Arithmetic. Our hope is that students will understand that number theory is not just a collection of tricks and isolated results but has a coherence fueled directly by a connected narrative that spans centuries.

The second belief is that mathematics students (and indeed professional mathematicians) come to the subject with different preferences and evolving strengths. Therefore, we have endeavored to present differing approaches to number theory. One way to achieve this is the obvious one of selecting material from both the algebraic and the analytic disciplines. Less obviously, in the early part of the book particularly, we sometimes present several different proofs of a single result. The aim is to try to capture the imagination of the reader and help her or him to discover his or her own taste in mathematics. The book is written under the assumption that students are being exposed to the power of analysis in courses such as complex variables, as well as the power of abstraction in courses such as algebra. Thus we use notions from finite group theory at several points to give alternative proofs. Often the resulting approaches simplify and promote generalization, as well as providing elegance. We also use this approach because we want to try to explain how different approaches to elementary results are worked out later in different approaches to the subject in general. Thus Euler's proof of the Fundamental Theorem of Arithmetic could be taken to prefigure the development of analytic number theory with its ingenious use of the Euler product Formula. When we move further into the analytic aspects of arithmetic, Euler's relatively simple observation may seem like a rather flimsy pretext. However, the view that many nineteenth-century mathematicians took of functions (complex functions particularly) was profoundly influenced by the Fundamental Theorem of Arithmetic. In their view, many functions are factorizable objects, and we will try to illustrate this in describing some of the great achievements of that century.

Having spoken of different approaches, it will surprise few readers that number theory has many streams. A major surprise is the fact that some of these meet again: Chapter 11 shows that many of the themes in Chapters 1–10 become reconciled further on. The classical class number formula reconciles the analytic stream of ideas with the algebraic. We also discuss – necessarily in general terms – the L-function associated with an elliptic curve and the conjectures of Birch and Swinnerton-Dyer, which draw together the elliptic, algebraic and analytic streams. The underlying motif is the theory of L-functions. As we enter a new millennium, it has become clear that one of the ways into the deepest parts of number theory requires a better understanding of these fundamental objects.

The third belief is that number theory is a living subject, even when studied at an elementary level. The onset of electronic computing gave the subject an enormous boost, and it is a pleasure to be able to record some recent developments. The language of arithmetical complexity has helped to change the way we think about numbers. Modern computers can carry out calculations with numbers that are almost unimaginably large. We recommend that any reader unfamiliar with modern number theory packages tries a few experiments using some of the excellent free software available from the internet. To start to think of the issues raised by large integer calculation can be no bad thing. Intellectually too, this computational topic illustrates an interesting point about the enduring nature of the paradigm. Our story begins over two millennia ago, yet it is the same questions that continue to fascinate us. What are the primes like? Where can they be found? How can the prime factors of an integer be computed? Whether these questions will endure awhile longer nobody can tell. The history of these problems already presents a fascinating story worth telling, and one that says a lot about one of the most important and beautiful narratives of enquiry in human history – mathematics.

One of the most striking and pleasurable aspects of number theory is the extent of time and range of cultures over which it has been studied. We do not go into a detailed history of the developments described here, but the names and places given in the list of "Dramatis Personae" should give some idea of how widely number theory has been studied. The names in this list are rather crudely Anglicized and the locations somewhat arbitrarily modernized. The many living mathematicians who have made significant contributions to the topics covered here have been omitted but may be found on the Web site in [113]. A densely written, comprehensive review of number theory up to about 1920 may be found in Dickson's history [42], [43], [44]; a discursive and masterly account of the four millennia ending in 1798 is provided by Weil [157].

Finally, we say something about the way this book could be used. It is based on three courses taught at the University of East Anglia on various aspects of number theory (analytic, algebraic/geometric, and computational), mostly at the final-year undergraduate level. We were motivated in part by G. A. and J. M. Jones' attractive book [84]. Their book sets out to deal with the subject as it is actually taught. Typically, third-year students will not have done a course in number theory and their experience will necessarily be fragmentary. Like [84], our book begins in quite an elementary way. We have found that the different years at a university do not equate neatly with different abilities: Students in their early years can often be stretched well beyond what seems possible, and upper-level students do not complain about beginning in simple ways. We will try to show how different chapters can be put together to make a course; the book can be used as a basis for two upper-level courses and one at an intermediate level.

We thank many people for contributing to this text. Notable among them are Christian Röttger, for writing up notes from an analytic number theory course at UEA; Sanju Velani, for making available notes from his analytic number theory course; several cohorts of UEA undergraduates for feedback on lecture courses; Neal Koblitz and Joe Silverman for their inspiring books; and Elena Nardi for help with the ancient Greek in Section 1.7.1. We thank Karim Belabas, Robin Chapman, Sue Everest, Gareth and Mary Jones, Graham Norton, David Pierce, Peter Pleasants, Christian Röttger, Alice Silverberg, Shaun Stevens, Alan and Honor Ward, and others for pointing out errors and suggesting improvements. Errors and solecisms that remain are entirely the authors' responsibility.

February 14, 2005   Graham Everest
Norwich, UK   Thomas Ward

NOTATION AND TERMINOLOGY

"Arithmetic" is used both as a noun and an adjective. The particular notation used is collected at the start of the index. The symbols $\mathbb{N}$, $\mathbb{P}$, $\mathbb{Z}$, $\mathbb{Q}$, $\mathbb{R}$, $\mathbb{C}$ denote the natural numbers $\{1, 2, 3, \dots\}$, prime numbers $\{2, 3, 5, 7, \dots\}$, integers, rational numbers, real numbers, and complex numbers, respectively. Any field with $q = p^r$ elements, $p \in \mathbb{P}$ and $r \in \mathbb{N}$, is denoted $\mathbb{F}_q$, and $\mathbb{F}_q^*$ denotes its multiplicative group; the field $\mathbb{F}_p$, $p \in \mathbb{P}$, is identified with the set $\{0, 1, \dots, p-1\}$ under addition and multiplication modulo p. For a complex number $s = \sigma + it$, $\Re(s) = \sigma$ and $\Im(s) = t$ denote the real and imaginary parts of s respectively. The symbol $\mid$ means "divides", so for $a, b \in \mathbb{Z}$, $a \mid b$ if there is an integer k with $ak = b$. For any set X, $|X|$ denotes the cardinality of X. The greatest common divisor of a and b is written $\gcd(a, b)$. Products are written using $\cdot$ as in $12 = 3 \cdot 4$ or $n! = 1 \cdot 2 \cdots (n-1) \cdot n$. The order of growth of functions f, g (usually these are functions $\mathbb{N} \to \mathbb{R}$) is compared using the following notation:

$$f \sim g \text{ if } \frac{f(x)}{g(x)} \longrightarrow 1 \text{ as } x \to \infty;$$

$$f = \mathrm{O}(g) \text{ if there is a constant } A > 0 \text{ with } f(x) \leqslant Ag(x) \text{ for all } x;$$

$$f = \mathrm{o}(g) \text{ if } \frac{f(x)}{g(x)} \longrightarrow 0 \text{ as } x \to \infty.$$

In particular, $f = \mathrm{O}(1)$ means that f is bounded. The relation $f = \mathrm{O}(g)$ will also be written $f \ll g$, particularly when it is being used to express the fact that two functions are commensurate, $f \ll g \ll f$. A sequence $a_1, a_2, \ldots$ will be denoted (a_n).

REFERENCES

The references are not comprehensive, and material that is not explicitly cited is nonetheless well-known. It is inevitable that we have borrowed ideas and used them inadvertently without citation; we apologize for any egregious instances of this. The general references that are likely to be most accessible without much background are as follows. For Chapter 2, [147]; for Chapters 3 and 4, [77], [96], [147], and [154]; for Chapters 5–7, [27] and [143]; for Chapters 8–10, [4], [75], and [81]; for Chapter 9, [6]; and for Chapter 12, [21], [22], [36], [90], and [66].

POSSIBLE COURSES

A course on analytic number theory could follow Chapters 1, 8, 9, and 10; one on Diophantine problems or elliptic curves could follow Chapters 1, 2, 5, 6, and 7. A lower-level course on algebraic number theory could be based on Chapters 1, 2, 3 and 4; one on complexity could be based on Chapters 1 and 12. (These could also be used for the complexity part of a course on cryptography.) The exercises are generally routine applications of the methods in the text, but exercises marked * are to be viewed as projects, some of them requiring further reading and research.

DRAMATIS PERSONAE

Person	Date	Country
Pythagoras of Samos	569 B.C.–475 B.C.	Greece, Egypt
Euclid of Alexandria	325 B.C.–265 B.C.	Greece, Egypt
Eratosthenes of Cyrene	276 B.C.–194 B.C.	Libya, Greece, Egypt
Diophantus of Alexandria	200–284	Greece, Egypt
Hypatia of Alexandria	370–415	Egypt
Sun Zi	400–460	China
Brahmagupta	598–670	India
Abu Ali al-Hasan ibn al-Haytham	965–1040	Iraq, Egypt
Bhaskaracharya	1114–1185	India
Leonardo Pisano Fibonacci	1170–1250	Italy
Qin Jiushao	1202–1261	China
Pietro Antonio Cataldi	1548–1626	Italy
Claude Gaspar Bachet de Méziriac	1581–1638	France
Marin Mersenne	1588–1648	France
Pierre de Fermat	1601–1665	France
James Stirling	1692–1770	Scotland
Leonhard Euler	1707–1783	Switzerland, Russia
Joseph–Louis Lagrange	1736–1813	Italy, France
Lorenzo Mascheroni	1750–1800	Italy, France
Adrien-Marie Legendre	1752–1833	France
Jean Baptiste Joseph Fourier	1768–1830	France
Johann Carl Friedrich Gauss	1777–1855	Germany
Siméon Denis Poisson	1781–1840	France
August Ferdinand Möbius	1790–1868	Germany
Niels Henrik Abel	1802–1829	Norway
Carl Gustav Jacob Jacobi	1804–1851	Germany
Johann Peter Gustav Lejeune Dirichlet	1805–1859	France, Germany
Joseph Liouville	1809–1882	France
Ernst Eduard Kummer	1810–1893	Germany
Evariste Galois	1811–1832	France
Karl Theodor Wilhelm Weierstrass	1815–1897	Germany
Pafnuty Lvovich Tchebychef	1821–1894	Russia
Georg Friedrich Bernhard Riemann	1826–1866	Germany, Italy
François Edouard Anatole Lucas	1842–1891	France
Jules Henri Poincaré	1854–1912	France
David Hilbert	1862–1943	Germany
Srinivasa Aiyangar Ramanujan	1887–1920	India, England
Louis Joel Mordell	1888–1972	USA, England
Carl Ludwig Siegel	1896–1981	Germany
Emil Artin	1898–1962	Austria, Germany
Kurt Mahler	1903–1988	Germany, UK, Australia
Derrick Henry Lehmer	1905–1991	USA
André Weil	1906–1998	France, USA

1

A Brief History of Prime

Most of the results in this book grow out of one theorem that has probably been known in some form since antiquity.

Theorem 1.1. [FUNDAMENTAL THEOREM OF ARITHMETIC] *Every integer greater than 1 can be expressed as a product of prime numbers in a way that is unique up to order.*

For the moment, we are using the term *prime* in its most primitive form – to mean an irreducible integer greater than one. Thus a positive integer p is prime if $p > 1$ and the factorization $p = ab$ into positive integers implies that either $a = 1$ or $b = 1$. The expression "up to order" means simply that we regard, for example, the two factorizations $6 = 2 \cdot 3 = 3 \cdot 2$ as the same.

Theorem 1.1, the Fundamental Theorem of Arithmetic, will reverberate throughout the text. The fact that the primes are the building blocks for all integers already suggests they are worth particular study, rather in the way that scientists study matter at an atomic level. In this case, we need a way of looking for primes and methods to construct them, identify them, and even quantify their appearance if possible. Some of these quests took thousands of years to fulfill, and some are still works in progress. At the end of this chapter, we will give a proof of Theorem 1.1, but for now we want to get on with our main theme.

1.1 Euclid and Primes

The first consequence of the Fundamental Theorem of Arithmetic for the primes is that there must be infinitely many of them.

Theorem 1.2. [EUCLID] *There are infinitely many primes.*

To emphasize the diversity of approaches to number theory, we will give several proofs of this famous result.

EUCLID'S PROOF IN MODERN FORM. If there are only finitely many primes, we can list them as $p_1, \ldots, p_r$. Let

$$N = p_1 \cdots p_r + 1 > 1.$$

By the Fundamental Theorem of Arithmetic, N can be factorized, so it must be divisible by some prime p_k of our list. Since p_k also divides $p_1 \cdots p_r$, it must divide the difference

$$N - p_1 \cdots p_r = 1,$$

which is impossible, as $p_k > 1$. □

EULER'S ANALYTIC PROOF. Assume that there are only finitely many primes, so they may be listed as $p_1, \ldots, p_r$. Consider the product

$$X = \prod_{k=1}^{r} \left(1 - \frac{1}{p_k} \right)^{-1}.$$

The product is finite since 1 is not a prime and by hypothesis there are only finitely many primes. Now expand each factor into a convergent geometric series,

$$\frac{1}{1 - \frac{1}{p}} = 1 + \frac{1}{p} + \frac{1}{p^2} + \frac{1}{p^3} + \cdots.$$

For any fixed K, we deduce that

$$\frac{1}{1 - \frac{1}{p}} \geqslant 1 + \frac{1}{p} + \frac{1}{p^2} + \cdots + \frac{1}{p^K}.$$

Putting this into the equation for X gives

$$X \geqslant \left(1 + \frac{1}{2} + \frac{1}{2^2} + \cdots + \frac{1}{2^K} \right) \cdot \left(1 + \frac{1}{3} + \frac{1}{3^2} + \cdots + \frac{1}{3^K} \right)$$

$$\cdot \left(1 + \frac{1}{5} + \frac{1}{5^2} + \cdots + \frac{1}{5^K} \right) \cdots \left(1 + \frac{1}{p_r} + \frac{1}{p_r^2} + \cdots + \frac{1}{p_r^K} \right)$$

$$= 1 + \frac{1}{2} + \frac{1}{3} + \frac{1}{4} + \cdots$$

$$= \sum_{n \in \mathcal{N}(K)} \frac{1}{n}, \tag{1.1}$$

where

$$\mathcal{N}(K) = \{ n \in \mathbb{N} \mid n = p_1^{e_1} \cdots p_r^{e_r}, e_i \leqslant K \text{ for all } i \}$$

denotes the set of all natural numbers with the property that each prime factor appears no more than K times. Notice that the identity (1.1) requires the Fundamental Theorem of Arithmetic. Given any number $n \in \mathbb{N}$, if K is large enough, then $n \in \mathcal{N}(K)$, so we deduce that

$$X \geqslant \sum_{n=1}^{\infty} \frac{1}{n}.$$

The series on the right-hand side (known as the *harmonic series*) diverges to infinity, but X is finite. Again we have reached a contradiction from the assumption that there are finitely many primes.   □

Let us recall why the harmonic series diverges to infinity. As with Theorem 1.2, there are many ways to prove this; the first is elementary, while the second compares the series with an integral.

ELEMENTARY PROOF. Notice that

$$1 + \frac{1}{2} \geqslant \frac{1}{2},$$
$$\frac{1}{3} + \frac{1}{4} \geqslant \frac{1}{2},$$
$$\frac{1}{5} + \frac{1}{6} + \frac{1}{7} + \frac{1}{8} \geqslant \frac{1}{2},$$

and so on. For any $k \geqslant 1$,

$$\frac{1}{2^k + 1} + \frac{1}{2^k + 2} + \cdots + \frac{1}{2^{k+1}} \geqslant 2^k \cdot \frac{1}{2^{k+1}} = \frac{1}{2}.$$

This means that

$$\sum_{n=1}^{2^{k+1}} \frac{1}{n} \geqslant \frac{k}{2} \text{ for all } k \geqslant 1,$$

and it follows that $\displaystyle\sum_{n=1}^{\infty} \frac{1}{n}$ diverges.   □

Hidden in the last argument is some indication of the *rate* at which the harmonic series diverges. Since the sum of the first 2^{k+1} terms exceeds $k/2$, the sum of the first N terms must be approximately $C \log N$ for some positive constant C. The second proof improves on this: Equation (1.2) gives a sharper lower bound as well as an upper bound.

Exercise 1.1. Try to prove that $\displaystyle\sum_{n=1}^{\infty} \frac{1}{n^2}$ diverges using the same technique of grouping terms together. Of course, this will not work since this series converges, but you will see something mildly interesting. In particular, can you use this to estimate the sum?

USING THE INTEGRAL TEST. Compare $\sum_{n=1}^{N} \frac{1}{n}$ with the integral

$$\int_{1}^{N} \frac{1}{x}\, dx = \log N.$$

Figure 1.1 shows $\sum_{n=1}^{6} \frac{1}{n}$ trapped between $\int_{0}^{6} \frac{1}{x+1}\, dx$ and $1 + \int_{1}^{6} \frac{1}{x}\, dx$; in general, it follows that

$$\log(N+1) \leqslant \sum_{n=1}^{N} \frac{1}{n} \leqslant 1 + \log N. \tag{1.2}$$

This shows again that the harmonic series diverges and that the partial sum of the first N terms is approximately $\log N$.

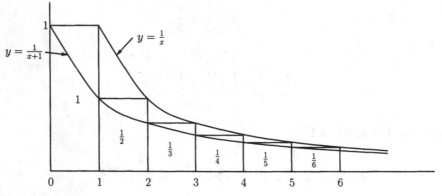

Figure 1.1. Graphs of $y = \frac{1}{x}$ and $y = \frac{1}{x+1}$ trapping the harmonic series.

□

This proof is a harbinger of more subtle results. Comparing series with integrals is a powerful technique; more generally, using *analytic* techniques to study properties of numbers has been one of the most important ideas in number theory.

Exercise 1.2. Extend the method illustrated in Figure 1.1 to show that the sequence (a_n) defined by

$$a_n = \sum_{m=1}^{n} \frac{1}{m} - \log n$$

is decreasing (that is, $a_{n+1} \leqslant a_n$ for all n) and nonnegative. Deduce that it converges to some number γ, and estimate γ to three digits. This number is known as the Euler–Mascheroni constant. It is not known if γ is rational, although it is expected not to be.

1.2 Summing Over the Primes

We begin this section with yet another proof that there are infinitely many primes. Recall that $\mathbb{P}$ denotes the set of prime numbers.

Theorem 1.3. *The series* $\displaystyle\sum_{p\in\mathbb{P}} \frac{1}{p}$ *diverges.*

Several proofs are offered; each one provides different insights. We adopt the convention that p always denotes a prime so, for example, $\displaystyle\sum_{p>N} a_p$ denotes $\displaystyle\sum_{p\in\mathbb{P}, p>N} a_p$.

Notice that Theorem 1.3 tells us something about the sequence (p_n) of primes that begins $p_1 = 2$, $p_2 = 3$, $p_3 = 5,\ldots$. For example, the sequence $(n^{1+\varepsilon}/p_n)$ cannot be bounded for any $\varepsilon > 0$.

FIRST PROOF OF THEOREM 1.3. We argue by contradiction: Assume that the series converges. Then there is some N such that

$$\sum_{p>N} \frac{1}{p} < \frac{1}{2}.$$

Let

$$Q = \prod_{p\leqslant N} p$$

be the product of all the primes less than or equal to N. The numbers

$$1 + nQ, \quad n \in \mathbb{N},$$

are never divisible by primes less than N because such primes do divide Q. Now consider

$$P = \sum_{t=1}^{\infty} \left(\sum_{p>N} \frac{1}{p}\right)^t < \sum_{t=1}^{\infty} \frac{1}{2^t} = 1.$$

We claim that

$$\sum_{n=1}^{\infty} \frac{1}{1+nQ} \leqslant \sum_{t=1}^{\infty} \left(\sum_{p>N} \frac{1}{p}\right)^t$$

because every term on the left-hand side appears on the right-hand side at least once. (Convince yourself of this claim by taking $N = 11$ and finding some terms on the right-hand side.) It follows that

$$\sum_{n=1}^{\infty} \frac{1}{1+nQ} \leqslant 1. \tag{1.3}$$

However, the series in Equation (1.3) diverges since

$$\sum_{n=1}^{K} \frac{1}{1+nQ} \geqslant \frac{1}{2Q} \sum_{n=1}^{K} \frac{1}{n}$$

for any K, and the right-hand side diverges as $K \to \infty$. This contradiction proves the theorem.   □

SECOND PROOF OF THEOREM 1.3. We will prove a stronger result, namely

$$\sum_{p \leqslant N} \frac{1}{p} > \log \log N - 2. \tag{1.4}$$

Fix N and let

$$\mathfrak{N}(N) = \{ n \in \mathbb{N} : \text{ all prime factors of } n \text{ are less than or equal to } N \}.$$

Then (just as in Euler's analytic proof of Theorem 1.2 on p. 8)

$$\sum_{n \in \mathfrak{N}(N)} \frac{1}{n} = \prod_{p \leqslant N} \left(1 + p^{-1} + p^{-2} + p^{-3} + \cdots \right)$$

$$= \prod_{p \leqslant N} \left(1 - p^{-1} \right)^{-1}.$$

If $n \leqslant N$, then certainly $n \in \mathfrak{N}(N)$, so

$$\sum_{n \leqslant N} \frac{1}{n} \leqslant \sum_{n \in \mathfrak{N}(N)} \frac{1}{n}.$$

It follows by Equation (1.2) that

$$\log N \leqslant \sum_{n \in \mathfrak{N}(N)} \frac{1}{n} = \prod_{p \leqslant N} \left(1 - p^{-1} \right)^{-1}. \tag{1.5}$$

In order to estimate the right-hand side of Equation (1.5), we need the following bound. For any $v \in [0, 1/2]$,

$$\frac{1}{1-v} \leqslant e^{v+v^2}. \tag{1.6}$$

To see why the bound (1.6) holds, let $f(v) = (1-v) \exp(v + v^2)$. Then

$$f'(v) = v(1 - 2v) \exp(v + v^2) \geqslant 0 \text{ for } v \in [0, \tfrac{1}{2}],$$

so the fact that $f(0) = 1$ implies that $f(v) \geqslant 1$ for all $v \in [0, 1/2]$.
    For any prime p, $v = \frac{1}{p} \leqslant \frac{1}{2}$, so by the bound (1.6)

$$\prod_{p\leqslant N} \left(1 - p^{-1}\right)^{-1} \leqslant \prod_{p\leqslant N} \exp\left(p^{-1} + p^{-2}\right).$$

Combining this with Equation (1.5) and taking logarithms gives

$$\log\log N \leqslant \sum_{p\leqslant N} \left(p^{-1} + p^{-2}\right). \tag{1.7}$$

Finally, we observe that

$$\sum_{p} \frac{1}{p^2} < \sum_{n=2}^{\infty} \frac{1}{n^2} < 1, \tag{1.8}$$

so the contribution to the right-hand side of Equation (1.7) from $\sum_{p\leqslant N} p^{-2}$ is bounded independently of N. This completes the second proof of Theorem 1.3. □

Exercise 1.3. Prove the second inequality in Equation (1.8) using the integral test: Show that

$$\sum_{n=2}^{N} \frac{1}{n^2} < \int_{2}^{N} \frac{1}{(x-1)^2}\, dx \leqslant 1 \quad \text{for all} \quad N \geqslant 2.$$

In fact, an estimate stronger than Equation (1.4) holds. Mertens showed that there is a constant A (approximately 0.261) such that

$$\sum_{p\leqslant N} \frac{1}{p} = \log\log N + A + O\left(\frac{1}{\log N}\right). \tag{1.9}$$

Exercise 1.4. Is it possible to prove Equation (1.9) with $O(1)$ in place of

$$A + O\left(\frac{1}{\log N}\right)$$

using only the methods of the second proof of Theorem 1.3?

The third proof of Theorem 1.3 extends the relationship between products such as $\prod_{p\in\mathbb{P}} \left(1 - p^{-1}\right)^{-1}$ and the harmonic series to a factorization of a *function* that will later turn out to have a starring role.

Definition 1.4. *The* Riemann zeta function *is defined by*

$$\zeta(\sigma) = \sum_{n=1}^{\infty} \frac{1}{n^\sigma}$$

wherever this makes sense.

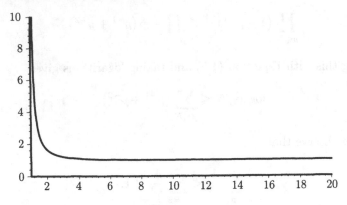

Figure 1.2. The graph of $\zeta(\sigma)$ for $1 < \sigma \leqslant 20$.

Understanding the properties of this function turns out to be the key to many deeper properties of the prime numbers. For now, we simply think of σ as being a real number and note that the series defining $\zeta(\sigma)$ converges by the integral test for $\sigma > 1$ to a positive sum and diverges at $\sigma = 1$. For $\sigma > 1$, $\zeta(\sigma)$ is a decreasing function of σ.

Viewed as a real function of a real variable, the zeta function does not look particularly subtle or useful. Figure 1.2 shows the graph of $\zeta(\sigma)$ for $1 < \sigma \leqslant 20$. Some indication of just how complicated this function really is appears when it is viewed as a complex-valued function of a complex variable. It is clear that the series defining the zeta function converges for $s = \sigma + it$ when $\sigma > 1$ (see p. 166 for more on this). Figure 1.3 shows the function $\Re(\zeta(\tfrac{3}{2} + it))$ for $0 \leqslant t \leqslant 60$, giving the first insight into the complex properties of the zeta function.

In Chapter 8, the Riemann zeta function is extended to a complex analytic function defined on the whole complex plane with the exception of a single pole, and this opens up the most mysterious aspect of the zeta function – its behavior along the line $\Re(s) = \tfrac{1}{2}$. Figure 9.1 on p. 186 gives some idea of how complicated this is.

Recall that p will be used to denote a prime number, so a product over the variable p means a product over $p \in \mathbb{P}$.

The first step in understanding the zeta function is the *Euler product representation*, which is a factorization of the zeta function into terms corresponding to primes. The idea of factorizing a function will be discussed again at the start of Chapter 9.

Theorem 1.5. [EULER PRODUCT REPRESENTATION] *For any $\sigma > 1$,*

$$\zeta(\sigma) = \prod_{p} \left(1 - p^{-\sigma}\right)^{-1}.$$

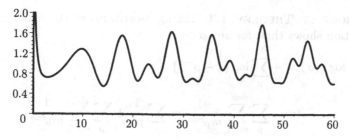

Figure 1.3. The graph of $\Re(\zeta(\frac{3}{2}+it))$ for $0 \leqslant t \leqslant 60$.

PROOF. For any $\sigma > 1$,

$$\left(1 - 2^{-\sigma}\right)\zeta(\sigma) = \sum_{n=1}^{\infty} \frac{1}{n^{\sigma}} - \sum_{n=1}^{\infty} \frac{1}{(2n)^{\sigma}}$$

$$= \sum_{n \text{ odd}} \frac{1}{n^{\sigma}}$$

$$= 1 + \sum_{p|n \Rightarrow p>2} \frac{1}{n^{\sigma}},$$

where the last sum is taken over those n with all prime factors greater than 2 (that is, the odd numbers greater than 2).

Now let P be a large prime and repeat the same argument with each of the primes $3, 5, \ldots, P$ in turn. This gives

$$\left(1 - 2^{-\sigma}\right)\left(1 - 3^{-\sigma}\right)\left(1 - 5^{-\sigma}\right) \cdots \left(1 - P^{-\sigma}\right)\zeta(\sigma) = 1 + \sum_{p|n \Rightarrow p>P} \frac{1}{n^{\sigma}}.$$

The last sum ranges over those n with the property that all the prime factors of n are greater than P. Thus the last sum is a subsum of the tail of the convergent series defining $\zeta(\sigma)$, and in particular it must tend to zero as P goes to infinity. It follows that

$$\lim_{P \to \infty} \left(1 - 2^{-\sigma}\right)\left(1 - 3^{-\sigma}\right)\left(1 - 5^{-\sigma}\right) \cdots \left(1 - P^{-\sigma}\right)\zeta(\sigma) = 1,$$

so

$$\zeta(\sigma) = \prod_{p} \left(1 - p^{-\sigma}\right)^{-1}.$$

□

Remark 1.6. An infinite product is defined to be convergent if the corresponding partial products form a convergent sequence, *that does not converge to zero*. The nonzero condition is imposed to allow us to take logarithms of infinite products, thereby connecting infinite products and infinite sums in a meaningful way.

THIRD PROOF OF THEOREM 1.3. Taking logarithms of the Euler product representation shows that, for any $\sigma > 1$,

$$\log \zeta(\sigma) = -\sum_p \log \left(1 - p^{-\sigma}\right)$$

$$= -\sum_p \sum_{m=1}^{\infty} \frac{-1}{mp^{m\sigma}} = \sum_p \frac{1}{p^\sigma} + \sum_p \sum_{m=2}^{\infty} \frac{1}{mp^{m\sigma}}. \qquad (1.10)$$

Notice that the series involved converge absolutely, so rearrangement is permissible. For any prime p,

$$1 - \frac{1}{p^\sigma} \geqslant \frac{1}{2},$$

so

$$\sum_p \sum_{m=2}^{\infty} \frac{1}{mp^{m\sigma}} < \sum_p \sum_{m=2}^{\infty} \frac{1}{p^{m\sigma}}$$

$$= \sum_p \frac{1}{p^{2\sigma}} \frac{1}{1 - p^{-\sigma}}$$

$$\leqslant 2 \sum_p \frac{1}{p^{2\sigma}} \leqslant 2\zeta(2\sigma) < 2\zeta(2),$$

which shows that the last double sum in Equation (1.10) is bounded. The bound $2\zeta(2)$ holds for any $\sigma \geqslant 1$, and the double sum converges for $\sigma > \frac{1}{2}$.

Thus

$$\log \zeta(\sigma) = \sum_p \frac{1}{p^\sigma} + O(1).$$

The left-hand side goes to infinity as σ tends to 1 from above, so the sum on the right-hand side must do the same. □

1.3 Listing the Primes

Early in the history of the subject, Eratosthenes[1] devised a kind of sieve for listing the primes. To illustrate his method – the *sieve of Eratosthenes* – we consider the problem of finding all the primes up to 50. First arrange all the integers between 1 and 50 in a grid.

---

[1] Eratosthenes of Cyrene (276 B.C.–194 B.C.) was born in what is now Libya. He made major contributions to many subjects, including finding surprisingly accurate estimates for the circumference of the Earth and the distances from the Earth to the Sun and the Moon.

1 2 3 4 5 6 7 8 9 10
11 12 13 14 15 16 17 18 19 20
21 22 23 24 25 26 27 28 29 30
31 32 33 34 35 36 37 38 39 40
41 42 43 44 45 46 47 48 49 50

Now do the sieving: Eliminate 1, then start with 2 and cross out all numbers greater than 2 and divisible by 2. Then take the next surviving number 3 and cross out all the multiples of 3 that are greater than 3. Repeat with the next surviving number and continue until the numbers divisible by 7 are crossed out.

Exercise 1.5. Why can you stop sieving once you get to 7?

The remaining numbers are the prime numbers below 50, as shown below.

□ 2 3 □ 5 □ 7 □ □ □
11 □ 13 □ □ □ 17 □ 19 □
□ □ 23 □ □ □ □ □ 29 □
31 □ □ □ □ □ 37 □ □ □
41 □ 43 □ □ □ 47 □ □ □

Understanding the patterns of the surviving numbers remains one of the great challenges facing mathematics two thousand years after Eratosthenes.

This method has great value, allowing people throughout history to rapidly create lists of primes. It fails to meet our longer-term objectives however. It elegantly and efficiently produces lists of primes without having to do trial divisions but does not help to decide if a given large number (with hundreds of digits, for example) is prime.

Table 1.1. Early prime hunters.

Name	Date	Bound
Pietro Cataldi	1588	750
T. Brancker	1688	100000
Felkel Kulik	1876	100330200
Derrick Henry Lehmer	1909	10006721

Table 1.1 is a short list of some of the calculations of prime tables in recent history; in each case all the primes up to the bound were listed. A rather different problem is to find exactly how many primes there are below a certain bound (without finding them all). Kulik listed the smallest factors of all the integers up to his bound and in particular found all the primes up to his bound. Lehmer's table was widely distributed and as a result was very influential (despite being shorter than Kulik's table).

1.3.1 Functions that Generate Primes.

In the seventeenth century attention turned to finding formulas that would generate the primes. Euler pointed out the following polynomial example.

Example 1.7. The polynomial $x^2 + x + 41$ yields prime values for $0 \leqslant x \leqslant 39$, but $x = 40, 41$ do not yield primes.

What is striking about this example is that it is prime for many values in succession relative to the size of the coefficients and the degree.

Exercise 1.6. (a) [GOLDBACH 1752] Prove that if $f \in \mathbb{Z}[x]$ has the property that $f(n)$ is prime for all $n \geqslant 1$, then f must be a constant.
(b) Extend your argument to show that if $f \in \mathbb{Z}[x]$ has the property that $f(n)$ is prime for all $n \geqslant N$ for some N, then f must be a constant.
(c) Let $P \in \mathbb{Z}[x_1, \ldots, x_k]$ be a polynomial in $k \geqslant 2$ variables with integer coefficients. Define a function f by $f(n) = P(n, 2^n, 3^n, \ldots, (k-1)^n)$, and assume that $f(n) \to \infty$ as $n \to \infty$. Show that $f(n)$ is composite for infinitely many values of n.

Remarkably, there is an explicit integral polynomial in several variables whose set of *positive* values as the variables run through the nonnegative integers coincides with the primes. This polynomial was discovered as a by-product of research into Hilbert's 10th Problem, which asked if there could be an algorithm to determine if a polynomial Diophantine[2] problem has a solution. However, once again, this is useless with regard to the aim of finding ways to generate primes efficiently.

There are ingenious "formulas" for the primes. Many of these require knowledge of the first $(n-1)$ primes to produce the nth prime, and none of them seem to be computationally useful. We will prove one striking result of this kind here, and two further results in Exercise 1.24 on p. 33 and in Exercise 8.9 on p. 163. The result proved here rests on Bertrand's Postulate, which is the first of many results that say something about how the prime numbers appear and how the next prime compares in size with the previous prime. The arguments below are intricate but elementary, and the basic contradiction arrived at in the proof of Theorem 1.9 is similar to one that will be used to prove Zsigmondy's Theorem (Theorem 1.15) in Section 8.3.1.

We need a lemma that says something about the growth in the product of all the primes up to n. As usual p will be used to denote a prime.

Lemma 1.8. *For any $n \geqslant 1$,*

$$\sum_{p \leqslant n} \log p < 2n \log 2. \tag{1.11}$$

---

[2] Diophantine problems are discussed in Chapter 2. The term is used to denote problems involving equations in which only integer solutions are sought.

PROOF. Let

$$M = \binom{2m+1}{m} = \frac{(2m+1)(2m)\cdots(m+2)}{m!}.$$

This is a binomial coefficient, so it is an integer (see Exercise 1.10 for a stronger form of this). The coefficient M appears twice in the binomial expansion of $2^{2m+1} = (1+1)^{2m+1}$, so $M < 2^{2m}$. If $m+1 < p \leqslant 2m+1$ for some prime p, then p divides the numerator of M but does not divide the denominator, so

$$\prod_{p \in A(m)} p \quad \text{divides} \quad M,$$

where $A(m)$ denotes the set of primes p with $m+1 < p \leqslant 2m+1$. It follows that

$$\sum_{p \leqslant 2m+1} \log p - \sum_{p \leqslant m+1} \log p = \sum_{p \in A(m)} \log p \leqslant \log M < 2m \log 2. \qquad (1.12)$$

We now prove Equation (1.11) by induction. It holds for $n \leqslant 2$, so suppose it holds for all $n \leqslant k - 1$. If k is even, then

$$\sum_{p \leqslant k} \log p = \sum_{p \leqslant k-1} \log p < 2(k-1) \log 2 < 2k \log 2$$

by the inductive hypothesis. If k is odd, write $k = 2m+1$ and then

$$\sum_{p \leqslant 2m+1} \log p = \sum_{p \leqslant 2m+1} \log p - \sum_{p \leqslant m+1} \log p + \sum_{p \leqslant m+1} \log p$$
$$< 2m \log 2 + 2(m+1) \log 2$$
$$= 2(2m+1) \log 2 = 2k \log 2,$$

since $m+1 < k$. Thus the inequality (1.11) holds for all n by induction. $\square$

Theorem 1.9. [BERTRAND'S POSTULATE] *If $n \geqslant 1$, then there is at least one prime p with the property that*

$$n < p \leqslant 2n. \qquad (1.13)$$

PROOF. For any real number x, let $\lfloor x \rfloor$ denote the integer part of x. Thus $\lfloor x \rfloor$ is the greatest integer less than or equal to x. Let p be any prime. Then

$$\left\lfloor \frac{n}{p} \right\rfloor + \left\lfloor \frac{n}{p^2} \right\rfloor + \left\lfloor \frac{n}{p^3} \right\rfloor + \cdots$$

is the largest power of p dividing $n!$ (see Exercise 8.7(a) on p. 162). Fix $n \geqslant 1$ and let

$$N = \prod_{p \leqslant 2n} p^{k(p)}$$

be the prime decomposition of $N = (2n)!/(n!)^2$. The number of times that a given prime p divides N is the difference between the number of times it divides $(2n)!$ and $(n!)^2$, so

$$k(p) = \sum_{m=1}^{\infty} \left(\left\lfloor \frac{2n}{p^m} \right\rfloor - 2 \left\lfloor \frac{n}{p^m} \right\rfloor \right), \tag{1.14}$$

and each of the terms in the sum is either 0 or 1, depending on whether $\left\lfloor \frac{2n}{p^m} \right\rfloor$ is odd or even. If $p^m > 2n$ the term is certainly 0, so

$$k(p) \leqslant \left\lfloor \frac{\log 2n}{\log p} \right\rfloor. \tag{1.15}$$

Now the proof of the theorem proceeds by a contradiction argument. Assume there is some $n \geqslant 1$ for which there is no prime satisfying the inequality (1.13), and let p be a prime factor of $N = (2n)!/(n!)^2$. Thus $p < n$ by our assumption, and $k(p) \geqslant 1$. If

$$\frac{2}{3}n < p \leqslant n$$

then

$$2p \leqslant 2n < 3p \quad \text{and} \quad p^2 > \frac{4}{9}n^2 > 2n,$$

so Equation (1.14) becomes

$$k(p) = \left\lfloor \frac{2n}{p} \right\rfloor - 2 \left\lfloor \frac{n}{p} \right\rfloor = 2 - 2 = 0.$$

We deduce that $p \leqslant \frac{2}{3}n$ for every prime factor p of N. It follows that

$$\sum_{p|N} \log p \leqslant \sum_{p \leqslant 2n/3} \log p \leqslant \frac{4}{3}n \log 2 \tag{1.16}$$

by Lemma 1.8. Now if $k(p) \geqslant 2$ then by the bound (1.15),

$$2 \log p \leqslant k(p) \log p \leqslant \log 2n,$$

so $p \leqslant \sqrt{2n}$ and thus there are at most $\sqrt{2n}$ possible values of p. Hence

$$\sum_{k(p) \geqslant 2} k(p) \log p \leqslant \sqrt{2n} \log 2n.$$

Together with the inequality (1.16), this shows that

$$\log N \leqslant \sum_{k(p)=1} \log p + \sum_{k(p)\geqslant 2} k(p) \log p$$

$$\leqslant \sum_{p|N} \log p + \sqrt{2n} \log 2n$$

$$\leqslant \frac{4}{3} \log 2 + \sqrt{2n} \log 2n. \tag{1.17}$$

Now N is the largest coefficient (namely the middle one) in the binomial expansion of

$$2^{2n} = (1+1)^{2n},$$

so

$$2^{2n} = 2 + \binom{2n}{1} + \binom{2n}{2} + \cdots + \binom{2n}{2n-1} \leqslant 2nN.$$

Substituting this estimate into the inequality (1.17) gives

$$2n \log 2 \leqslant \frac{4}{3} n \log 2 + \log 2n + \sqrt{2n} \log 2n. \tag{1.18}$$

It is clear that the inequality (1.18) cannot hold for large values of n; a simple calculation shows that (1.18) implies that n does not exceed 500.

It follows that if $n > 500$, then there is a prime satisfying the inequality (1.13). A calculation confirms that (1.13) also holds for all $n \leqslant 500$, completing the proof of the theorem. □

Notice that a consequence of Equation (1.13) is that if the primes are listed in order as $p_1, p_2, \ldots$, then

$$p_{n+1} < 2p_n \quad \text{for all } n \geqslant 1. \tag{1.19}$$

It is clear that Theorem 1.9 gives another proof that there must be infinitely many primes. In each interval of the form $(n, 2n]$ there is at least one. This gives us a bound for the prime counting function

$$\pi(X) = |\{p \leqslant X \mid p \in \mathbb{P}\}|.$$

The proof of Euclid's Theorem 1.2 already says a little more than the purely qualitative statement that $\pi(X) \to \infty$ as $X \to \infty$: from the proof of Theorem 1.2 we see that

$$p_{n+1} \leqslant p_1 p_2 \cdots p_n + 1.$$

This tells us something about $\pi(X)$. Define a sequence (u_n) by setting $u_1 = 2$ and $u_{n+1} = u_1 \cdots u_n + 1$ for $n \geqslant 1$. Then

$$\pi(X) \geqslant \min\{n \mid u_n \geqslant X\}.$$

This is an extremely slowly growing sequence, and the bound obtained for $\pi(X)$ is very far from the truth.

Theorem 1.9 says more: there are at least N primes in the interval

$$(1, 2^N] = (1, 2] \cup (2, 4] \cup (4, 8] \cup \cdots \cup (2^{N-1}, 2^N],$$

so $\pi(2^N) > N$. It follows that $\pi(X)$ is larger than $C \log(X)$ for some positive constant C, infinitely often. Something closer to the truth about the asymptotic behavior of $\pi(X)$ is the Prime Number Theorem (Theorem 8.1). Finding more refined estimates for $\pi(X)$ generally involves deep problems in analytic number theory. An exception is the result of Tchebychef, described in Exercise 8.7 on p. 162, which uses elementary methods to give better bounds for $\pi(X)$.

Bertrand's Postulate is enough to exhibit a striking but impractical formula for the primes. More importantly, the bound (1.13) immediately motivates the question of whether the upper estimate $2n$ could be reduced, perhaps for all large n only, and this is the subject of ongoing research.

Corollary 1.10. *There exists a real number θ with the property that*

$$\left\lfloor 2^{2^{2^{\cdot^{\cdot^{\cdot^\theta}}}}} \right\rfloor$$

is a prime number for any number of iterations of the exponential.

PROOF. Let q_1 be any prime, and choose a sequence of primes (q_n) with the property that

$$2^{q_n} < q_{n+1} < 2^{q_n+1}. \tag{1.20}$$

This is possible by Bertrand's Postulate. Now define functions $f^{(1)}, f^{(2)}, \ldots$ by $f^{(1)}(x) = \log_2(x)$ and $f^{(n+1)}(x) = \log_2(f^{(n)}(x))$ for $n \geqslant 1$. Define sequences (u_n) and (v_n) by

$$u_n = f^{(n)}(q_n) \quad \text{and} \quad v_n = f^{(n)}(q_n + 1).$$

By the inequality (1.20),

$$q_n < f^{(1)}(q_{n+1}) < f^{(1)}(q_{n+1} + 1) < q_n + 1,$$

so by applying the increasing function $f^{(n)}$ we have

$$u_n < u_{n+1} < v_{n+1} < v_n.$$

It follows that the sequence (u_n) is increasing and bounded above, so it converges. Let

$$\theta = \lim_{n \to \infty} u_n.$$

Define functions $g^{(n)}$ by $g^{(1)}(x) = 2^x$ and $g^{(n+1)}(x) = 2^{g^{(n)}(x)}$ for all $n \geqslant 1$. Then

$$g^{(n)}(u_n) < g^{(n)}(\theta) < g^{(n)}(v_n),$$

so

$$q_n < g^{(n)}(\theta) < q_n + 1 \text{ for all } n \geqslant 1$$

as required.   □

Exercise 1.7. [MILLS] A deep result of Ingham improves Equation (1.13) to say that there is a constant C such that

$$p_{n+1} - p_n < Cp_n^{5/8}.$$

Assuming this result, modify the proof of Corollary 1.10 to show that there is a real number θ with the property that $\lfloor \theta^{3^n} \rfloor$ is a prime for all $n \geqslant 1$.

Exercise 1.8. [RICHERT] Use Theorem 1.9 to show that every integer greater than 6 is a sum of distinct primes. (Hint: Show this is true for the numbers 7 to 19, then use Theorem 1.9 to see that we can keep adding new primes to the set of sums obtained without missing out any integers).

Exercise 1.9. [DRESSLER] (a) Modify the proof of Theorem 1.9 to show that

$$p_{n+1} < 2p_n - 10 \text{ for all } n > 6.$$

(Hint: Assume there is an integer $n \geqslant 1000$ for which no prime p has the property $n < p < 2n - 10$, and consider the primes dividing $N = \binom{2n-10}{n-10}$.) (b)*Use your result to prove that every positive integer apart from 1, 2, 4, 6 and 9 can be written as a sum of distinct odd primes.

1.3.2 Mersenne Primes

Mersenne[3] noticed that $2^2 - 1 = 3$, $2^3 - 1 = 7$, $2^5 - 1 = 31$, and $2^7 - 1 = 127$ are all primes. He suggested on the basis of experiments that $2^p - 1$ would be a prime whenever p is a prime that exceeds by 3 or less an even power of 2.

Lemma 1.11. *If* $2^n - 1$ *is prime, then* n *is prime.*

PROOF. We prove the contrapositive statement that n being composite forces $2^n - 1$ to be composite. If $n = ab$ with $a, b > 1$, then

$$2^n - 1 = (2^a - 1)(2^{n-a} + 2^{n-2a} + \cdots + 2^a + 1),$$

so $2^n - 1$ is composite.   □

The list of primes noticed by Mersenne does not continue uninterrupted because $2^{11} - 1$ is composite. A prime of the form $2^p - 1$ is known as a *Mersenne*

---

[3] Marin Mersenne (1588–1648) was a French friar in the religious order of the Minims. He defended Descartes and Galileo against their theological critics and worked to undermine alchemy and astrology. He wrote on music as part of his studies in physics and mathematics.

prime. The next few Mersenne primes are $2^{13} - 1$, $2^{17} - 1$ and $2^{19} - 1$. It is not known if there are infinitely many Mersenne primes. That $2^{19} - 1$ is prime was known to Cataldi in 1588, and this was the largest known prime for 150 years. Fermat discovered that $2^{23} - 1$ is not prime in 1640; in 1732 Euler knew that $2^{29} - 1$ is not prime but that $2^{31} - 1$ is prime.

It is worth pausing to say something about how this knowledge, which potentially requires the factorization of ten-digit numbers, accrued. Generally this involved a mixture of improving technique with congruences, some guile, and some heroic calculations. The first of several theoretical advances was discovered by Fermat and is now known as Fermat's Little Theorem.

Theorem 1.12. [FERMAT'S LITTLE THEOREM] *For any prime p and any integer a,*

$$a^p \equiv a \pmod{p}.$$

In keeping with our philosophy about differing approaches, we present two proofs of Fermat's Little Theorem.

COMBINATORIAL PROOF. It is enough to prove the statement when a is a positive integer, so we use induction. The result is true for $a = 1$ because both sides are 1. Assume it is true for $a = b$. Now

$$(b+1)^p = b^p + pb^{p-1} + \cdots + pb + 1 = \sum_{j=0}^{p} \binom{p}{j} b^j$$

by the Binomial Theorem. For $0 < j < p$, $\binom{p}{j} = \frac{p!}{j!(p-j)!}$ has a numerator divisible by p and denominator not divisible by p; the Fundamental Theorem of Arithmetic then shows that $\binom{p}{j}$ is divisible by p for $j = 1, \ldots, p-1$. So

$$(b+1)^p \equiv b^p + 1 \equiv b + 1 \pmod{p}$$

by the inductive hypothesis. Thus Fermat's Little Theorem is proved.   □

Exercise 1.10. Prove that the product of any n successive integers is divisible by $n!$.

A second, and often more useful, version of Fermat's Little Theorem can be written as follows. Integers a and b are said to be *coprime* if $\gcd(a, b) = 1$. For all $a \in \mathbb{Z}$ that are coprime to p,

$$a^{p-1} \equiv 1 \pmod{p}. \tag{1.21}$$

This form is easily seen to be equivalent to Theorem 1.12 as follows:

$$a^p - a = a(a^{p-1} - 1),$$

so when p does not divide a the Fundamental Theorem of Arithmetic shows that $p | (a^{p-1} - 1)$ if and only if $p | (a^p - a)$.

The second proof of Fermat's Little Theorem proves the congruence (1.21) and uses slightly more sophisticated ideas from group theory. The virtue of this second proof is that it is quicker and (as we shall see) is better suited to generalization. It does require some properties of modular arithmetic (see Exercise 1.28 on p. 38).

PROOF USING GROUP THEORY. Work in the group $G = (\mathbb{Z}/p\mathbb{Z})^*$ of nonzero residues modulo p under multiplication. The residue of a generates a cyclic subgroup of G whose order must divide that of G by Lagrange's Theorem. Since the order of G is $(p-1)$, we deduce Equation (1.21).   □

This proof is something of an anachronism: Lagrange's Theorem generalized Fermat's Little Theorem. However, thinking of residues using group theory is a powerful tool and gives rise to many more results, so it is useful to begin thinking in those terms now. Exercise 3.6 on p. 62 gives a good example where a proof using group theory can be favourably compared with a proof that only uses congruences.

Exercise 1.11. Fermat's Little Theorem says that, for any prime p, $2^{p-1} - 1$ is divisible by p. It sometimes happens that $2^{p-1} - 1$ is divisible by p^2. Find all the primes p with this property for $p < 10^6$. Such primes are called *Wieferich primes*, and it is not known if there are infinitely many of them.

Exercise 1.12. *A pair of congruences that arises in the Catalan problem (see p. 57) for odd primes p, q is

$$p^{q-1} \equiv 1 \pmod{q^2} \quad \text{and} \quad q^{p-1} \equiv 1 \pmod{p^2}. \tag{1.22}$$

A pair of odd primes satisfying Equation (1.22) is called a *Wieferich pair*. Find all the Wieferich pairs with $p, q < 10^4$.

Exercise 1.13. An integer n is called a *perfect number* if it is equal to the sum of its proper divisors. Thus $6 = 1 + 2 + 3$ is a perfect number.
(a) If $q = 2^p - 1$ is a Mersenne prime, prove that $2^{p-1}q$ is a perfect number.
(b) Prove that if n is an *even* perfect number, then n has the form $2^{p-1}(2^p - 1)$ for some prime of the form $2^p - 1$.

It is not known if there are any odd perfect numbers, but there are certainly no odd perfect numbers smaller than 10^{400}.

Write $M_n = 2^n - 1$ for the nth Mersenne number. The Mersenne numbers have special properties that make them particularly suitable for primality testing. The next result is the first of a series of results showing that divisors of M_n are quite prescribed when n is prime.

Lemma 1.13. *Suppose p is a prime and q is a nontrivial prime divisor of M_p. Then $q \equiv 1$ modulo p.*

Again, we give two proofs.

PROOF USING THE EUCLIDEAN ALGORITHM. The condition that q divides M_p amounts to

$$2^p \equiv 1 \pmod{q}.$$

By Fermat's Little Theorem, $2^{q-1} \equiv 1$ modulo q. Let $d = \gcd(p, q - 1)$. If $d = p$, then $p \mid (q - 1)$ as required. The only other possibility is $d = 1$ since p is prime. By Theorem 1.23 (see p. 35), in this case there are integers a and b with $1 = pa + (q - 1)b$. Notice that one of a and b must be negative. Now

$$2 \equiv 2^1 \equiv 2^{pa+(q-1)b} \equiv (2^p)^a (2^{(q-1)})^b \equiv 1^a 1^b \equiv 1 \pmod{q}, \qquad (1.23)$$

which is impossible as $q > 1$, so the result is proved. $\qquad\qquad\square$

In the preceding argument, we have made use of negative exponents of expressions modulo q, but only in the form

$$1^{-a} \equiv 1 \pmod{q} \text{ for } a > 0. \qquad (1.24)$$

PROOF USING GROUP THEORY. Work in the group G of nonzero residues modulo q. In this group 2 generates a cyclic subgroup whose order divides p since $2^p - 1 \equiv 0$ modulo q. Since 2 is not the identity and p is prime, the order of 2 must be p. Again, by Lagrange's Theorem, this order must divide the order of the group G, which is $(q - 1)$. $\qquad\qquad\square$

Example 1.14. Lemma 1.13 is a significant help in factorizing M_n. To see how this works, we present Fermat's proof from 1640 that $2^{23} - 1$ is not prime. If q is a prime dividing $2^{23} - 1$, then $q \equiv 1$ modulo 23. Now $23n + 1$ is a prime smaller than $\sqrt{2^{23} - 1}$ only for

$$n = 2, 12, 20, 26, 30, 36, 42, 44, 50, 56, 60, 62, 72, 84, 86, 102, 104, 110.$$

Trial division shows that M_{23} is divisible by the first of the resulting numbers, 47. In general, there is no reason to expect the smallest possible candidate to be a divisor, but even if the largest were the first such divisor, only 18 trial divisions are involved.

In 1876, Lucas discovered a test for proving the primality of Mersenne numbers. Using this test, he proved that

$$2^{127} - 1 = 170141183460469231731687303715884105727$$

is prime, but $2^{67} - 1$ is not. This disproved the suggestion of Mersenne.

The latter number occupies a special place in the history (and folklore) of mathematics. First, Lucas showed it is not prime but was not able to exhibit a nontrivial factor, which might seem a remarkable idea. In fact, it is something we will encounter again in the computational number theory sections. Second, this number was the subject of a famous talk given by Prof. F. N. Cole to the American Mathematical Society in 1903 entitled "On the Factorization of Large Numbers." On one blackboard, he wrote out the decimal expansion of $2^{67} - 1$ and on another he proceeded to compute the product of 193707721 and 761838257287, thereby showing them to be equal. The legend goes that after this silent lecture he sat down to "prolonged applause."

The specific arithmetic properties of Mersenne numbers mean that results on the primality of later terms in the sequence sometimes predated results on earlier terms. For example, $2^{127} - 1$ was shown to be prime in 1876 while $2^{89} - 1$ and $2^{107} - 1$ were shown to be prime in 1914.

Exercise 1.14. *[LUCAS–LEHMER TEST] Define an integer sequence by

$$S_1 = 4 \quad \text{and} \quad S_{n+1} = S_n^2 - 2 \quad \text{for} \quad n \geqslant 2.$$

Let p be an odd prime. Prove that $M_p = 2^p - 1$ is a prime if and only if $S_{p-1} \equiv 0$ modulo M_p.

1.3.3 Zsigmondy's Theorem

Although the proof of the conjecture that there are infinitely many Mersenne primes seems a long way off, it is known that the sequence starts to produce *new prime factors* very quickly. A prime p is a *primitive divisor* of M_n if p divides M_n but does not divide M_m for any $m < n$. Table 1.2 shows the prime factorization of M_n for $2 \leqslant n \leqslant 24$, with primitive divisors shown in bold.

The pattern that seems to emerge from Table 1.2 turns out to reflect something genuine. Sequences such as the Mersenne sequence, after a few initial terms, always have primitive divisors.

Theorem 1.15. [ZSIGMONDY] *Let $M_n = 2^n - 1$. Then for every $n \neq 6, n > 1$, the term M_n has a primitive divisor.*

As seen in Table 1.2, M_6 does not have a primitive divisor, so this result is optimal. The proof of Theorem 1.15 is presented in Section 8.3.1 on p. 167, after we have proved the Möbius inversion formula (Theorem 8.15). A basic result that will be needed for the proof can be proved now, using the Binomial Theorem. Notice that this result, proved as the next exercise, already shows that the divisors of the sequence (M_n) have a special structure.

Exercise 1.15. Let p denote a prime, and for any integer N, define $\mathrm{ord}_p(N)$ to be the exact power of p that divides N. Thus $\mathrm{ord}_p(N) = a$ means $p^a | N$ but $p^{a+1} \nmid N$.
(a) Prove that ord_p behaves like a logarithm in the sense that

$$\mathrm{ord}_p(xy) = \mathrm{ord}_p(x) + \mathrm{ord}_p(y)$$

for all integers x, y.
(b) Prove that if $p \mid M_n$ then $\mathrm{ord}_p(M_{kn}) = \mathrm{ord}_p(M_n) + \mathrm{ord}_p(k)$.
(c) Deduce that $\gcd(M_n, M_m) = M_{\gcd(n,m)}$ for all m, n.

Table 1.2. Primitive divisors of (M_n).

n	M_n	Factorization
2	3	**3**
3	7	**7**
4	15	$3 \cdot$ **5**
5	31	**31**
6	63	$3^2 \cdot 7$
7	127	**127**
8	255	$3 \cdot 5 \cdot$ **17**
9	511	$7 \cdot$ **73**
10	1023	$3 \cdot 11 \cdot 31$
11	2047	**23 · 89**
12	4095	$3 \cdot 5 \cdot 7 \cdot$ **13**
13	8191	**8191**
14	16383	$3 \cdot$ **43** $\cdot 127$
15	32767	$7 \cdot 31 \cdot$ **151**
16	65535	$3 \cdot 5 \cdot 17 \cdot$ **257**
17	131071	**131071**
18	262143	$3^3 \cdot 7 \cdot$ **19** $\cdot 73$
19	524287	**524287**
20	1048575	$3 \cdot 5^2 \cdot 11 \cdot 31 \cdot$ **41**
21	2097151	$7 \cdot 127 \cdot$ **337**
22	4194303	$3 \cdot 23 \cdot 89 \cdot$ **683**
23	8388607	**47 · 178481**
24	16777215	$3 \cdot 5 \cdot 7 \cdot 13 \cdot 17 \cdot$ **241**

Exercise 1.16. (a) Show that if q is a prime then every prime divisor of M_q is a primitive divisor.

(b) If M_n does not have a primitive divisor show that M_n divides the quantity

$$n \prod_{\substack{p \mid n, \\ p < n}} M_{n/p}.$$

(c) Deduce that for $n > 6$, every term M_n has a primitive divisor if n has only two distinct prime divisors. (Hint: take logarithms of the quantities in (b) and compare the growth rates of both sides.)

(d) What can you deduce if n has three distinct prime divisors?

Zsigmondy's Theorem holds in greater generality, though we will not prove the following result here.

Theorem 1.16. [ZSIGMONDY] *Let* $a_n = c^n - d^n$, *where* $c > d$ *are positive coprime integers. Then* a_n *always has a primitive divisor unless*

(1) $c = 2, d = 1$ *and* $n = 6$*; or*
(2) $c + d = 2^k$ *and* $n = 2$.

Exercise 1.17. Find some nontrivial examples of case (2) of the theorem.

A more general result is considered in Exercise 8.19 on p. 169.

Exercise 1.18. Prove that the sequence (u_n) does not satisfy a Zsigmondy Theorem in each of the following cases. This means that for every N there is a term u_n, $n > N$, which does not have a primitive divisor.
(a) $u_n = an + b$ for integers a and b;
(b) $u_n = n^2 + an + b$ for integers a and b with the property that the zeros of $x^2 + ax + b$ are integers;
(c)*$u_n = n^2 + an + b$ for integers a and b.

Exercise 1.19. *Can any polynomial $u_n = n^d + a_{d-1}n^{d-1} + \cdots + a_0$ for integers $a_0, \ldots, a_{d-1}$ have the property that the sequence (u_n) satisfies a Zsigmondy Theorem?

1.3.4 Mersenne Primes in the Computer Age

The arrival of electronic computers extended the limits of large Mersenne prime-hunting dramatically.

Table 1.3 is a short list showing how the size of the largest known Mersenne prime has grown over recent years; $\#M_p$ denotes the number of decimal digits in M_p. In 1978, Nickol and Noll were 18-year-old students. We do not distinguish here between a Mersenne prime that is the largest known at the time from a Mersenne prime for which all smaller Mersenne primes are known; see the references for a more detailed discussion. In Table 1.3, (G) denotes GIMPS and (P) denotes PrimeNet; these are distributed computer searches using idle time on many thousands of computers all over the world. Because of the special properties of Mersenne numbers (and related numbers of special shape), it has usually been the case that the largest explicitly known prime number is a Mersenne prime.

1.4 Fermat Numbers

Fermat noticed that the expression $F_n = 2^{2^n} + 1$ takes prime values for the first few values of n:

$$F_0 = 3, \quad F_1 = 5, \quad F_2 = 17, \quad F_3 = 257, \quad \text{and} \quad F_4 = 65537.$$

He believed the sequence might always take prime values. Euler in 1732 gave the first counterexample, when he showed that $641 | F_5$.

Euler, in common with Fermat and many others, was able to perform these impressive calculations through a good use of technique to minimize the amount of calculation required. Since Euler's time, many other Fermat numbers have been investigated and shown to be composite. No prime values

Table 1.3. Largest known prime values of M_p (from Caldwell's Prime Pages [25]).

p	$\#M_p$	Date	Discoverer
17	6	1588	Cataldi
19	6	1588	Cataldi
31	10	1772	Euler
61	19	1883	Pervushin
89	27	1911	Powers
107	33	1914	Powers
127	39	1876	Lucas
521	157	1952	Robinson
607	183	1952	Robinson
1279	386	1952	Robinson
2203	664	1952	Robinson
2281	687	1952	Robinson
3217	969	1957	Riesel
4253	1281	1961	Hurwitz
4423	1332	1961	Hurwitz
9689	2917	1963	Gillies
9941	2993	1963	Gillies
11213	3376	1963	Gillies
19937	6002	1971	Tuckerman
21701	6533	1978	Nickol and Noll
23209	6987	1979	Noll
44497	13395	1979	Nelson and Slowinski
86243	25962	1982	Slowinski
110503	33265	1988	Colquitt and Welsh
132049	39751	1983	Slowinski
216091	65050	1985	Slowinski
756839	227832	1992	Slowinski and Gage
859433	258716	1994	Slowinski and Gage
1257787	378632	1996	Slowinski and Gage
1398269	420921	1996	Armengaud, Woltman et al. (G)
2976221	895932	1997	Spence, Woltman et al. (G)
3021377	909526	1998	Clarkson, Woltman, Kurowski et al. (G, P)
6972593	2098960	1999	Hajratwala, Woltman, Kurowski et al. (G, P)
13466917	4053946	2001	Cameron, Woltman, Kurowski et al. (G, P)
20996011	6320430	2003	Shafer, Woltman, Kurowski et al. (G, P)
24036583	7235733	2004	Findley, Woltman, Kurowski et al. (G)

of F_n with $n > 4$ have been discovered, and it is generally expected that only finitely many terms of the sequence (F_n) are prime.

To begin, we return to Euler's result that 641 divides F_5. First, notice that $640 = 5 \cdot 2^7 \equiv -1$ modulo 641 so working modulo 641,

$$1 = (-1)^4 \equiv (5 \cdot 2^7)^4 = 5^4 \cdot 2^{28}.$$

Now $5^4 = 625 \equiv -16$ modulo 641 and $16 = 2^4$. Hence

$$1 \equiv -2^{32} \equiv -2^{2^5} \pmod{641}.$$

Of course, this elegant argument is useful only once we suspect that 641 is a factor of F_5. Euler also used some cunning to reach that point.

Lemma 1.17. *Suppose p is a prime with $p|F_n$. Then $p = 2^{n+1}k + 1$ for some $k \in \mathbb{N}$.*

Example 1.18. When $n = 5$, Lemma 1.17 shows that if p is a prime dividing F_5, then $p = 2^6k + 1 = 64k + 1$ for some k. Thus the list of possible divisors is greatly reduced. We only have to test F_5 for divisibility by

$$65, 129, 193, 257, 321, 385, 449, 513, 577, 641, \ldots,$$

of which $65, 129, 321, 385, 513, \ldots$ are not primes. Therefore we only have to test $193, 257, 449, 577, 641, \ldots$ and so on. At the fifth attempt, we find that $641|F_5$.

PROOF OF LEMMA 1.17. Suppose p is a prime with $p|F_n$, so $2^{2^n} \equiv -1$ modulo p and p is odd. Hence

$$2^{2^{n+1}} = (2^{2^n})^2 \equiv (-1)^2 \equiv 1 \pmod{p}.$$

Let $d = \gcd(2^{n+1}, p-1)$, and write $d = 2^{n+1}a + (p-1)b$ for integers a and b using Theorem 1.23. Just as in Equation (1.23) one of a and b will be negative, so we again use Equation (1.24) to argue that

$$2^d = 2^{2^{n+1}a+(p-1)b} \equiv (2^{2^{n+1}})^a (2^{p-1})^b \equiv 1 \pmod{p}.$$

Since $d|2^{n+1}$, $d = 2^c$ for some $0 \leqslant c \leqslant n+1$ so

$$2^{2^c} = 2^d \equiv 1 \pmod{p}.$$

However, $2^{2^n} \equiv -1$ modulo p and $-1 \not\equiv 1$ modulo p, so the smallest possibility for c is $(n+1)$. Hence $d = 2^{n+1}$. On the other hand, $d|(p-1)$ so $p-1 = k2^{n+1}$ as claimed. $\qquad\square$

Exercise 1.20. Strengthen Lemma 1.17 by showing that any prime p dividing F_n must have the form $2^{n+2}k + 1$ for some $k \in \mathbb{N}$.

1.5 Primality Testing

We have covered enough ground to take a first look at the challenges thrown up by primality testing. Given a small integer, one can determine if it is prime by testing for divisibility by known small primes. This method becomes totally unfeasible very quickly. We are really trying to factorize. The ability to rapidly factorize large integers remains the Holy Grail of computational number theory. Later we will look at some more sophisticated techniques and estimate the range of integers for which they are applicable.

For now, we concentrate on properties of primes that can be used to help determine primality. Fermat's Little Theorem is an example, although it does not give a necessary and sufficient condition for primality, just a necessary one. The next result does give a necessary and sufficient condition; it is known as Wilson's Theorem because of a remark to this effect allegedly made by John Wilson in 1770 to the mathematician Edward Waring. An early proof was published by Lagrange in 1772. The theorem first seems to have been noted by al-Haytham[4] some 750 years before Wilson.

Theorem 1.19. *An integer $n > 1$ is prime if and only if*

$$(n - 1)! \equiv -1 \pmod{n}.$$

PROOF OF 'ONLY IF' DIRECTION. We prove that the congruence is satisfied when n is prime and leave the converse as an exercise. Assume that $n = p$ is an odd prime. (The congruence is clear for $n = 2$.)

Each of the integers $1 < a < p - 1$ has a unique multiplicative inverse distinct from a modulo p (see Corollary 1.25). Uniqueness is obvious; for distinctness, note that $a^2 \equiv 1$ modulo p implies $p | (a+1)(a-1)$, forcing $a \equiv \pm 1$ modulo p by primality. Thus in the product

$$(p - 1)! = (p - 1)(p - 2) \cdots 3 \cdot 2 \cdot 1,$$

all the terms cancel out modulo p except the first and the last. Their product is clearly -1 modulo p. □

Exercise 1.21. Prove the converse: If $n > 1$ and $(n - 1)! \equiv -1$ modulo n, then n is prime.

Exercise 1.22. [GAUSS] Prove the following generalization of Theorem 1.19. Let

$$P_n = \prod_{\substack{m < n, \\ \gcd(m,n)=1}} m$$

be the product of all positive integers less than n and coprime to n. Then $P_n + 1$ is divisible by n if n is equal to 4, p^k, or $2p^k$ for some odd prime p, and $P_n - 1$ is divisible by n if n is not of that form.

---

[4] Abu Ali al-Hasan ibn al-Haytham (964–1040) lived in Persia and Egypt. He is most famous for *Alhazen's Problem*: Find the point on a spherical mirror where a light will be reflected to an observer. In number theory, in addition to proving what we often call Wilson's Theorem, al-Haytham worked on perfect numbers (see Exercise 1.13).

Exercise 1.23. [CLEMENT] (a) Use al-Haytham's Theorem (Theorem 1.19) to prove that, for $n > 1$, n and $n + 2$ are both prime if and only if

$$4\left((n-1)! + 1\right) + n \equiv 0 \quad (\mathrm{mod}\ n(n+2)).$$

(b) Prove that, for $n > 13$, the triple n, $n + 2$, and $n + 6$ are all prime if and only if

$$4320\left(4\left((n-1)! + 1\right) + n\right) + 361n(n+2) \equiv 0 \ \mathrm{mod}\ \left(n(n+2)(n+6)\right).$$

(c) Find a similar characterization of prime triples of the form n, $n + 4$, and $n + 6$.

Primes p for which $p + 2$ is also a prime are called *twin primes*, and it is a long-standing conjecture that there are infinitely many twin primes. A remarkable result of Brun from 1919 is that the reciprocals of the twin primes (whether there are infinitely many or not) are summable:

$$\sum_{p,p+2\in\mathbb{P}} \frac{1}{p} = B < \infty. \tag{1.25}$$

Numerical estimation of *Brun's constant B* is very difficult.

Exercise 1.24. Theorem 1.19 gives another 'formula' for the primes. Show that $(n-2)!$ is congruent to 1 or 0 modulo n depending on whether n is prime or not, for $n \geqslant 3$.
(a) Deduce that the prime counting function $\pi(X) = |\{p \in \mathbb{P} \mid p \leqslant X\}|$ may be written

$$\pi(X) = 1 + \sum_{j=3}^{X} \left((j-2)! - j\left\lfloor \frac{(j-2)!}{j}\right\rfloor\right), \quad X \geqslant 3,$$

with $\pi(1) = 0$, $\pi(2) = 1$.
(b) Define a function f by $f(x, x) = 0$ and

$$f(x, y) = \frac{1}{2}\left(1 + \frac{x-y}{|x-y|}\right) \quad \text{for } x \neq y.$$

Use Theorem 1.9 to prove that

$$p_n = 1 + \sum_{j=1}^{2^n} f(n, \pi(j)).$$

In principle, Theorem 1.19 seems to offer a general primality test because the condition is necessary and sufficient. The problem is that in practice it is impossible to compute $(n-1)!$ modulo n in a reasonable amount of time for any integer that is not quite small. In Chapter 12 we will seek to give a better understanding of what counts as "small" or "large" in terms of modern computing.

Fermat's Little Theorem offers another hope. Taking $a = 2$, Fermat's Little Theorem implies that

$$2^{p-1} \equiv 1 \pmod{p} \text{ whenever } p \text{ is prime.} \tag{1.26}$$

At various times in history, it has been thought that a kind of converse might be true: If n is odd and $2^{n-1} \equiv 1$ modulo n, might it follow that n is prime? Calculations tend to support this, and for $n < 341$ this does indeed successfully detect primality.

Example 1.20. Testing the congruence $2^{n-1} \equiv 1$ modulo n fails to detect the fact that $n = 341 = 11 \cdot 31$ is composite. By Fermat's Little Theorem, $2^{10} \equiv 1$ modulo 11 so $2^{340} \equiv 1^{34} \equiv 1$ modulo 11. Also $2^5 = 32 \equiv 1$ modulo 31, so

$$2^{340} = (2^5)^{68} \equiv 1^{68} = 1 \pmod{31}.$$

Thus $2^{340} - 1$ is divisible by the coprime numbers 11 and 31, and hence by their product 341, so $2^{340} \equiv 1$ modulo 341.

However, Fermat's Little Theorem says more than Equation (1.26): It gives the congruence

$$a^{p-1} \equiv 1 \pmod{p}$$

for *any base* a, not just $a = 2$. Taking $a = 3$ in Example 1.20, we quickly find

$$3^{340} \equiv 56 \pmod{341},$$

which contradicts Fermat's Little Theorem with $a = 3$, showing that 341 cannot be prime. Notice the recurrence of a phenomenon encountered before: Using $a = 3$, we have shown that a number is not prime without exhibiting a nontrivial factor.

This method suggests the following as a primality test. Given an integer n, choose numbers a at random with $1 < a < n$ and test to see if $a^{n-1} \equiv 1$ modulo n. If not, then n is definitely composite. If the congruence is satisfied for several such a, we might view this as compelling evidence that n must be prime. Unfortunately, this also fails as a primality test.

Exercise 1.25. Prove that $n = 561$ is a composite number that satisfies Fermat's Little Theorem for every possible base by showing that $a^{560} \equiv 1$ modulo 561 for every a, $1 < a < n$ with $\gcd(a, 561) = 1$. (Hint: Use Fermat's Little Theorem on each of the factors 3, 11, and 17 of 561.)

A composite integer that satisfies the congruence of Fermat's Little Theorem for all bases coprime to itself is known as a *Carmichael number*; these will be discussed in more detail in Section 12.5. It was not known whether there are infinitely many Carmichael numbers until 1994, when Alford, Granville, and Pomerance not only proved that there are infinitely many but gave some measure of how many there are asymptotically. The existence of infinitely many Carmichael numbers renders the test based on Fermat's Little Theorem test too unreliable. Later, we will see however that a more sophisticated version is salvageable as a primality test.

1.6 Proving the Fundamental Theorem of Arithmetic

We uncover Euclid's real genius once we try to prove the Fundamental Theorem of Arithmetic. There are two parts to it: existence and uniqueness. The existence part is not difficult. Let $n > 1$ be an integer, and choose r with $2^r > n$. If n itself is not divisible by any a with $1 < a < n$, then nothing else needs to be said. Otherwise, we can write $n = ab$ with $1 < a, b < n$. Again, if a and b cannot be factorized, further then we are done. If this is not the case then at least one of them can be factorized. Once we have done this r times, we have $n = a_1 \cdots a_r$ with each $1 < a_i < n$. This implies $n \geqslant 2^r$, giving a contradiction. Thus n must be a product of no more than r prime factors.

It is when we come to the uniqueness part of the proof that we uncover a subtlety – namely, that the definition of prime as an irreducible element is not really adequate to prove the Fundamental Theorem of Arithmetic. Suppose we try to argue as follows: Consider two factorizations for n into primes, say

$$p_1 \cdots p_r = n = q_1 \cdots q_s.$$

We would like to say that because p_1 divides the right-hand side, it must divide one of the q_i. However, if we are working with the definition of *prime* as *irreducible*, then we need a result that tells us that being irreducible forces this divisibility property. Such a result may be found using the Euclidean Algorithm.

Later, we will see examples in rings that are closely related to $\mathbb{Z}$ whose elements have genuinely different factorizations into irreducibles.

Exercise 1.26. Let

$$A = \{n \in \mathbb{N} \mid n \equiv 1 \pmod 4\},$$

and call $n \neq 1$ an *A-prime* if the only divisors of n in A are 1 and n.
(a) Show that every element of A except 1 factorizes as a finite product of A-primes.
(b) Show that this factorization into A-primes is not unique.

1.6.1 The Euclidean Algorithm

Given $a, b > 0$ in $\mathbb{Z}$, we can always find q and r with $a = bq + r$ and $0 \leqslant r < b$. Indeed, for q we can simply take the integer part $\lfloor a/b \rfloor$ of a/b and then show that by defining $r = a - bq$ we must have $0 \leqslant r < b$.

Something very interesting happens when we iterate this process. It will help to define $q = q_1$ and $r = r_1$ and continue to find quotients and remainders as follows:

$$a = bq_1 + r_1, \qquad 0 \leqslant r_1 < b$$
$$b = r_1 q_2 + r_2, \qquad 0 \leqslant r_2 < r_1$$
$$\vdots \qquad\qquad\qquad \vdots$$
$$r_{n-3} = r_{n-2} q_{n-1} + r_{n-1}, \quad 0 \leqslant r_{n-1} < r_{n-2}$$
$$r_{n-2} = r_{n-1} q_n + r_n, \qquad 0 \leqslant r_n < r_{n-1}$$
$$r_{n-1} = r_n q_{n+1} + 0.$$

The sequence of remainders is decreasing and each term is nonnegative, so the sequence must terminate. We have written r_n for the last nonzero remainder, so $r_n | r_{n-1}$. We claim that r_n is the greatest common divisor of a and b.

Example 1.21. Let $a = 17$ and $b = 11$. Then the Euclidean Algorithm gives the equations

$$17 = 11 \cdot 1 + 6,$$
$$11 = 6 \cdot 1 + 5,$$
$$6 = 5 \cdot 1 + 1,$$
$$5 = 1 \cdot 5 + 0.$$

The last nonzero remainder is the greatest common divisor of 17 and 11, which is clearly 1.

To prove that $r_n = \gcd(a, b)$, we need a better notion of greatest common divisor than the intuitive one.

Definition 1.22. *If a and b in $\mathbb{Z}$ are not both zero, d is said to be a greatest common divisor of a and b if*

(1) *$d | a$ and $d | b$; and*
(2) *if d' is any number with $d' | a$ and $d' | b$, then $d' | d$.*

The first condition says d is a common divisor of a and b, while the second says it is the greatest such divisor.

Note that we say "a" greatest common divisor rather than "the" greatest common divisor because if d satisfies this condition then $-d$ will also satisfy the definition. If we work in $\mathbb{N}$, then the greatest common divisor will be unique. The notation $\gcd(a, b)$ denotes the unique nonnegative greatest common divisor of a and b. If $\gcd(a, b) = 1$, then we will call a and b *coprime*.

Exercise 1.27. Using Definition 1.22, show that $r_n = \gcd(a, b)$. (Hint: Work your way up and then down the chain of equations to verify the two properties.)

The next result is fundamental to the structure of the integers; it is an easy consequence of the Euclidean Algorithm and is sometimes referred to as Bezout's Lemma.

Theorem 1.23. *If $d = \gcd(a, b)$ with $a, b \in \mathbb{Z}$ not both zero, then there are numbers $x, y \in \mathbb{Z}$ with*

$$d = ax + by. \tag{1.27}$$

PROOF. The idea is to work your way up the chain of equations in the Euclidean Algorithm, always expressing the remainder in terms of the previous two remainders. Writing $*$ for an integer, we get

$$\gcd(a, b) = r_n = r_{n-2} - r_{n-1}q_n$$
$$= r_{n-2}(1 + q_n q_{n-1}) - r_{n-3}q_n$$
$$= r_{n-3} \cdot * + r_{n-4} \cdot *$$
$$\vdots$$
$$= b \cdot * + r_1 \cdot *$$
$$= a \cdot * + b \cdot *.$$

$\square$

Example 1.24. Using the equations from Example 1.21 we find that

$$1 = 6 - 5$$
$$= 6 - (11 - 6)$$
$$= 2 \cdot 6 - 11$$
$$= 2(17 - 11) - 11$$
$$= 2 \cdot 17 - 3 \cdot 11.$$

Corollary 1.25. *Let $n > 1$ and a denote elements of $\mathbb{Z}$. Then a and n are coprime if and only if there exists x with*

$$ax \equiv 1 \pmod{n}.$$

That is, $\gcd(a, n) = 1$ if and only if a is invertible modulo n.

PROOF. The congruence is equivalent to the existence of an integer y with

$$ax + ny = 1.$$

If a and n have a factor in common then that factor will also divide 1, so the congruence implies a and n are coprime. Conversely, if a and n are coprime then 1 is a greatest common divisor of a and n so we can use Theorem 1.23 to see that there are integers x and y with $ax + ny = 1$, which translates into the congruence. $\square$

Exercise 1.28. Let p be a prime. Prove that the set $(\mathbb{Z}/p\mathbb{Z})^*$ of nonzero elements in $\mathbb{Z}/p\mathbb{Z}$ forms a group under multiplication modulo p.

One of the remarkable things about the Euclidean Algorithm is that it finds the greatest common divisor of two integers without factorizing either of them. We will see later how this has been exploited in powerful ways by computational number theory in recent years.

Exercise 1.29. Prove the Fundamental Theorem of Arithmetic using Theorem 1.23. (Hint: This is done in greater generality on p. 47.)

1.6.2 An Inductive Proof of Theorem 1.1

We wish to prove that any natural number n has a decomposition $n = p_1 \cdots p_r$ into primes uniquely up to rearrangement of the prime factors.

For $n = 2$, the theorem is clearly true. We proceed by induction. Suppose that the Fundamental Theorem of Arithmetic holds for all natural numbers strictly less than some $a > 1$. We want to deduce the Fundamental Theorem of Arithmetic for a. Let
$$D = \{d \mid d > 1, d|a\}$$
denote the set of non-identity divisors of a. The set D is nonempty since it contains a, so it has a smallest element, which we denote p. This smallest element must be a prime because if it had a nontrivial divisor that would be a smaller element of D. Thus we have a decomposition

$$a = pb, p \text{ prime}, b < a.$$

Since $b < a$, by the inductive hypothesis, the Fundamental Theorem of Arithmetic holds for b, so there is a prime decomposition

$$b = p_1 \cdots p_s$$

into primes uniquely up to rearrangement. It follows that

$$a = p \cdot p_1 \cdots p_s$$

is a prime decomposition of a, and a has no other prime decomposition *involving the prime p.*

Suppose that a has another prime decomposition,

$$a = q_1 \cdots q_r,$$

in which the prime p does not appear. In particular, $q_1 \neq p$. Moreover, by the definition of p, $q_1 > p$ since $q_1 \in D$, $1 \leqslant q_1 - p < q_1$. Let $c = q_2 \cdots q_r$, and define
$$a_0 = a - pc = p(b - c) = (q_1 - p)c. \tag{1.28}$$

Now $1 \leqslant a_0 < a$ and the divisors $(b-c)$, $(q_1 - p)$, and c are all less than a. By the inductive hypothesis, the numbers a_0, $(b-c)$, $(q_1 - p)$, and c all have unique prime decompositions. By Equation (1.28), the prime p must appear in any prime decomposition of a_0 and therefore (by uniqueness) must also appear in the decomposition of $(q_1 - p)$ or that of c.

Now p cannot appear in a prime decomposition of $(q_1 - p)$ because that would require $p|q_1$, which is impossible, as p and q_1 are distinct primes. Nor can p appear in a prime decomposition of $c = q_2 \cdots q_r$ by assumption. Thus the assumption of a second prime decomposition for a leads to a contradiction, completing the proof of the Fundamental Theorem of Arithmetic.

1.7 Euclid's Theorem Revisited

In this section, three further proofs of Theorem 1.2 are given, each interesting and suggestive in its own right.

1.7.1 What Did Euclid Really Prove?

First, we return to the master's proof. The following is a translation of Euclid's proof taken from Joyce's Web translation of Euclid's *Elements*. In Euclid's time, numbers were thought of as relatively concrete lengths of line segments. Thus, for example, a number A *measures* a number B if a stick of length A could be used to fit into a stick of length B a whole number of times. In modern terminology, A divides B. We start with Euclid's Theorem in (an approximation of) Euclid's language:

> Οἱ πρῶτοι ἀριθμοὶ πλείους εἰσὶ παντὸς τοῦ
> προτεθέντος πλήθους πρώτων ἀριθμῶν.

A translation of this is the following theorem, which is Proposition 20 of Book IX in Euclid's *Elements*.

Theorem 1.26. *The prime numbers are more than any assigned multitude of prime numbers.*

PROOF. Let A, B, and C be the assigned prime numbers. I say that there are more prime numbers than A, B, and C. Take the least number DE measured by A, B, and C. Add the unit DF to DE.

Then EF is either prime or not.

First, let it be prime. Then the prime numbers A, B, C, and EF have been found, which are more than A, B, and C.

Next, let EF not be prime. Therefore, it is measured by some prime number. Let it be measured by the prime number G. I say that G is not the same as any of the numbers A, B, and C.

If possible, let it be so.

Now A, B, and C measure DE, and therefore G also measures DE. But it also measures EF. Therefore G, being a number, measures the remainder, the unit DF, which is absurd.

Therefore G is not the same as any one of the numbers A, B, and C, and by hypothesis it is prime. Therefore, the prime numbers A, B, C, and G have been found, which are more than the assigned multitude of A, B, and C. Therefore, prime numbers are more than any assigned multitude of prime numbers.   □

There is little between this argument and Euclid's proof in modern form on p. 8. Euclid did not have our modern notion of infinity, so he proved that there are more primes than any prescribed number. He also often stated proofs using examples (in this case, what he really proves is that there are more than three primes), but it is clear he understood the general case. It is possible that part of the reason for this is the notational difficulties involved in dealing with arbitrarily large finite lists of objects.

1.7.2 A Topological Proof of Theorem 1.2

In 1955, Furstenberg gave a completely different type of proof of the infinitude of the primes using ideas from topology.

FURSTENBERG'S TOPOLOGICAL PROOF OF THEOREM 1.2. Define a topology on the integers $\mathbb{Z}$ by taking as a basis the arithmetic progressions. For each prime p, let S_p denote the arithmetic progression $p\mathbb{Z}$. Since

$$S_p = \mathbb{Z} \backslash \big((p\mathbb{Z} + 1) \cup \cdots \cup (p\mathbb{Z} + (p-1)) \big),$$

the set S_p is the complement of an open set, and thus is closed. Let $S = \bigcup_p S_p$ be the union of all the sets S_p as p varies over the primes. If there are only finitely many primes, then S is a finite union of closed sets, and thus is closed. However, every integer except ± 1 is in some S_p, so the complement of S is $\{1, -1\}$, which is clearly not open. It follows that S cannot be closed and therefore cannot be a finite union, so there must be infinitely many primes. □

In contrast with the other proofs of Theorem 1.2, this is qualitative – all it tells us about the prime counting function is that $\pi(X) \to \infty$ as $X \to \infty$.

1.7.3 Goldbach's Proof

Goldbach showed how one may use a sequence of integers with the property that an infinite subsequence are pairwise coprime to give a different proof.

GOLDBACH'S PROOF OF THEOREM 1.2. We claim that the Fermat numbers $F_n = 2^{2^n} + 1$ are pairwise coprime:.

$$m \neq n \implies \gcd(F_m, F_n) = 1. \qquad (1.29)$$

The first step is to show by induction that

$$F_m - 2 = F_0 F_1 \cdots F_{m-1} \text{ for all } m \geqslant 1. \tag{1.30}$$

To see why this is true, first note that $F_1 - 2 = F_0$ and assume that Equation (1.30) holds for $m \leqslant k$. Then

$$\begin{aligned}
F_0 F_1 \cdots F_{k-1} F_k &= (F_k - 2) F_k \\
&= \left(2^{2^k} - 1\right)\left(2^{2^k} + 1\right) \\
&= 2^{2^{k+1}} - 1 = F_{k+1} - 2,
\end{aligned}$$

showing Equation (1.30) by induction. Thus for $m > n$,

$$d \big| F_m, d \big| F_n \implies d \big| F_m - 2 \implies d \big| 2,$$

which forces d to be 1 since all the F_n are odd numbers. This proves Equation (1.29).

This in turn means there must be infinitely many primes. By Theorem 1.1, each F_n has a prime factor p_n, say, and by Equation (1.29) these are all distinct. □

The proof using Fermat numbers actually does a little more than prove there are infinitely many primes. It also gives some insight into how many primes there are that are smaller than a given number. By the time we reach the number F_n, we must have seen at least n different primes, so

$$\pi(X) \geqslant \frac{1}{\log 2} \log\left(\frac{\log(X-1)}{\log 2}\right),$$

which is approximately proportional to $\log \log X$. This is far weaker than the remark on p. 21.

NOTES TO CHAPTER 1: The exact history of Theorem 1.1 is not clear, and it is likely that it was known and used long before it was explicitly stated. The earliest precise formulation and proof seems to be due to Gauss [67], but it could be argued that Euclid certainly knew that if a prime p divides a product ab, then p must divide a or b, and that his geometrical formalism and approach to exposition did not require him to consider products of more than three terms (see Section 1.7.1). Many of the proofs of Euclid's Theorem are featured in the Prime Pages Web site [25]; Ribenboim's book [125] describes no fewer than 11 proofs. Example 1.7 is related to subtle problems in algebraic number theory; see Ribenboim's book [125] for a discussion and detailed references. That the positive values of a polynomial in several variables could coincide with the primes is essentially a by-product of Matijasevič's solution to one of Hilbert's famous problems. Some of the history and references and two explicit polynomials are given in accessible form in the paper [85] of Jones, Sato, Wada and Wiens. The proofs of Lemma 1.8 and Theorem 1.9 are those of

Erdös [51] and Kalmar, and may be found in Hardy and Wright [75]; that of Corollary 1.10 follows a survey paper of Dudley [46]. Bertrand's Postulate (Theorem 1.9) was first proved by Tchebychef [151, Tome I, pp. 49–70, 63]. He also proved that for any $e > \frac{1}{5}$, there is a prime between x and $(1 + e)x$ for x sufficiently large. The deep result of Ingham [80] has been improved a great deal — for example, Baker, Harman and Pintz [8] have shown that there is a prime in the interval $[x - x^{0.525}, x]$ for x sufficiently large. Exercise 1.7 is due to Mills [107]. Exercise 1.8 comes from a paper of Richert [127]; Exercise 1.9 from a paper of Dressler [45]. Further material on Mersenne primes – and on large primes in general – may be found on Caldwell's Prime Pages Web site [25]; Table 1.3 is taken from his Web site. A recent account of the GIMPS record-breaking prime is in Ziegler's short article [167]. Zsigmondy's Theorems 1.15 and 1.16 appeared first in his paper [168]; a more accessible proof may be found in a short paper by Roitman [132]. Deep recent work has extended this to a larger class of sequences: Bilu, Hanrot and Voutier have shown that for $n > 30$ the nth term of any Lucas or Lehmer sequence has a primitive divisor in their paper [15]. The current status of Fermat numbers and their factorization may be found on Keller's Web site [88]. Parts of the intricate connection between group theory and the origins of modern number theory, and in particular a discussion of how Gauss used group-theoretic concepts long before they were formalized, are in a paper of Wußing [164]. For more on the very special numbers found in Exercise 1.11 see Ribenboim's popular article [123]. The inductive proof of Theorem 1.1 in Section 1.6.2 is taken from Hasse's classic text [76] and is attributed there to Zermelo. Hasse's text is also the source of the statement of Euclid's Theorem in Greek in Section 1.7.1. We thank David Joyce for permission to use the translation in Section 1.7 from his Web site [86]; this Web site is based on several translations of Euclid's work, but the primary and most accessible source remains the translation by Heath [53]. Exercise 1.24 is taken from Hardy and Wright [75]. Furstenberg's proof of Euclid's Theorem appeared in [63]. Exercise 1.23 is taken from Clement's paper [31]. Brun's result in Equation (1.25) appeared originally in his paper [24]; a modern proof may be found in the book of LeVeque [100]. Finally, we make some remarks concerning Section 1.7.2. Using topology in this setting might seem odd, but perhaps Euler's proof using the harmonic series seemed odd when it first appeared. We don't wish to stretch the point, but it could just be that Furstenburg's proof points forward to new ways of looking at arithmetic in just the same way as Euler's did. Profound structures in the integers have certainly been uncovered using methods from ergodic theory, combinatorics, functional analysis, and Fourier analysis; see a survey paper of Bergelson [11], the book by Furstenberg [64], and a new approach in a paper of Gowers [72] for some of these startling results. In a similar vein, Green and Tao [73] have recently proved the deep result that the primes contain arbitrarily long arithmetic progressions.

2

Diophantine Equations

Diophantine equations are equations (very often involving polynomials with integer coefficients) in which the solutions are required to be integers. They have been studied since antiquity and are mathematically both challenging and attractive because of the great diversity of methods that are needed to understand them.

2.1 Pythagoras

In this chapter, we are going to explore the relationship between the Fundamental Theorem of Arithmetic and the study of polynomial Diophantine problems. We begin with an equation handed down from antiquity,

$$x^2 + y^2 = z^2. \tag{2.1}$$

We know that an equation of this kind is related to a right-angled triangle with side lengths x, y, and z. Right-angled triangles have been studied and used for four thousand years (at least). Equation (2.1) is called the *Pythagorean equation* to honor Pythagoras for his result connecting Equation (2.1) to right-angled triangles. We seek to identify all the integral solutions; that is, to find all triples of integers (x, y, z) that satisfy Equation (2.1). The main point in the first three sections of this chapter is to emphasize the symbiosis between properties of numbers and solutions of equations.

To motivate what follows, rearrange the equation to read

$$x^2 = z^2 - y^2 = (z + y)(z - y). \tag{2.2}$$

If we knew that $\gcd(z + y, z - y) = 1$, then we could apply the Fundamental Theorem of Arithmetic to argue that both $(z + y)$ and $(z - y)$ must themselves be squares and use the resulting equations to parametrize all triples of solutions.

To refine the proof, we resort to a congruence argument. First, we may assume that the triple (x, y, z) contains no common prime factor – otherwise we may divide through by the square of that factor. A triple (x, y, z) is called a *primitive solution* of Equation (2.1) if x, y, and z have no common factor. Second, we may assume that only one of the three is even because if two are then the third must be, contrary to the primitive condition. Now the even one out (so to speak) cannot be z because

$$x^2 + y^2 \equiv 0 \pmod 4$$

is impossible with x and y being odd. Thus we may suppose one of x or y is even. Without loss of generality, suppose it is x that is even. Write $x = 2x'$ and substitute into Equation (2.2) to give

$$x'^2 = \left(\frac{z+y}{2}\right)\left(\frac{z-y}{2}\right).$$

Notice that each of $(z \pm y)/2$ must be an integer because z and y are both odd. More than that, they must be coprime because any common factor of any two of x, y, and z must divide the third. Hence, any common divisor of $(z \pm y)/2$ will also divide their sum and their difference, z and y, and we are assuming the triple (x, y, z) is primitive.

Thus at last we may apply the Fundamental Theorem of Arithmetic to deduce that $(z \pm y)/2$ are both squares, say

$$z + y = 2m^2, z - y = 2n^2, m > n. \tag{2.3}$$

We are assuming z and y are positive so $z + y > z - y$, giving the inequality between m and n. Solving Equation (2.3) for z and y and then using Equation (2.1) to find x gives the following characterization of *primitive Pythagorean triples*.

Theorem 2.1. *The primitive integral solutions of the Pythagorean equation*

$$x^2 + y^2 = z^2$$

with even x are given by

$$x = 2mn, \ y = m^2 - n^2, \ z = m^2 + n^2$$

with $m > n$ coprime integers, not both odd.

The integers $m > n$ are said to *parametrize* the solutions of the equation.

Exercise 2.1. For any primitive solution of Equation (2.1) show that one of x, y, or z is divisible by 3, one by 4, and one by 5.

Exercise 2.2. Finding integral solutions to Equation (2.1) is equivalent to finding rational solutions to $x^2 + y^2 = 1$. Find the second point of intersection with the circle $x^2 + y^2 = 1$ of the line with slope t through the point $(1, 0)$, and show that letting t run through all rationals gives all rational solutions to $x^2 + y^2 = 1$.

Using geometry to construct new rational solutions of Diophantine equations from old ones is a powerful idea that will be taken up again in Section 5.1.

2.2 The Fundamental Theorem of Arithmetic in Other Contexts

In the integers, the Fundamental Theorem of Arithmetic is a direct consequence of the existence of the Euclidean Algorithm. In certain rings, the two properties are not equivalent. For example, the Fundamental Theorem of Arithmetic holds in the ring of integer polynomials $\mathbb{Z}[x]$, even though this ring does not have a Euclidean Algorithm. Nonetheless, in many arithmetic contexts, the Fundamental Theorem of Arithmetic can be proven easily because one has a Euclidean Algorithm. We will consider only commutative rings with a multiplicative identity, written 1.

Definition 2.2. *A commutative ring R is* Euclidean *if there is a function*

$$N : R \backslash \{0\} \to \mathbb{N}$$

with the following properties:

(1) $N(ab) = N(a)N(b)$ *for all $a, b \in R$, and*
(2) *for all $a, b \in R$, if $b \neq 0$, then there exist $q, r \in R$ such that*

$$a = bq + r \text{ and } r = 0 \text{ or } N(r) < N(b).$$

Such a function is called a norm *on R.*

Much of what follows can be done with weaker conditions. In particular, one does not need such a strong property as (1). However, in many cases, the norm does have this property, so we assume it to allow a speedier and more natural development of the argument.

Example 2.3. The following are examples of Euclidean rings.

(1) Let $R = \mathbb{Z}[i]$ denote the Gaussian integers, so

$$R = \{x + iy \mid x, y \in \mathbb{Z}\},$$

where $i^2 = -1$. Setting $N(x + iy) = x^2 + y^2$ shows that R is a Euclidean ring.

(2) Let $\mathbb{F}$ denote any field and let $R = \mathbb{F}[x]$ be the ring of polynomials with coefficients in $\mathbb{F}$. Define $N(f) = 2^{\deg(f)}$, where $\deg(f)$ is the degree of f in $\mathbb{F}[x]$, which is defined for all nonzero elements of R.

We prove the first of these; the second is an exercise.

PROOF THAT $\mathbb{Z}[i]$ IS EUCLIDEAN. Condition (1) of Definition 2.2 is easily verified by direct computation. For property (2), let $a, b \neq 0 \in R$ and write $ab^{-1} = p + iq$ with $p, q \in \mathbb{Q}$. Now define $m, n \in \mathbb{Z}$ by

$$m \in [p - 1/2, p + 1/2), \ n \in [q - 1/2, q + 1/2).$$

Let $q = m + in \in R$ and $r = a - b(m + in)$. For $r \neq 0$,

$$\begin{aligned} N(r) &= N((ab^{-1} - m - in)b) \\ &= N(p + iq - m - in)N(b) \\ &= N(p - m + i(q - n))N(b) \leqslant \left(\tfrac{1}{4} + \tfrac{1}{4}\right)N(b) < N(b), \end{aligned}$$

showing property (2).   □

Exercise 2.3. When $R = \mathbb{Z}$, for any fixed a and b, the values of q and r in Definition 2.2(2) are uniquely determined. Is the same true when $R = \mathbb{Z}[i]$?

In any ring, we define greatest common divisors in exactly the same way as before. A greatest common divisor is defined up to multiplication by units (invertible elements). In any Euclidean ring, the function N can be used to define a Euclidean Algorithm, which can be used to find the greatest common divisor just as for the integers.

Definition 2.4. *In a ring R,*

(1) α *divides* β, *written* $\alpha | \beta$, *if there is an element* $\gamma \in R$ *with* $\beta = \alpha\gamma$;
(2) u *is a* unit *if u divides* 1;
(3) π *(not equal to zero nor to a unit) is* prime *if for all* $\alpha, \beta \in R$,

$$\pi | \alpha\beta \implies \pi | \alpha \text{ or } \pi | \beta;$$

(4) *a non-unit μ is* irreducible *if*

$$\mu = \alpha\beta \implies \alpha \text{ or } \beta \text{ is a unit.}$$

Notice that $u \in R$ is a unit if and only if there is some μ with $u\mu = 1$. We write $U(R)$ or R^* for the units in the commutative ring R; this is an Abelian group under multiplication. If the recent clutch of definitions are new to you, we recommend the following exercise.

Exercise 2.4. (a) Show that, in any commutative ring, every prime element is irreducible.

(b) Show that, in a Euclidean ring, u is a unit if and only if $N(u) = 1$.

(c) Show that there are infinitely many units in $\mathbb{Z}[\sqrt{3}]$.

(d) Show that $3 + \sqrt{-2}$ is an irreducible element of $\mathbb{Z}[\sqrt{-2}]$.

(e) Let $\xi = \frac{-1+\sqrt{-3}}{2}$ and $R = \mathbb{Z}[\xi]$. Prove that R is a Euclidean domain with respect to the norm $N(a + b\xi) = a^2 - ab + b^2 = (a + b\xi)(a + b\bar{\xi})$ and find all the units in R.

Exercise 2.5. Prove the *Remainder Theorem*: For a polynomial $f \in \mathbb{F}[x]$, $\mathbb{F}$ a field, $f(a) = 0$ if and only if $(x - a)|f(x)$.

Exercise 2.6. Give a different proof of Lemma 1.17 on p. 31 using group theory by considering the multiplicative group of units $U(\mathbb{Z}/F_n\mathbb{Z}) = (\mathbb{Z}/F_n\mathbb{Z})^*$.

Exercise 2.7. Prove that $\mathbb{Z}[x]$ does not have a Euclidean Algorithm by showing that the equation $2f(x) + xg(x) = 1$ has no solution for $f, g \in \mathbb{Z}[x]$, but 2 and x have no common divisor in $\mathbb{Z}[x]$.

Despite the conclusion of Exercise 2.7, the ring $\mathbb{Z}[x]$ does have unique factorization into irreducibles.

We will say that a ring has the Fundamental Theorem of Arithmetic if either of the following properties hold.

(FTA1) Every irreducible element is prime.

(FTA2) Every nonzero non-unit can be factorized uniquely up to order and multiplication by units.

Theorem 2.5. *Every Euclidean ring has the Fundamental Theorem of Arithmetic.*

PROOF. Clearly, every irreducible μ has $N(\mu) \geqslant 2$. Arguing as we did in $\mathbb{Z}$ shows we cannot keep factorizing into irreducibles forever, so the existence part is easy. To complete the argument, we just need to show that every irreducible is prime. This follows easily from Theorem 1.23. Let μ be an irreducible and suppose that μ divides $\alpha\beta$ but μ does not divide α. Clearly, the greatest common divisor of μ and α is 1 because μ admits only itself and units as divisors and μ does not divide α, so we can write

$$\mu x + \alpha y = 1$$

for some $x, y \in R$ by Theorem 1.23. Multiply through by β to obtain

$$\mu x \beta + \alpha \beta y = \beta.$$

Since μ divides both terms on the left-hand side, it must divide the right-hand side, and this completes the proof. $\square$

2.3 Sums of Squares

The resolution of the Pythagorean equation (Equation (2.1)) is an elementary and well-known result. We are now going to show how the Fundamental Theorem of Arithmetic in other contexts can yield solutions to less tractable Diophantine equations. Consider the following problem: Which integers can be represented as the sum of two squares? That is, what are the solutions to the Diophantine problem

$$n = x^2 + y^2?$$

When n is a prime, experimenting with a few small values suggests the following.

Theorem 2.6. *The prime p can be written as the sum of two squares if and only if $p = 2$ or p is congruent to 1 modulo 4.*

To prove this, we are going to use the Fundamental Theorem of Arithmetic in the ring of Gaussian integers $R = \mathbb{Z}[i]$ with norm function $N : R \to \mathbb{N}$ defined by $N(x + iy) = x^2 + y^2$ as in Example 2.3(1).

Lemma 2.7. *If p is 2 or a prime congruent to 1 modulo 4, then the congruence*

$$T^2 + 1 \equiv 0 \pmod{p}$$

is solvable in integers.

PROOF. This is clear for $p = 2$ so suppose $p = 4n + 1$ for some integer $n > 0$. Using al-Haytham's Theorem (Theorem 1.19),

$$(p - 1)! = (p - 1)(p - 2) \cdots 3 \cdot 2 \cdot 1 \equiv -1 \pmod{p}.$$

Now

$$4n = p - 1 \equiv -1 \pmod{p},$$
$$4n - 1 = p - 2 \equiv -2 \pmod{p},$$
$$\vdots$$
$$2n + 1 = p - 2n \equiv -2n \pmod{p}.$$

It follows that

$$(-1)(-2) \cdots (-2n)(2n)(2n - 1) \cdots 3 \cdot 2 \cdot 1 = (2n)!(-1)^{2n} \equiv -1 \pmod{p}.$$

Thus $T = (2n)!$ has $T^2 + 1 \equiv 0$ modulo p, proving the lemma. □

PROOF OF THEOREM 2.6. The case $p = 2$ is trivial. The case when p is congruent to 3 modulo 4 is also dealt with easily; no integer that is congruent to 3 modulo 4 can be the sum of two squares because squares are 0 or 1 modulo 4.

Assume that p is a prime congruent to 1 modulo 4. By Lemma 2.7, we can write

$$cp = T^2 + 1 = (T + \mathrm{i})(T - \mathrm{i}) \text{ in } R = \mathbb{Z}[\mathrm{i}]$$

for some integers T and c.

Suppose (for a contradiction) that p is irreducible in R. Then since $\mathbb{Z}[\mathrm{i}]$ has the Fundamental Theorem of Arithmetic, p is prime. Hence p must divide one of $T \pm \mathrm{i}$ in R since it divides their product, and this is impossible because p does not divide the coefficient of i. It follows that p cannot be irreducible in R, so

$$p = \mu\nu$$

is a product of two non-units in R. Taking the norm of both sides shows that

$$p^2 = N(\mu\nu) = N(\mu)N(\nu).$$

This is an equation in $\mathbb{Z}$, so by the Fundamental Theorem of Arithmetic there are three possibilities.

1. $N(\mu) = 1$ and $N(\nu) = p^2$, which is impossible since μ is not a unit;
2. $N(\nu) = 1$ and $N(\mu) = p^2$, which is impossible since ν is not a unit;
3. $N(\mu) = N(\nu) = p$, which must be the case, and this means there is a nontrivial solution to the equation $x^2 + y^2 = p$.

$\square$

What is being witnessed here is a symbiotic relationship between certain Diophantine equations and the structure of an associated ring. To illustrate this, we now give a theorem that characterizes the primes of $\mathbb{Z}[\mathrm{i}]$.

Theorem 2.8. *The primes of $R = \mathbb{Z}[\mathrm{i}]$ are of three types,*

(1) $1 + \mathrm{i}$,
(2) *integer primes $p \equiv 3$ modulo 4,*
(3) *factors $x \pm \mathrm{i}y$ of the integer primes $p \equiv 1$ modulo 4,*

together with all multiples of these types by units.

Exercise 2.8. Prove Theorem 2.8. (Hint: Show that any prime in $\mathbb{Z}[\mathrm{i}]$ divides a prime in $\mathbb{Z}$.)

Exercise 2.9. Prove that if a prime p is a sum of two squares, $p = a^2 + b^2$, then this representation is unique (apart from the obvious changes).

Exercise 2.10. Prove that the positive integer n is a sum of two squares if and only if every prime p with $p \equiv 3$ modulo 4 that divides n does so to an even exponent.

2.3.1 Lagrange's Four Squares Theorem

One of the many classical results of elementary number theory extends Theorem 2.6 to all integers – at the expense of allowing more squares to be added together. Bachet conjectured the result, and Diophantus stated it; Fermat may have had a proof. The first published proof was that of Lagrange in 1770, which we now present.

Lemma 2.9. *Let p be an odd prime. Then there are integers a and b with*

$$a^2 + b^2 + 1 \equiv 0 \pmod{p}.$$

PROOF. Define the sets

$$A = \left\{ a^2 \mid 0 \leqslant a \leqslant \frac{p-1}{2} \right\}$$

and

$$B = \left\{ -b^2 - 1 \mid 0 \leqslant b \leqslant \frac{p-1}{2} \right\}.$$

No two elements of A are congruent modulo p, and no two elements of B are congruent modulo p. It follows that each of the sets A and B contains $\frac{p+1}{2}$ elements modulo p, so by the pigeonhole principle[1] there must be an element of A that is equal to an element of B modulo p since there are only p distinct integers modulo p. Thus there are integers a and b with

$$a^2 + b^2 + 1 \equiv 0 \pmod{p}$$

as required.   □

Theorem 2.10. [LAGRANGE] *Every positive integer is a sum of four integer squares.*

PROOF. The first step is to note the Euler four-square identity,

$$(a^2 + b^2 + c^2 + d^2)(w^2 + x^2 + y^2 + z^2) = (aw + bx + cy + dz)^2$$
$$+ (ax - bw - cz + dy)^2$$
$$+ (ay + bz - cw - dx)^2$$
$$+ (az - by + cx - dw)^2,$$

which may be proved simply by expanding the right-hand side. This identity means that the property of being written as a sum of four squares is preserved under products. By the Fundamental Theorem of Arithmetic, it is therefore

---

[1] The 'pigeonhole' principle states that if $(Q + 1)$ letters are placed in Q pigeonholes, one pigeonhole must contain more than one letter. It is readily proved by contradiction.

sufficient to prove that any prime is a sum of four integer squares. It is clear that $2 = 1^2 + 1^2 + 0^2 + 0^2$ is a sum of four integer squares, so it is enough to prove that any odd prime is a sum of four integer squares.

Let p be an odd prime. By Lemma 2.9, there are integers a, b, c, d and m with

$$mp = a^2 + b^2 + c^2 + d^2. \tag{2.4}$$

If $m = 1$ then we are done, so assume that $m > 1$. The proof proceeds by finding an expression for $m'p$ as a sum of four squares, with $0 < m' < m$. This can be repeated, reducing the size of m each time, until we eventually must find an expression for the prime p itself as a sum of four squares.

Now notice that if an even integer $2n$ is a sum of two squares, $2n = x^2 + y^2$, then the integers x and y are either both even or both odd. It follows that the identity

$$n = \left(\frac{x+y}{2}\right)^2 + \left(\frac{x-y}{2}\right)^2 \tag{2.5}$$

expresses n as a sum of two integer squares. Returning to Equation (2.4), if m is even, then either none, two, or four of the numbers a, b, c, d are even. Thus we can use Equation (2.5) twice to deduce that $(\frac{m}{2})p$ is a sum of four squares. In this case we have halved the size of m.

If m is odd, write

$$w \equiv a \pmod{m}$$
$$x \equiv b \pmod{m}$$
$$y \equiv c \pmod{m}$$
$$z \equiv d \pmod{m}$$

with $-\frac{m}{2} < w, x, y, z < \frac{m}{2}$. Then

$$w^2 + x^2 + y^2 + z^2 < m^2$$

and

$$w^2 + x^2 + y^2 + z^2 \equiv 0 \pmod{m}.$$

It follows that

$$w^2 + x^2 + y^2 + z^2 = km$$

for some k, $0 < k < m$. Now in Euler's four-square identity

$$(a^2 + b^2 + c^2 + d^2)(w^2 + x^2 + y^2 + z^2) = (aw + bx + cy + dz)^2$$
$$+ (ax - bw - cz + dy)^2$$
$$+ (ay + bz - cw - dx)^2$$
$$+ (az - by + cx - dw)^2 \tag{2.6}$$

the left-hand side is km^2p. By our choice of w, x, y, z we see that $ax \equiv bw$ and $dy \equiv cz$ modulo m, so $(ax - bw - cz + dy)^2$ is divisible by m^2. A similar argument shows that

$$(ay + bz - cw - dx)^2$$

and

$$(az - by + cx - dw)^2$$

are also divisible by m^2. For the first term,

$$aw + bx + cy + dz \equiv w^2 + x^2 + y^2 + z^2 \equiv 0 \pmod{m},$$

so the right-hand side of Equation (2.6) is divisible by m^2. It follows that the identity (2.6) can be divided through by m^2, resulting in an expression for kp as a sum of four squares, with $0 < k < m$.

Repeating this reduction a finite number of times will reduce m to 1, resulting in an expression for the odd prime p as a sum of four squares, completing the proof. $\square$

Exercise 2.11. *[LEGENDRE] Show that every integer not of the form

$$4^n(8k + 7)$$

is a sum of three integer squares.

Exercise 2.12. Suppose a prime p is a sum of four squares. Is it true that the representation is unique? What if p is a sum of three squares?

2.4 Siegel's Theorem

In this section, we show how a direct application of the Fundamental Theorem of Arithmetic in rings that are larger than the integers, for example the Gaussian integers $\mathbb{Z}[i]$, can yield all the integral solutions to certain cubic equations. In the first example, we use the Fundamental Theorem of Arithmetic only in $\mathbb{Z}$.

Theorem 2.11. *The only integral solution of the equation*

$$y^2 = x^3 + x \tag{2.7}$$

is $x = 0, y = 0$.

PROOF. Let x and y be integers with $y^2 = x^3 + x$. Write the right-hand side of the equation as $x^3 + x = x(x^2 + 1)$. Any factor of x will divide x^2, so any factor common to x and $x^2 + 1$ will also divide 1. Thus x and $x^2 + 1$ must be coprime and hence, by the Fundamental Theorem of Arithmetic, both must be squares (since their product is y^2). Writing $z^2 = x^2 + 1$, we see that

$$1 = z^2 - x^2 = (z + x)(z - x).$$

By the Fundamental Theorem of Arithmetic in $\mathbb{Z}$, $(z + x)$ and $(z - x)$ must both be 1 or both be -1.

Solving for x and z shows that $x = 0$ in both cases. $\square$

Theorem 2.12. *The only integral solution of the equation*

$$y^2 = x^3 - 1 \tag{2.8}$$

is $x = 1, y = 0$.

For this equation, it looks as if we should factorize the right-hand side over $\mathbb{Z}$, but it does not seem easy to get to the proof that way. Instead we factorize over a bigger ring that is also known to satisfy the Fundamental Theorem of Arithmetic.

PROOF OF THEOREM 2.12. Rewrite the equation as

$$y^2 + 1 = x^3$$

and then factorize the left-hand side as $(y+\mathrm{i})(y-\mathrm{i})$ in $\mathbb{Z}[\mathrm{i}]$. We claim that the two factors $y \pm \mathrm{i}$ must be coprime. To see why, let $\delta = \gcd(y+\mathrm{i}, y-\mathrm{i})$; δ must divide the difference $y + \mathrm{i} - (y - \mathrm{i}) = 2\mathrm{i}$. However, we claim that no factor of 2 can divide $y \pm \mathrm{i}$. This is because x must be odd; if x is even then $x^3 \equiv 0$ modulo 8, which means that $y^2 + 1 \equiv 0$ modulo 8 and this congruence has no solutions. We deduce that δ must be a unit, and the two factors $y \pm \mathrm{i}$ are coprime in $\mathbb{Z}[\mathrm{i}]$.

Applying the Fundamental Theorem of Arithmetic in $\mathbb{Z}[\mathrm{i}]$, we deduce that each factor $y + \mathrm{i}$, $y - \mathrm{i}$ must be a unit multiple of a cube. Since all units are themselves cubes, we deduce that each of $y \pm \mathrm{i}$ is a cube in $\mathbb{Z}[\mathrm{i}]$, so assume

$$y + \mathrm{i} = (a + b\mathrm{i})^3, a, b \in \mathbb{Z}.$$

Equating imaginary parts gives

$$1 = 3a^2b - b^3 = b(3a^2 - b^2).$$

By the Fundamental Theorem of Arithmetic in $\mathbb{Z}$, the solutions are greatly restricted: $b = (3a^2 - b^2) = \pm 1$. If $b = 1$, then $3a^2 - 1 = 1$, which is impossible as no integer a has $3a^2 = 2$. The only alternative is $b = -1$, in which case $a = 0$, yielding the unique solution $y = 0$ and $x = 1$. $\qquad\square$

Exercise 2.13. Use the preceding method in the ring $\mathbb{Z}[\sqrt{-2}]$ to prove that the only integral solutions of

$$y^2 = x^3 - 2$$

are $x = 3$, $y = \pm 5$.

Later we will be thinking of the set of solutions to equations such as these *geometrically*, so we will describe the solutions as points (x, y) in the plane.

Now consider the example

$$y^2 = x^3 - 3. \tag{2.9}$$

Experimentation with small integers suggests that there will be no integral solutions, but we encounter a difficulty when we try to prove this using the preceding methods. The reason is that the Fundamental Theorem of Arithmetic does not hold in the ring $\mathbb{Z}[\sqrt{-3}]$ (see Exercise 3.17 on p. 73.) On the other hand, the Fundamental Theorem of Arithmetic *does* hold in the bigger ring $\mathbb{Z}[\omega]$, where $\omega = e^{2\pi i/3}$ is a nontrivial cube root of unity.

This suggests that we might try to find all the solutions (x, y) over the ring $\mathbb{Z}[\omega]$ as a precursor to finding all the solutions over the smaller ring $\mathbb{Z}$. This might seem audacious but historically this is just what happened in the general case.

Theorem 2.13. [SIEGEL'S THEOREM] *Suppose $a, b, c \in \mathbb{Q}$. Then there are only finitely many integer pairs (x, y) with*

$$y^2 = x^3 + ax^2 + bx + c, \tag{2.10}$$

provided the cubic polynomial $x^3 + ax^2 + bx + c$ has no repeated zeros.

This theorem will not be proved here – see the notes at the end of the chapter for references where complete proofs may be found. The curve described by an equation of the shape Equation (2.10) is known as an *elliptic curve* provided the right-hand side has no repeated zeros. In order for Siegel's Theorem to hold, some condition about the cubic polynomial is clearly needed because, for example, the equation $y^2 = x^3$ has infinitely many integral solutions. We will devote considerable space to studying the remarkable properties of elliptic curves.

Exercise 2.14. Prove that the polynomial $x^3 + ax + b$ has no repeated zero if and only if $4a^3 + 27b^2 \neq 0$.

The genius of people such as Siegel is that they are willing to take an imaginative step up from particular cases, and are in addition able to supply the guile needed to complete the proof. In fact, he gave two different proofs of Theorem 2.13. In his second proof Siegel showed that there are only finitely many solutions (x, y) with x and y lying inside a suitably large ring containing $\mathbb{Z}$ in which the Fundamental Theorem of Arithmetic holds. The rings in which Siegel proposed to work typically contain infinitely many units, in contrast with the integers $\mathbb{Z}$. We can appreciate some of the technical difficulties he had to overcome by considering the techniques that went into his second proof in some special cases. His second proof turned out to be very important: He first reduced the given equation to a finite number of linear equations over a finitely generated group. Subsequently, methods were developed in Diophantine Approximation that applied to these linear equations and allowed, ultimately, a practical method for finding all the integral solutions of the equation in Theorem 2.13.

Exercise 2.15. Fix a square-free integer $d > 1$, and assume that $\mathbb{Z}[\sqrt{d}]$ satisfies the Fundamental Theorem of Arithmetic. Show that the equation

$$y^2 = x^3 + d$$

has only finitely many integral solutions. (Hint: You may assume that the units of the ring $\mathbb{Z}[\sqrt{d}]$ are all of the form $\pm u^n$ for some unit $u > 1$.)

The rings Siegel worked with are obtained by inverting certain chosen primes. This technique provides us with a new class of rings to study. As a simple illustrative example, let S denote the set $\{2\}$ and let $\mathbb{Z}_S$ denote the ring $\mathbb{Z}[\frac{1}{2}]$ consisting of all rational numbers with a denominator consisting of a power of 2. Given any nonzero $q \in \mathbb{Q}$, write $q = 2^r q'$, where $r \in \mathbb{Z}$ and the numerator and denominator of q' are odd. Define the S-norm of q to be $|q|_S = |q'|$. The ring R has infinitely many units, consisting of the rational numbers $\pm 2^k$ for $k \in \mathbb{Z}$. The ring R is sometimes called the ring of S-integers of $\mathbb{Z}$, and its units are known as S-units.

Exercise 2.16. Prove that the ring $\mathbb{Z}_S$ is a Euclidean ring with respect to $|.|_S$.

The next exercise will provide a further illustration of some of the techniques needed to prove Siegel's Theorem. We have already seen examples where the Fundamental Theorem of Arithmetic fails in some quadratic rings. We overcame that failure in $\mathbb{Z}[\sqrt{-3}]$ by working in the bigger ring $R = \mathbb{Z}[\omega]$, where ω is a nontrivial cube root of unity. Letting $S = \{2\}$ as before, R is a subring of an even bigger ring $R_S = \mathbb{Z}[\sqrt{-3}, \frac{1}{2}]$.

Exercise 2.17. Define a norm function on $R_S = \mathbb{Z}[\sqrt{-3}, \frac{1}{2}]$ with the property that R_S is a Euclidean ring. Find all solutions to Equation (2.9) in the ring R_S. Again, this exercise shows there are only finitely many solutions to a specific cubic equation in a ring with infinitely many units.

Theorem 2.14 below is quite deep and we will not prove it. The proof requires Theorem 4.14 from Chapter 3. The notes at the end of the chapter reference a proof in the literature. It shows that the Fundamental Theorem of Arithmetic in $\mathbb{Z}[\sqrt{d}]$ can be recovered by inverting a finite list of primes.

Theorem 2.14. *Let d be a nonsquare integer. There is a finite list of primes*

$$p_1, \ldots, p_r$$

with the property that $\mathbb{Z}[\sqrt{d}, \frac{1}{p_1}, \ldots, \frac{1}{p_r}]$ has the Fundamental Theorem of Arithmetic.

Combining the techniques learned thus far allows a special case of Siegel's Theorem to be proved. An integer is called *square-free* if it is not divisible by the square of any integer greater than 1.

Exercise 2.18. Suppose $d < 0$ is a square-free integer with the property that $\mathbb{Z}[\sqrt{d}, \frac{1}{p}]$ has the Fundamental Theorem of Arithmetic for some prime p. Show that the equation

$$y^2 = x^3 + d \qquad (2.11)$$

has only finitely many integral solutions.

When explicit approaches such as this succeed, they allow the determination of all the integral solutions. Determining all the integral solutions predicted by Siegel's Theorem is generally quite a difficult problem and requires powerful methods from transcendence theory. It was not until late in the twentieth century that these methods were sufficiently well advanced to allow for a practical method of solving a given equation.

An *S-unit equation* is one of the form

$$a_1 x_1 + \cdots + a_n x_n = 1$$

with a_i fixed constants in some field $\mathbb{K}$, and the solutions x_i are sought in a *finitely generated* subgroup of $\mathbb{K}^*$. For the cubic equations studied here, Siegel reduced the problem of finding all the integral solutions to finding the solutions of a finite number of S-unit equations all having $n = 2$. He then showed that such an equation has only finitely many solutions. In general S-unit equations turn out to lie behind many other Diophantine equations and they have come to be studied as important in their own right.

2.5 Fermat, Catalan, and Euler

Finally we mention three famous Diophantine problems, all of which have recently been solved. There are detailed references in the notes at the end of the chapter.

2.5.1 Fermat

Fermat's Last Theorem, now proved by Wiles, states that the equation

$$x^n + y^n = z^n, \quad n \geqslant 3, \qquad (2.12)$$

has no nontrivial solutions. (A solution is trivial if one of x, y or z is zero.) Clearly, it is only necessary to prove this in the case when $n = p$ is a prime. A startling aspect of the solution is that it depends on deep results concerning the arithmetic of elliptic curves: If $a^p + b^p + c^p = 0$ for a prime p and integers a, b, c, then the elliptic curve with equation

$$y^2 = x(x - a^p)(x + b^p)$$

turns out to have properties that Wiles was able to show were impossible. We will be studying the arithmetic of elliptic curves in Chapters 5 and 6.

Exercise 2.19. *Prove that Equation (2.12) has no nontrivial solutions with n equal to $3, 4$, or 5.

Exercise 2.20. *Prove that Equation (2.12) has no nontrivial solutions in Gaussian integers with $n = 4$.

Exercise 2.21. *Prove that Equation (2.12) has no solutions x, y, z in positive integers with n a Gaussian integer.

2.5.2 Catalan

The Catalan equation is

$$u^x - v^y = 1, \quad u, v, x, y \in \mathbb{N}, \quad u, v, x, y \geqslant 2. \tag{2.13}$$

A solution is $3^2 - 2^3 = 1$; the Catalan problem is to show that there are no others, and this has recently been proved.

2.5.3 Euler

Euler conjectured that an nth power cannot be written as the sum of fewer than n nontrivial nth powers for $n \geqslant 3$. Lander and Parkin made a computer search for nontrivial solutions to the Diophantine equation

$$\sum_{i=1}^{n} x_i^5 = y^5, \quad n \leqslant 6.$$

Among the solutions, they found a counterexample to Euler's conjecture for $n = 5$. Their resulting announcement matches the famous seminar of Cole described on p. 27 for its brevity and drama: The entire text of their paper is as follows.

"A direct search on the CDC 6600 yielded

$$27^5 + 84^5 + 110^5 + 133^5 = 144^5$$

as the smallest instance in which four fifth powers sum to a fifth power. This is a counterexample to a conjecture by Euler [see L. E. Dickson, History of the theory of numbers, Vol. 1, p. 648, Chelsea, New York, 1952] that at least n nth powers are required to sum to an nth power, $n > 2$."

In addition, it was shown that the case $n = 4$, namely the Diophantine equation

$$u^4 + v^4 + w^4 = x^4, \tag{2.14}$$

has no solutions in positive integers with $x < 220000$.

In a dramatic development, Elkies used a mixture of sophisticated theory and a computer search to find a solution to Equation (2.14),

$$2682440^4 + 15365639^4 + 18796760^4 = 20615673^4. \qquad (2.15)$$

Following this, Roger Frye found that the minimal solution to Equation (2.14) is

$$95800^4 + 217519^4 + 414560^4 = 422481^4$$

and showed that there are no other solutions with $u \leqslant v \leqslant w < x < 1000000$.

NOTES TO CHAPTER 2: Much of the material in this chapter is part of algebraic number theory. Stewart's book [147] is an accessible introduction at this level; for more advanced treatments, see the books of Hasse [76], Janusz [83] or Lang [96]. A sophisticated text on related topics is Serre's classic book [137]. Barbeau's book [10] discusses Pell's equation in detail and requires very little background. A proof of Theorem 2.14 can be found in Lang [96, Chapter I, Proposition 17]. The seminal finiteness results on S-unit equations mentioned at the end of Section 2.4 may be found in the papers of Evertse [60], Schlickewei [134], and van der Poorten and Schlickewei [120]. These results have found wide application; a surprising connection to ergodic theory is shown in a paper of Schmidt and Ward [135]. For attractive accounts of Fermat's Last Theorem, see the popular accounts of Ribenboim [126] and van der Poorten [119]; a serious introduction at a high level to the mathematics behind Wiles' extraordinary proof [162] may be found in the proceedings [35] of an instructional conference edited by Cornell, Silverman and Stevens. Exercise 2.20 comes from a short note by Cross [38]; Exercise 2.21 comes from a paper of Zuehlke [169] and uses some transcendence theory. The Catalan problem Equation (2.13) was initially reduced to a finite calculation and then solved completely by Mihăilescu; see the paper of Metsänkylä [106] for an account and the monograph [58, p. 159] by Everest, van der Poorten, Shparlinski and Ward for an overview of related questions. An accessible account of the Catalan problem before its final solution may be found in the book of Ribenboim [124]. The results of Lander and Parkin appeared in their paper [92]; their dramatic announcement quoted in Section 2.5.3 is [91]. The state of Euler's problem in 1967 is surveyed in a paper of Lander, Parkin and Selfridge [93]. Equation (2.15) of Elkies is in [49].

3

Quadratic Diophantine Equations

Attempts to go beyond the Pythagorean Diophantine equation quickly lead to general questions about quadratic Diophantine problems. Apparently simple questions seem to require an excursion into the theory of finite fields. For example, we prove that any finite field has a primitive root in order to develop the classical theory of the Legendre symbol and the Quadratic Reciprocity Law. Some general theory of quadratic rings and quadratic forms is established, up to the finiteness of the class number for quadratic forms.

3.1 Quadratic Congruences

Suppose we now seek to generalize our earlier results and understand the Diophantine equation

$$x^2 + 2y^2 = p \tag{3.1}$$

when p is a prime and x and y are integers. We can do this by using properties of the ring $\mathbb{Z}[\sqrt{-2}]$, but we also need a better understanding of the arithmetic of the integers modulo p when p is a prime.

Exercise 3.1. Let $R = \mathbb{Z}[\sqrt{-2}]$.
(a) Show that the function $N : R \to \mathbb{N}$ defined by

$$N(x + y\sqrt{-2}) = x^2 + 2y^2$$

satisfies $N(\alpha\beta) = N(\alpha)N(\beta)$ for all $\alpha, \beta \in R$.
(b) Determine all the units in R.
(c) Show that R is Euclidean with respect to N.

Following our earlier method, we now expect to use unique factorization in R together with some knowledge of congruences to understand Equation (3.1). The relevant congruence to study for this equation is

$$T^2 + 2 \equiv 0 \pmod{p}. \tag{3.2}$$

Exercise 3.2. Compute the list of primes $p < 1000$ for which the congruence (3.2) has a solution with $T \in \mathbb{Z}$.

It is becoming clear that we need some tool that will guarantee the existence of a solution for certain congruences and rule out a solution for others. For example, your computations in Exercise 3.2 should suggest that for primes $p \equiv 1$ or 3 modulo 8 there is a solution, while there is no solution for primes $p \equiv 5$ or 7 modulo 8. (The prime $p = 2$ does give a solution.) Our earlier approach suggests that the area we need to look at is the arithmetic of $\mathbb{Z}/p\mathbb{Z}$. Previously we used al-Haytham's Theorem in a crucial way, and here we have no obvious analog. It turns out that the property we need is directly related to a natural concept in group theory.

Definition 3.1. *An element a of $\mathbb{Z}/p\mathbb{Z}$ is a* primitive root modulo p *if the powers of a generate all the nonzero residues modulo p.*

Example 3.2. It is easy to prove that the powers of 2 yield all the nonzero residues modulo 5: $2^0 \equiv 1$, $2^1 \equiv 2$, $2^2 \equiv 4$, $2^3 \equiv 3$ modulo 5. Thus 2 is a primitive root modulo 5. Similarly, 3 is a primitive root modulo 7, but 2 is not since no power of 2 is congruent to 3 modulo 7.

The set of residues modulo p forms a field: The existence of a primitive root a modulo p is the same as the statement that the multiplicative group $(\mathbb{Z}/p\mathbb{Z})^*$ of the field $\mathbb{Z}/p\mathbb{Z}$ is cyclic, generated by a. We will use freely other equivalent ways of saying this. If G denotes a finite Abelian group with n elements, written multiplicatively, then a generates G if and only if any of the following equivalent conditions hold:

1. $a^m = 1, 1 < m \leqslant n \implies m = n$;
2. the order of a is n;
3. $a^m = 1, 1 < m \implies n|m$.

Theorem 3.3. *The multiplicative group of any finite field is cyclic.*

This is an important result, and we will spend some time proving it. When we have done this, we can return to our equations. The proof of Theorem 3.3 involves an important example of an arithmetic function.

Definition 3.4. *An* arithmetic function *is any function* $\mathrm{f} : \mathbb{N} \to \mathbb{C}$. *An arithmetic function with* $\mathrm{f}(1) \neq 0$ *and*

$$\mathrm{f}(mn) = \mathrm{f}(m)\mathrm{f}(n)$$

whenever m and n are coprime is called multiplicative. *(Note that this implies* $\mathrm{f}(1) = 1$.) *If f has this property not only for coprime m, n, but for all $m, n \in \mathbb{N}$, then f is called* completely multiplicative.

Multiplicative arithmetic functions will be discussed further in Section 8.2. One of the most important arithmetic functions is

$$\phi(n) = |\{1 \leqslant a \leqslant n \mid \gcd(a, n) = 1\}|,$$

called the *Euler phi-function*.

Exercise 3.3. Let p be a prime. Show that $\phi(p^e) = p^{e-1}(p-1)$ for any $e \geqslant 1$.

Lemma 3.5. *The Euler phi-function is multiplicative.*

We will postpone the proof slightly to note an immediate corollary of Lemma 3.5 and Exercise 3.3.

Corollary 3.6. *If n is factorized into powers of distinct primes, $n = \prod_p p^{e_p}$, then*

$$\phi(n) = \prod_{p|n} (p-1)p^{e_p-1} = n \prod_{p|n} \frac{p-1}{p}.$$

Exercise 3.4. Give an example to show that ϕ is not completely multiplicative.

Exercise 3.5. (a) Find all values of $n \in \mathbb{N}$ with $\phi(n) = \frac{1}{2}n$.
(b) Find all values of $n \in \mathbb{N}$ with $\phi(n) = \phi(2n)$.
(c) Find all six values of $n \in \mathbb{N}$ with $\phi(n) = 12$.
(d) Find the smallest $n \in \mathbb{N}$ for which $\frac{\phi(n)}{n} < \frac{1}{4}$.
(e) Find a sequence of integers (n_j) for which $\frac{\phi(n_j)}{n_j} \to 0$ as $j \to \infty$.

The proof of Lemma 3.5 depends on the following result.

Theorem 3.7. [CHINESE REMAINDER THEOREM] *Suppose $m, n \in \mathbb{N}$ are co-prime. Then the simultaneous congruences*

$$x \equiv a \pmod{m},$$
$$x \equiv b \pmod{n},$$

have a solution $x \in \mathbb{N}$ for any $a, b \in \mathbb{Z}$, and the solution is unique modulo mn.

The Chinese Remainder Theorem was discovered by Chinese mathematicians in the fourth century A.D. The first appearance seems to have been in a work of Sun-Zi, and a general treatment was given by Qin[1] Jiushao. Special

---

[1] Also transliterated as Ch'in Chiu-Shao. Jiushao seems to have been both a rogue and a mathematical genius. His work *Shushu Jiuzhang (Mathematical Treatise in Nine Sections)* appeared in 1247 and contained many important and novel results and methods. The so-called Chinese Remainder Theorem is among these, attributed to experts in astronomy and calenders. It has been suggested that the theorem does not bear his name because in the form 'Chin' it was too easily confused with 'Chinese'.

results of the same sort were used by Fibonacci in Italy and al-Haytham in Iraq. We will see it again in Chapter 12 (see p. 256) in greater generality.

PROOF OF THE CHINESE REMAINDER THEOREM. The coprimality condition guarantees that there exist m', n' such that

$$mm' = 1 \pmod{n} \text{ and } nn' = 1 \pmod{m} \tag{3.3}$$

by Corollary 1.25. Then $x = bmm' + ann'$ satisfies both the required congruences.

If, on the other hand, x and y satisfy both congruences, then $(x - y)$ is divisible by m and by n. Since m and n are coprime, $(x - y)$ must be divisible by mn.   □

Example 3.8. Solve the simultaneous congruences $x \equiv 2$ modulo 17 and $x \equiv 8$ modulo 11. We find $m' = 2$ and $n' = 14$ in the proof of the Chinese Remainder Theorem. Then

$$x = 8 \cdot (17 \cdot 2) + 2 \cdot (11 \cdot 14) = 580$$

satisfies the two congruences. (The smallest solution is the remainder of 580 divided by $11 \cdot 17$, namely 19.)

PROOF OF LEMMA 3.5. Let m and n be coprime. Define a map

$$\Phi : \mathbb{Z}/mn\mathbb{Z} \to \mathbb{Z}/m\mathbb{Z} \times \mathbb{Z}/n\mathbb{Z}$$

by

$$x \mapsto (x \pmod{m}, x \pmod{n}),$$

where we think of the elements of $\mathbb{Z}/mn\mathbb{Z}$ as $\{0, 1, \ldots, mn-1\}$. By the Chinese Remainder Theorem, Φ is a bijection. (In fact, Φ is an isomorphism of rings.) Now define

$$(\mathbb{Z}/n\mathbb{Z})^* = \{1 \leqslant a \leqslant n : \gcd(a, n) = 1\}$$

and likewise for n and mn. Since x is coprime to mn if and only if it is coprime both to m and n, Φ restricts to these subsets:

$$\Phi : (\mathbb{Z}/mn\mathbb{Z})^* \to (\mathbb{Z}/m\mathbb{Z})^* \times (\mathbb{Z}/n\mathbb{Z})^*.$$

Here Φ is still a bijection. (In fact, the set $(\mathbb{Z}/k\mathbb{Z})^*$ is the set of units $U(\mathbb{Z}/k\mathbb{Z})$ of $\mathbb{Z}/k\mathbb{Z}$, and Φ is an isomorphism of (multiplicative) groups.) By definition, the cardinality of $(\mathbb{Z}/m\mathbb{Z})^*$ is just $\phi(m)$ and likewise for n and mn, which completes the proof of Lemma 3.5.   □

The next exercise is a generalization of Fermat's Little Theorem (Theorem 1.12), called the Euler–Fermat Theorem.

Exercise 3.6. Given $n > 1$ in $\mathbb{N}$, show that for any $a \in \mathbb{Z}$ with $\gcd(a, n) = 1$

$$a^{\phi(n)} \equiv 1 \pmod{n}. \tag{3.4}$$

Exercise 3.6 is a pretty standard one found in most texts that deal with the ϕ-function. It is a good test case for our earlier remarks about how different approaches can yield different benefits. It is possible to prove Equation (3.4) using congruences modulo p^r for each prime power p^r dividing n, together with the Binomial Theorem. Another, slicker, proof simply uses Lagrange's Theorem on the group $U(\mathbb{Z}/n\mathbb{Z}) = (\mathbb{Z}/n\mathbb{Z})^*$.

Theorem 3.9. *For any* $n \in \mathbb{N}$,

$$\sum_{d|n} \phi(d) = n. \tag{3.5}$$

PROOF. First check the equality when $n = p^r$ is a prime power. The left-hand side is

$$1 + \sum_{i=1}^{r}(p-1)p^{i-1} = 1 + (p^r - 1) = n$$

by summing the geometric progression or noticing that it is a telescoping sum. Next, observe that both sides of Equation (3.5) are multiplicative arithmetic functions. For the left-hand side, this follows from

$$\sum_{d|mn} \phi(d) = \sum_{d_1|m}\sum_{d_2|n} \phi(d_1 d_2) = \sum_{d_1|m} \phi(d_1)\sum_{d_2|n}\phi(d_2)$$

for any pair of coprime integers (m, n). Note that d divides mn if and only if there exist divisors d_1 of m and d_2 of n such that $d = d_1 d_2$, so it is enough to check the prime power case. $\qquad\square$

We can now prove Theorem 3.3. In the proof, we will be working with a general finite field. Such a field can always be explicitly presented using polynomials; however, nowhere will we need an explicit presentation. This suggests that more abstract methods might also be applicable to prove the theorem. Indeed, a proof can be given that only uses the theory of finite Abelian groups.

PROOF OF THEOREM 3.3. Let $\mathbb{F}$ be a finite field with q elements. We are going to prove that if g is any element of $\mathbb{F}^*$, then g^j has the same order as g if and only if $\gcd(j, q-1) = 1$. This will allow us to find how many elements there are of each order, showing in particular that there are $\phi(q-1)$ distinct generators in total.

Example 3.10. The distinct powers of 3 in $\mathbb{F}_7^*$ are

$$3^0 \equiv 1, 3^1 \equiv 3, 3^2 \equiv 2, 3^3 \equiv 6, 3^4 \equiv 4, 3^5 \equiv 5.$$

The only values of j, $1 \leqslant j \leqslant 6$ with $\gcd(j, 6) = 1$ are 1 and 5. Since $3^5 \equiv 5$ modulo 7, 5 is another generator of $\mathbb{F}_7^*$. Similarly, $\mathbb{F}_{11}^* = \langle 2 \rangle$ (the multiplicative group generated by 2). The values of j between 1 and 10 for which $\gcd(j, 10) = 1$ are $j = 1, 3, 7, 9$ so there are four possibilities for generators of $\mathbb{F}_{11}^*$, namely $2^1 \equiv 2, 2^3 \equiv 8, 2^7 \equiv 7, 2^9 \equiv 6$.

Exercise 3.7. Prove that in any field, a polynomial of degree d has no more than d zeros. (Hint: Use Exercise 2.5 on p. 47).

Returning to the proof of Theorem 3.3, suppose $d|(q-1)$ and a is an element of $\mathbb{F}^*$ of order d (if one exists). Then

$$a^d = 1 \text{ in } \mathbb{F} \text{ and } a^m = 1 \text{ with } 0 \leqslant m < d \text{ implies } m = 0.$$

The elements $1, a, a^2, \cdots, a^{d-1}$ are all distinct, otherwise $a^i = a^j$ would imply that $a^k = 1$ with some $0 < k < d$. We claim that if an element a of order d exists, then the other elements of order d in $\mathbb{F}^*$ are precisely those powers a^j with $1 \leqslant j < d$ and $\gcd(j,d) = 1$. Thus if there is an element of order d, then there will be precisely $\phi(d)$ of them. If a does have order d, then the only other elements of order d must lie among the powers a^j above since any element of order d satisfies the equation

$$x^d - 1 = 0$$

in $\mathbb{F}$, this equation has at most d roots by Exercise 3.7, and each of the powers $a^j, 0 \leqslant j < d$ satisfies the equation. Thus all the elements of order d must lie among these powers. But which of the powers have order d? We now prove our claim that a^j has order d, $1 \leqslant j < d$, if and only if $\gcd(j,d) = 1$.

If $1 < \gcd(j,d) = d' < d$ then $1 < d/d' < d$ and

$$(a^j)^{d/d'} = (a^d)^{j/d'} = 1^{j/d'} = 1,$$

so a^j does not have order d (since $d/d' < d$).

Conversely, suppose that $\gcd(j,d) = 1$ and a^j has order d'' with

$$1 < d'' \leqslant d.$$

Then $a^{jd''} = 1$, so $d|jd''$ since a has order d. However, $\gcd(d,j) = 1$, which forces $d|d''$. On the other hand, $d'' \leqslant d$, so we must have $d = d''$. This completes the proof that a^j has order d if and only if $\gcd(j,d) = 1$.

Each of the $(q-1)$ elements of $\mathbb{F}^*$ has order dividing $(q-1)$, so by Theorem 3.9,

$$\sum_{d|(q-1)} \phi(d) = q - 1.$$

Thus, for every d dividing $(q-1)$, we must have $\phi(d)$ elements (not none) of order d. In particular, we have $\phi(q-1) \geqslant 1$ elements of order $(q-1)$.   □

Notice that we have proved a little more than Theorem 3.3. The proof shows how many elements of $\mathbb{F}_q$ there are of each possible order, finding in particular that there must be at least one element of order $(q-1)$, which is therefore a primitive root.

Exercise 3.8. Verify that 2 is a primitive root for the prime $p = 19$. Find all the elements of order 6 under multiplication modulo 19, expressed as integers between 1 and 18.

Despite the seemingly complete knowledge provided by the proof of Theorem 3.3, several closely related questions turn out to be extremely difficult. The following is a famous conjecture of Artin which remains an open problem.

Conjecture 3.11. [ARTIN] Any integer that is not a square or -1 is a primitive root modulo p for infinitely many primes p.

An apparently less ambitious question is to ask, given an explicitly presented finite field, whether there is an algorithm for determining a primitive root. For example, if p is a given prime, can we determine a primitive root for p? The most obvious thing to try is checking the integers $2, 3, 5, 6 \ldots$ (not 4 of course!) in the hope that a primitive root will soon be found. Thus one seeks an upper bound on the smallest primitive root, and this too is difficult. The smallest primitive root modulo p can be shown – conditionally – to be bounded by a constant multiple of $(\log p)^6$, a result of Shoup from 1992. However this result relies upon a hard unproven hypothesis stated in Section 12.7.1. This might not sound very satisfactory, but it turns out to have great practical value.

3.2 Euler's Criterion

Many problems concerning quadratic congruences can be reduced to solving the simplest such congruence, namely $x^2 \equiv a$ modulo p for a prime p and given a.

Definition 3.12. *Let p be an odd prime and a an integer. The Legendre symbol is defined by*

$$\left(\frac{a}{p}\right) = \begin{cases} 0 & \text{if } p \mid a, \\ 1 & \text{if } p \nmid a \text{ and } x^2 \equiv a \pmod{p} \text{ has a solution,} \\ -1 & \text{otherwise.} \end{cases}$$

If $a \neq 0$ and $\left(\frac{a}{p}\right) = -1$ then a is a *quadratic nonresidue* modulo p; otherwise a is a *quadratic residue* modulo p.

Some elementary properties of the Legendre symbol will be used without comment. In particular, if $a \equiv b$ modulo p, then $\left(\frac{a}{p}\right) = \left(\frac{b}{p}\right)$ and $\left(\frac{a^2}{p}\right) = 1$ for any $a \neq 0$.

Theorem 3.13. [EULER'S CRITERION] *Let p be an odd prime. Then*

$$\left(\frac{a}{p}\right) \equiv a^{(p-1)/2} \pmod{p}. \qquad (3.6)$$

PROOF. The statement is obvious if $a \equiv 0$ modulo p, so assume that a is coprime to p. Notice that the only square roots of 1 modulo p are congruent to ± 1 since $x^2 - 1 = (x - 1)(x + 1)$ in any field.

Now

$$(a^{(p-1)/2})^2 = a^{p-1} \equiv 1 \pmod{p},$$

so

$$a^{(p-1)/2} \equiv \pm 1 \pmod{p}.$$

Let g denote a generator of the cyclic group $(\mathbb{Z}/p\mathbb{Z})^*$. Then $a \equiv g^j$ modulo p for some j, and a is a quadratic residue if and only if j is even. Suppose a is a quadratic residue, so $j = 2j'$ for some integer j'. It follows that

$$a^{(p-1)/2} \equiv (g^j)^{(p-1)/2} = g^{j'(p-1)} = (g^{p-1})^{j'} \equiv 1 \pmod{p}.$$

Thus $\left(\frac{a}{p}\right) = 1$ implies that $a^{(p-1)/2} \equiv 1$ modulo p.

Conversely, if $a^{(p-1)/2} \equiv 1$ modulo p, then $g^{j(p-1)/2} \equiv 1$ modulo p. However, g has order $(p - 1)$ modulo p, so

$$(p - 1) \mid j(p - 1)/2, \text{ which implies } 2(p - 1) \mid j(p - 1).$$

Canceling $(p - 1)$ from both sides shows that j is even. Thus $a^{(p-1)/2} \equiv 1$ modulo p implies that $\left(\frac{a}{p}\right) = 1$.   □

Corollary 3.14. *The Legendre symbol satisfies*

$$\left(\frac{ab}{p}\right) = \left(\frac{a}{p}\right)\left(\frac{b}{p}\right).$$

That is, the Legendre symbol viewed as an arithmetic function

$$\left(\frac{\cdot}{p}\right) : \mathbb{Z} \to \{0, \pm 1\}$$

is *completely multiplicative.*

The proof follows immediately from Theorem 3.13 because the right-hand side of Equation (3.6) is completely multiplicative.

Exercise 3.9. Suppose that $p, q > 0$ are odd primes with $q = 4p + 1$. Prove that 2 is a primitive root modulo q. It follows that Artin's conjecture (on p. 65) for $a = 2$ would be proved if we knew there are infinitely many primes q of the form $4p + 1$ where p is a prime.

Exercise 3.10. Prove Corollary 3.14 using concepts from group theory. (Hint: The set of squares in the group $G = (\mathbb{Z}/p\mathbb{Z})^*$ forms a subgroup. The index of this subgroup in G is of order 2 if p is odd; see Exercise 3.12 below.)

3.3 The Quadratic Reciprocity Law

The main result on quadratic residues is a *reciprocity law*. Gauss did many calculations with quadratic residues and in particular studied whether there might be a relation between p being a quadratic residue modulo q and q being a quadratic residue modulo p when p and q are primes. Based on his extensive calculations, he conjectured and then proved (in several ways) the following: When one of p or q is congruent to 1 modulo 4, either both of the congruences

$$x^2 \equiv q \pmod p, \quad y^2 \equiv p \pmod q,$$

are solvable or both are not. If both p and q are congruent to 3 modulo 4, then one is solvable if and only if the other is not. This surprising result is of great importance.

Theorem 3.15. *Let p and q denote odd primes. If $p \equiv q \equiv 3$ modulo 4, then*

$$\left(\frac{q}{p}\right) = -\left(\frac{p}{q}\right).$$

If at least one of p or q is 1 modulo 4, then the symbols are equal.

Theorem 3.15 can be stated as a neater formula, and this is what we will prove. If p and q are odd primes, then

$$\left(\frac{q}{p}\right) = (-1)^{(p-1)/2 \cdot (q-1)/2} \left(\frac{p}{q}\right). \tag{3.7}$$

The even prime 2 has to be treated separately: The theorem below will be proved on p. 68.

Theorem 3.16. *If p is an odd prime, then*

$$\left(\frac{2}{p}\right) = 1 \text{ if and only if } p \equiv \pm 1 \pmod 8.$$

Exercise 3.11. (a) Show that Theorem 3.16 can be written in the form

$$\left(\frac{2}{p}\right) = (-1)^{(p^2-1)/8}$$

for an odd prime p.
(b) Prove that $\left(\frac{-1}{p}\right) = (-1)^{(p-1)/2}$.

Exercise 3.12. A Diophantine equation with solutions in $\mathbb{Z}$ must have solutions modulo p (that is, in $\mathbb{Z}/p\mathbb{Z}$) for all primes p.
(a) Show that the converse does not hold by proving that

$$(x^2 - 2)(x^2 - 3)(x^2 - 6) = 0$$

has a solution modulo p for every prime p but no integral solution.
(b) Show that $x^8 - 16 = 0$ has a solution modulo p for every prime p but no integral solution.

Exercise 3.13. (a) Show that Equation (3.7) is equivalent to Theorem 3.15.
(b) Show that if p is an odd prime, then

$$\left(\frac{-2}{p}\right) = 1 \text{ if and only if } p \equiv 1 \text{ or } 3 \pmod 8.$$

(c) Use the arithmetic of $\mathbb{Z}[\sqrt{-2}]$ to show that the prime p can be written

$$p = x^2 + 2y^2, \text{ with } x, y \in \mathbb{Z},$$

if and only if $p \equiv 1$ or 3 modulo 8.

Exercise 3.14. (a) Show that if $p > 3$ is a prime, then

$$\left(\frac{-3}{p}\right) = 1 \text{ if and only if } p \equiv 1 \pmod 3.$$

(b) Show that the map $x \mapsto x^3 + 2$ is a bijection on $\mathbb{Z}/p\mathbb{Z}$ for any odd prime p congruent to 2 modulo 3. Deduce that the equation $(x^2 + 3)(x^3 + 2) = 0$ has a solution modulo q for any prime q but has no integral solutions.

Exercise 3.15. *Show that a monic polynomial $f \in \mathbb{Z}[x]$ of degree 4 or less that has a solution modulo q for every prime q has an integral solution.

The proof of Theorem 3.16 acts as a dummy run for the proof of Theorem 3.15. The proofs given here are due to Serre.

PROOF OF THEOREM 3.16. The prime p is odd, so $p^2 - 1 \equiv 0$ modulo 8. Let $\mathbb{F}$ denote the field with p^2 elements. Then $\mathbb{F}^*$ is a cyclic group of order $p^2 - 1$ by Theorem 3.3. Since $p^2 - 1$ is divisible by 8, this implies that $\mathbb{F}^*$ contains an element of order 8. Let ζ denote such an element. Let

$$G = \zeta - \zeta^3 - \zeta^5 + \zeta^7. \tag{3.8}$$

Now $(\zeta^4)^2 = \zeta^8 = 1$, so $\zeta^4 = -1$ (ζ has order 8, so we cannot have $\zeta^4 = 1$) and therefore

$$\zeta^5 = -\zeta \text{ and } \zeta^7 = -\zeta^3.$$

Therefore

$$G = 2(\zeta - \zeta^3),$$

so $G^2 = 4(\zeta - \zeta^3)^2 = 4(\zeta^2 + \zeta^6 - 2\zeta^4)$. But $\zeta^4 + 1 = 0$ implies that $\zeta^6 + \zeta^2 = 0$. Therefore

$$G^2 = 8.$$

Recall that we are working in the field $\mathbb{F}$ so that 8 denotes not only the integer 8 but also the sum $1_\mathbb{F} + \cdots + 1_\mathbb{F}$ (seven additions), where $1_\mathbb{F}$ is the multiplicative identity in $\mathbb{F}^*$.

The proof of the theorem depends on finding two distinct expressions for G^p.

First expression for G^p:

$$
\begin{aligned}
G^p &= GG^{p-1} \\
&= G(G^2)^{(p-1)/2} \\
&= G8^{(p-1)/2} \text{ because } G^2 = 8 \\
&= G\left(\frac{8}{p}\right) \text{ by Euler's criterion} \\
&= G\left(\frac{2}{p}\right) \text{ by Corollary 3.14.}
\end{aligned}
$$

Second expression for G^p:

Define a function $f : \mathbb{Z} \to \{0, \pm 1\}$ to be 0 when j is even and $(-1)^{(j^2-1)/8}$ when j is odd. Notice that

$$f(j) = 1 \text{ if and only if } j \equiv \pm 1 \pmod 8.$$

The second expression for G^p is

$$G^p = f(p)G. \tag{3.9}$$

Equate the two expressions for G^p to obtain

$$G\left(\frac{2}{p}\right) = f(p)G.$$

Now G is not zero in $\mathbb{F}$ (because $G^2 = 8$), so cancelling gives

$$\left(\frac{2}{p}\right) = f(p) = 1 \text{ if and only if } p \equiv \pm 1 \pmod 8.$$

The field $\mathbb{F}$ has characteristic p, so $(a + b)^p = a^p + b^p$ in $\mathbb{F}$ because all binomial coefficients apart from the end ones are divisible by p. (A similar argument was used in the proof of Fermat's Little Theorem on p. 24.) Similarly, by induction,

$$(a_1 + \cdots + a_n)^p = a_1^p + \cdots + a_n^p.$$

Using Equation (3.8) and the definition of f,

$$
\begin{aligned}
G &= f(1)\zeta + f(3)\zeta^3 + f(5)\zeta^5 + f(7)\zeta^7 \\
&= f(0) + f(1)\zeta + f(2)\zeta^2 + \cdots + f(7)\zeta^7 \\
&= \sum_{j=0}^{7} f(j)\zeta^j.
\end{aligned}
$$

Thus

$$G^p = \left(\sum_{j=0}^{7} f(j)\zeta^j\right)^p = \sum_{j=0}^{7} f(j)\zeta^{jp}. \tag{3.10}$$

Note that $f(j)$ does not need to be raised to the power p because $f(j)^p = f(j)$.

Lemma 3.17. *For all $j \in \mathbb{Z}$, $f(p)f(jp) = f(j)$.*

Assuming this lemma for the moment, Equation (3.10) gives

$$G^p = \sum_{j=0}^{7} f(j)\zeta^{jp} = \sum_{j=0}^{7} f(p)f(jp)\zeta^{jp} = f(p)\sum_{j=0}^{7} f(jp)\zeta^{jp}.$$

Now, for fixed p, jp modulo 8 runs through $0,\ldots,7$ as j does, so this shows that $G^p = f(p)G$, proving Equation (3.9).

All we need to do now is prove Lemma 3.17, which states that

$$f(p)f(pj) = f(j).$$

Clearly, this is true if j is even, so suppose that j is odd. The statement is true for any odd pair j and p. This can be checked by examining all the possibilities for j and p modulo 16. Alternatively, notice that

$$(-1)^{((jp)^2-1)/8} = (-1)^{((jp)^2-p^2+p^2-1)/8}$$
$$= ((-1)^{p^2})^{(j^2-1)/8}(-1)^{(p^2-1)/8}$$
$$= (-1)^{(j^2-1)/8}(-1)^{(p^2-1)/8}.$$

This shows that $f(jp) = f(j)f(p)$, and Lemma 3.17 follows by multiplying both sides by $f(p)$ (whose square is 1). $\qquad\qquad\square$

Finally, we come to the proof of the Quadratic Reciprocity Law (Theorem 3.15). Theorem 3.3 will again play a pivotal role.

PROOF OF THEOREM 3.15. Consider the field $\mathbb{F}$ with p^{q-1} elements. Then $\mathbb{F}^*$ is a cyclic group with order $p^{q-1} - 1$ by Theorem 3.3. By Fermat's Little Theorem, $p^{q-1} \equiv 1$ modulo q. Thus there is an element ζ in $\mathbb{F}^*$ whose order is q. Define

$$G = \sum_{j=1}^{q-1} \left(\frac{j}{q}\right)\zeta^j. \qquad (3.11)$$

The sum G is called a *Gauss sum* because Gauss seems to have been the first person to systematically study sums such as these.

The proof works as before by finding two different expressions for G^p. We claim first that

$$G^2 = (-1)^{(q-1)/2}q. \qquad (3.12)$$

Using this, we can derive our first expression for G^p.
First expression for G^p:

$$G^p = GG^{p-1} = G(G^2)^{(p-1)/2} = G((-1)^{(q-1)/2}q)^{(p-1)/2}$$

$$= G(-1)^{(q-1)/2 \cdot (p-1)/2}q^{(p-1)/2} = G(-1)^{(q-1)/2 \cdot (p-1)/2}\left(\frac{q}{p}\right).$$

Second expression for G^p:

We claim that

$$G^p = \left(\frac{p}{q}\right) G. \tag{3.13}$$

Equating the two expressions gives

$$G(-1)^{(q-1)/2 \cdot (p-1)/2} \left(\frac{q}{p}\right) = \left(\frac{p}{q}\right) G.$$

We can cancel G because it is not zero in $\mathbb{F}$ (since its square is $(-1)^{(q-1)/2}q$, which is not zero in $\mathbb{F}$); the Quadratic Reciprocity Law follows at once.

The next step is to show Equation (3.13). By the Binomial Theorem,

$$G^p = \left(\sum_{j=1}^{q-1} \left(\frac{j}{q}\right) \zeta^j \right)^p = \sum_{j=1}^{q-1} \left(\frac{j}{q}\right) \zeta^{jp}$$

because $\left(\frac{j}{q}\right)^p = \left(\frac{j}{q}\right)$. By the multiplicativity of the Legendre symbol (Corollary 3.14), the right-hand side is

$$\left(\frac{p}{q}\right) \sum_{j=1}^{q-1} \left(\frac{jp}{q}\right) \zeta^{jp}$$

since $\left(\frac{p}{q}\right)$ is ± 1. Now jp modulo q runs through $1, \ldots, q-1$ as j does, so the second expression for G^p can be written as in Equation (3.13).

The only tricky part of this proof is to evaluate G^2. Expanding the product for G^2 gives

$$G^2 = \sum_{j=1}^{q-1} \left(\frac{j}{q}\right) \zeta^j \sum_{k=1}^{q-1} \left(\frac{-k}{q}\right) \zeta^{-k},$$

noting that as k runs through $1, \ldots, q-1$, so does $-k$ modulo q.

By the multiplicativity of the Legendre symbol, $\left(\frac{-k}{q}\right) = \left(\frac{-1}{q}\right)\left(\frac{k}{q}\right)$. Pulling the factor $\left(\frac{-1}{q}\right)$ out to the front and replacing k by jk in the second sum gives

$$G^2 = \left(\frac{-1}{q}\right) \sum_{j=1}^{q-1} \sum_{k=1}^{q-1} \left(\frac{j}{q}\right) \left(\frac{jk}{q}\right) \zeta^{j(1-k)}.$$

By the multiplicativity of the Legendre symbol $\left(\frac{j}{q}\right)\left(\frac{jk}{q}\right) = \left(\frac{k}{q}\right)$, so

$$G^2 = \left(\frac{-1}{q}\right) \sum_{j=1}^{q-1} \sum_{k=1}^{q-1} \left(\frac{k}{q}\right) \zeta^{j(1-k)}.$$

Next we add zero to both sides of this equation in a special form. On the right-hand side, add

$$0 = \sum_{k=1}^{q-1} \left(\frac{k}{q}\right) \zeta^{0(1-k)}. \tag{3.14}$$

This expression is zero because half of the nonzero residues modulo q are squares, so half of the values of the symbol are 1 and the other half are -1. Thus

$$G^2 = \left(\frac{-1}{q}\right) \sum_{j=0}^{q-1} \sum_{k=1}^{q-1} \left(\frac{k}{q}\right) \zeta^{j(1-k)}.$$

This double sum can be rearranged to give

$$G^2 = \left(\frac{-1}{q}\right) \sum_{k=1}^{q-1} \left(\frac{k}{q}\right) \sum_{j=0}^{q-1} \zeta^{j(1-k)}. \tag{3.15}$$

By Euler's criterion (Theorem 3.13), the term with $k = 1$ contributes

$$\left(\frac{-1}{q}\right) q = (-1)^{(q-1)/2} q$$

to G^2.

We claim that all the other terms (those with $k \neq 1$) in Equation (3.15) contribute nothing. Assume that $k \neq 1$, and write $\eta = \zeta^{1-k}$. Then η is a nontrivial qth root of 1. We claim that

$$S = 1 + \eta + \cdots + \eta^{q-1} = 0.$$

To see this, notice that

$$\eta S = \eta + \eta^2 + \cdots + \eta^{q-1} + \eta^q = 1 + \eta + \cdots + \eta^{q-1} = S,$$

which shows that $S = 0$ since $\eta \neq 1$.   □

Apart from being a very beautiful result, the Quadratic Reciprocity Law is important in that it allows the Legendre symbol to be rapidly computed. This is useful in many areas, including primality testing (see, for example, Section 12.6).

Example 3.18. Compute the Legendre symbol $\left(\frac{91}{167}\right)$ using the Quadratic Reciprocity Law. First notice that

$$\left(\frac{91}{167}\right) = \left(\frac{7}{167}\right) \left(\frac{13}{167}\right) = -\left(\frac{167}{7}\right) \left(\frac{167}{13}\right) = -\left(\frac{6}{7}\right) \left(\frac{11}{13}\right).$$

The problem has become more manageable and is readily finished by noting that

$$\left(\frac{11}{13}\right) = \left(\frac{13}{11}\right) = \left(\frac{2}{11}\right) = -1$$

and

$$\left(\frac{6}{7}\right) = \left(\frac{2}{7}\right)\left(\frac{3}{7}\right) = \left(\frac{3}{7}\right) = -\left(\frac{7}{3}\right) = -\left(\frac{1}{3}\right) = -1.$$

It follows that $\left(\frac{91}{167}\right) = -1$.

Exercise 3.16. Evaluate the Legendre symbols $\left(\frac{11}{37}\right)$, $\left(\frac{19}{31}\right)$, $\left(\frac{1003}{111}\right)$.

3.4 Quadratic Rings

It is tempting to conclude that we are now in a position to characterize those primes p that can be written in the form

$$p = x^2 + dy^2,$$

with $x, y \in \mathbb{Z}$ for a given $d \in \mathbb{Z}$. Unfortunately, this problem is a little more complicated than it first appears. The methods of this chapter are applicable only if the ring $\mathbb{Z}[\sqrt{-d}]$ is Euclidean, and this is not always the case. The structure of $\mathbb{Z}[\sqrt{-d}]$ is quite subtle, and some basic questions about these rings are still open.

Exercise 3.17. Show that the Fundamental Theorem of Arithmetic does not hold in the ring $\mathbb{Z}[\sqrt{-3}]$ by considering the two factorizations

$$2 \cdot 2 = 4 = (1 + \sqrt{-3})(1 - \sqrt{-3})$$

of 4. (Hint: Show that 2 cannot be a prime in this ring.)

Example 3.19. Consider the equation

$$x^2 + 5y^2 = p.$$

In order to understand this, we expect to use the Quadratic Reciprocity Law to solve

$$T^2 + 5 \equiv 0 \pmod{p}$$

for T.

Exercise 3.18. Show that $\left(\frac{-5}{p}\right) = 1$ if and only if $p \equiv 1, 3, 7,$ or 9 modulo 20.

In particular, the congruence $T^2 + 5 \equiv 0$ modulo 7 has a solution: it is easily found that $T = 3$ is a solution. However, the equation

$$x^2 + 5y^2 = 7$$

has no solution in integers.

Exercise 3.19. Show that the Fundamental Theorem of Arithmetic does not hold in the ring $\mathbb{Z}[\sqrt{-5}]$.

Exercise 3.20. Show that there are infinitely many rings $\mathbb{Z}[\sqrt{-d}]$, where d is a positive square-free integer, in which the Fundamental Theorem of Arithmetic does not hold.

The Quadratic Reciprocity Law is a useful tool for understanding when quadratic congruences have no solutions. For example, Exercise 3.18 shows that we will never obtain a solution to the equation $x^2 + 5y^2 = p$ if p is any prime that is congruent to 11 modulo 20. More than that, it can predict the existence of solutions when the equation cannot be checked easily by hand.

Exercise 3.21. (a) Show that $\mathbb{Z}[\sqrt{2}]$ is Euclidean with respect to the norm

$$N(x + y\sqrt{2}) = x^2 - 2y^2.$$

(b) Show that if p is an odd prime, then the equation

$$x^2 - 2y^2 = p$$

has a solution whenever $p \equiv \pm 1$ modulo 8 but has no solutions when $p \equiv \pm 3$ modulo 8. This is (by now) a routine use of the Quadratic Reciprocity Law together with the Euclidean property of $\mathbb{Z}[\sqrt{2}]$.

Exercise 3.22. When $d > 1$ is square-free, the ring $\mathbb{Z}[\sqrt{d}]$ has infinitely many units. Deduce that if

$$x^2 - dy^2 = p$$

has a solution in integers, then it has infinitely many solutions. (The first part of the exercise will be covered in the next section, but try to find a proof yourself.)

The statement in the first part of the exercise is not easy. The equation

$$x^2 - dy^2 = 1$$

is often called Pell's Equation after the seventeenth-century mathematician John Pell. This is now thought to be a misattribution. Brahmagupta seems to have known how to solve the equation long before Pell. In the twelfth century, Bhaskaracharya discovered the simplest of the infinitely many nontrivial solutions when $d = 61$, namely

$$x = 1766319049, \quad y = 226153980.$$

3.5 Units in $\mathbb{Z}[\sqrt{d}], d > 0$

For $d < 0$, the ring $R = \mathbb{Z}[\sqrt{d}]$ has only finitely many units, so we assume in this section that $d > 0$ is a fixed square-free integer. Write $\{t\}$ for the fractional part of a real number t.

Lemma 3.20. *There are infinitely many coprime pairs of integers p and $q > 0$ with*

$$|q\sqrt{d} - p| < \frac{1}{q}.$$

PROOF. Let $Q > 1$ denote an integer. Divide the interval $[0,1)$ into Q subintervals $[0, 1/Q), [1/Q, 2/Q), \ldots$ and consider the $(Q+1)$ numbers

$$0, \{\sqrt{d}\}, \{2\sqrt{d}\}, \ldots, \{Q\sqrt{d}\}.$$

There are $(Q+1)$ of them since $\sqrt{d}$ is irrational, so at least two must lie in a single one of the Q intervals of the form $[a/Q, (a+1)/Q)$ by the pigeonhole principle Thus there must be integers q_1, q_2 with $0 \leqslant q_1 < q_2 \leqslant Q$ such that

$$\left| \{q_2\sqrt{d}\} - \{q_1\sqrt{d}\} \right| < 1/Q.$$

Unwinding the definition of the fractional part, this means that there are integers p_1 and p_2 with

$$\left| q_2\sqrt{d} - p_2 - q_1\sqrt{d} + p_1 \right| = \left| (q_2 - q_1)\sqrt{d} - (p_2 - p_1) \right| < 1/Q.$$

The proof is now finished by choosing $Q \geqslant q = q_2 - q_1 > 0$ and $p = p_2 - p_1$. $\square$

This was originally proved by Dirichlet and is the starting point for a deep subject known as *Diophantine approximation*. This subject has to do with how well an irrational number can be approximated by rational numbers.

Exercise 3.23. Show that there is a constant $C > 0$ such that

$$\frac{C}{q} < |q\sqrt{d} - p|$$

for all integers p and $q > 0$.

Exercise 3.24. More generally, show that if α is algebraic of degree $k > 1$ (that is, α satisfies an irreducible polynomial of degree k with integer coefficients), then there is a constant $C(\alpha) > 0$ such that

$$\frac{C(\alpha)}{q^{k-1}} < |q\alpha - p|$$

for all integers p and $q > 0$.

Theorem 3.21. *If $d > 1$ is a square-free integer, then*

$$x^2 - dy^2 = 1$$

has infinitely many solutions in integers (x, y). Moreover, each solution corresponds to a unit in $R = \mathbb{Z}[\sqrt{d}]$ with norm 1. Any unit with norm 1 has the form $\pm u^n$ for $n \in \mathbb{Z}$, where u is a fixed unit with norm 1.

PROOF. Using Lemma 3.20, choose $p, q > 0$ with

$$|q\sqrt{d} - p| < \frac{1}{q}. \tag{3.16}$$

Then

$$p - \frac{1}{q} < q\sqrt{d} < p + \frac{1}{q},$$

so

$$|q\sqrt{d} + p| < 2q\sqrt{d} + \frac{1}{q}. \tag{3.17}$$

Multiplying the inequalities (3.16) and (3.17) shows that

$$|p^2 - dq^2| < 1 + 2\sqrt{d}.$$

We would like to show that the left-hand side is 1 for infinitely many pairs (p, q). We cannot deduce this at once, but notice that the right-hand side is a uniform bound (independent of p and q), so there must be an integer e with

$$1 \leqslant e < 1 + 2\sqrt{d}$$

such that for infinitely many pairs p and q,

$$p^2 - dq^2 = e.$$

There also must be infinitely many distinct pairs (p, q) and (p', q') such that $p \equiv p'$ modulo e, $q \equiv q'$ modulo e, and

$$pp' - dqq' \equiv p^2 - dq^2 \equiv 0 \pmod{e}.$$

Given such a distinct pair, write

$$pp' - dqq' = xe \quad \text{and} \quad pq' - q'p = ye$$

for integers x and y. Then

$$\begin{aligned}
x^2 - dy^2 &= \left(\frac{pp' - dqq'}{e}\right)^2 - d\left(\frac{pq' - qp'}{e}\right)^2 \\
&= \frac{1}{e^2}\left(p^2(p'^2 - dq'^2) - dq^2(p'^2 - dq'^2)\right) \\
&= \frac{1}{e}(p^2 - dq^2) = 1,
\end{aligned}$$

so there are infinitely many solutions.

To prove the claim about the structure of the unit group, consider the map

$$L : U(R) \rightarrow \mathbb{R}^2$$

defined by

$$L(x + y\sqrt{d}) = (\log(x + y\sqrt{d}), \log(x - y\sqrt{d})).$$

The image is a nontrivial discrete subgroup (see Exercise 3.25 below) of $\mathbb{R}^2$, so it must have rank 1 or 2 by Exercise 3.26.

On the other hand, $x^2 - dy^2 = 1$ implies that

$$\log(x - y\sqrt{d}) + \log(x + y\sqrt{d}) = 0,$$

so the image of L lies in a one-dimensional subspace of $\mathbb{R}^2$ and therefore the rank must be 1. This is enough to prove the claim: The image set must be $\{n(v, -v) \mid n \in \mathbb{Z}\}$ for some nonzero $v \in \mathbb{R}$ and the claim follows with u satisfying $L(u) = (v, -v)$.   □

Exercise 3.25. Explain why the image of L is a discrete subgroup of $\mathbb{R}^2$ in the proof of Theorem 3.21.

Exercise 3.26. Prove that a discrete subgroup of $\mathbb{R}^n$ has rank less than or equal to n.

Finding u is in general a nontrivial problem. In some books you will see the method of continued fractions used, which does give an algorithm. The method used here, which is a first step into the subject called geometry of numbers, was chosen for two reasons. First, using a generalization of this argument, one can go on to analyze the units of the ring of algebraic integers (see p. 84) inside a number field. This always turns out to be finitely generated with a rank that is easily computed from basic data about the number field. The method using continued fractions does not generalize. Second, the geometry of numbers, when worked out fully, really represents an application of topological ideas. If you ask what kind of shapes in space must contain lattice points, you quickly find yourself resorting to ideas such as compactness and connectedness as well as convexity.

A beautiful fact about the solutions of the equation in Theorem 3.21 is that they form a group. Moreover, the multiplication law on elements $x + y\sqrt{d}$ can be expressed in terms of polynomial functions on the coordinates x and y as follows:

$$(x_1 + y_1\sqrt{d})(x_2 + y_2\sqrt{d}) = (x_1x_2 + dy_1y_2) + (x_1y_2 + x_2y_1)\sqrt{d}.$$

In Chapter 5, we will encounter a whole family of Diophantine equations in two variables whose solutions form groups, and for which the multiplication law can be expressed in terms of *rational* functions on the coordinates.

3.6 Quadratic Forms

The subject of quadratic forms is a large one, and we will merely introduce it here via a classical proof of a result due to Lagrange and some comment on Gauss' Theorem. Consider the Diophantine equation

$$ax^2 + bxy + cy^2 = n \qquad (3.18)$$

in which we seek an integral solution (x, y) for given $a, b, c, n \in \mathbb{Z}$. The *discriminant* Δ of the quadratic form $ax^2 + bxy + cy^2$ is defined to be

$$\Delta = b^2 - 4ac.$$

Just as with the Pythagorean equation, there are several elementary reductions to be made.

First, if $\gcd(a, b, c) = d > 1$, then d must also divide n, and Equation (3.18) becomes

$$(a/d)x^2 + (b/d)xy + (c/d)y^2 = (n/d),$$

so without loss of generality we may assume that $\gcd(a, b, c) = 1$.

Second, if $\gcd(x, y) = e > 1$, then

$$a(x/e)^2 + b(x/e)(y/e) + c(y/e)^2 = (n/e^2),$$

so we may assume without loss of generality that x and y are coprime. As in the Pythagorean case, call solutions (x, y) with $\gcd(x, y) = 1$ *primitive*.

Third, if the discriminant Δ is a square, then the equation

$$at^2 + bt + c = 0$$

has rational solutions that may be written u_1/v_1 and u_2/v_2 in lowest terms, with v_1 and v_2 positive, so Equation (3.18) may be written as

$$a(v_1 x - u_1 y)(v_2 x - u_2 y) = n v_1 v_2.$$

This is not really a quadratic equation, but a pair of linear ones. For each integral pair (r, s) with $ars = n v_1 v_2$, solve the equations

$$v_1 x - u_1 y = r,$$
$$v_2 x - u_2 y = s.$$

Integral solutions to this pair of equations – if there are any – solve Equation (3.18).

Exercise 3.27. Let $\rho = \dfrac{1 - (-1)^b}{2}$. Show that $\dfrac{\Delta - \rho}{4}$ is an integer.

Theorem 3.22. [LAGRANGE] *Let Δ be a nonsquare integer. Then there is a quadratic form $ax^2 + bxy + cy^2$ of discriminant Δ with a primitive solution to*

$$ax^2 + bxy + cy^2 = n$$

if and only if the congruence

$$z^2 + \rho z - \left(\frac{\Delta - \rho}{4}\right) \equiv 0 \pmod{n} \tag{3.19}$$

has a solution z.

PROOF. Assume that (α, β) is a primitive integral solution to Equation (3.18). By Theorem 1.23, there are integers γ, δ with

$$\alpha\gamma + \beta\delta = 1.$$

Let

$$\begin{bmatrix} x \\ y \end{bmatrix} = \begin{bmatrix} \alpha & -\delta \\ \beta & \gamma \end{bmatrix} \begin{bmatrix} X \\ Y \end{bmatrix}.$$

Notice that $\det \begin{bmatrix} \alpha & -\delta \\ \beta & \gamma \end{bmatrix} = 1$, so this matrix is an invertible transformation on $\mathbb{Z}^2$.

Let

$$r = a\alpha\delta + c\beta\delta + \frac{b + \rho}{2}$$

and

$$s = a\delta^2 + b\delta\gamma + c\gamma^2.$$

Notice that by our choice of ρ, both r and s are integers. Now express Equation (3.18) in the variables X and Y to obtain

$$a(\alpha X - \delta Y)^2 + b(\alpha X - \delta Y)(\beta X - \gamma Y) + c(\beta X + \gamma Y)^2$$

$$= X^2(a\alpha^2 + ba\beta + c\beta^2) + XY(2a\alpha\delta - b(\alpha\gamma + \beta\delta) + 2c\beta\gamma) + Y^2(a\delta^2 + b\delta\gamma + c\gamma^2)$$

$$= nX^2 + (2r + \rho)XY + sY^2 = n.$$

The equation

$$nX^2 + (2r + \rho)XY + sY^2 = n \tag{3.20}$$

has the solution $X = 1, Y = 0$, corresponding to

$$\begin{bmatrix} \alpha \\ \beta \end{bmatrix} = \begin{bmatrix} \alpha & -\delta \\ \beta & \gamma \end{bmatrix} \begin{bmatrix} 1 \\ 0 \end{bmatrix}.$$

The discriminant of Equation (3.20) is

$$(2r + \rho)^2 - 4sn = \Delta, \tag{3.21}$$

so

$$r^2 + \rho r - \left(\frac{\Delta - \rho}{4}\right) = sn,$$

showing that r is a solution of the congruence (3.19).

Conversely, assume that r is a solution to the congruence (3.19). Then solving Equation (3.21) gives an integer s and hence the integer solution $X = 1, Y = 0$ to Equation (3.20). Changing back to the variables x, y using

$$\begin{bmatrix} X \\ Y \end{bmatrix} = \begin{bmatrix} \gamma & \delta \\ -\beta & \alpha \end{bmatrix} \begin{bmatrix} x \\ y \end{bmatrix}$$

gives an integral solution to the equation

$$nx^2 + (2r + \rho)xy + s^2 = n$$

that has discriminant Δ.   □

Example 3.23. Let $a = 1, b = 0, c = 5$, and $n = 7$, so $\rho = 0$ and $\Delta = -20$. Theorem 3.22 applies to say that there is a quadratic form representing 7 with discriminant -20 if and only if

$$z^2 + 5 \equiv 0 \pmod{7}$$

has a solution. We know that -5 is a quadratic residue modulo 7, so there is such a form. The proof constructs the form

$$7x^2 + 6xy + 2y^2,$$

and of course this represents 7 when $x = 1$ and $y = 0$.

Exercise 3.28. Prove that any odd prime congruent to 1 modulo 4 is a sum of two integer squares using Theorem 3.22 (cf. Theorem 2.6 where this was proved using different methods).

The next exercises explore the change of variables (X, Y) to (x, y) used in the proof of Theorem 3.22. An integer n is said to be *represented* by an integral quadratic form Q if there are integers x and y with $Q(x, y) = n$.

Exercise 3.29. Let

$$P(x, y) = ax^2 + bxy + cy^2$$

and

$$Q(x, y) = AX^2 + BXY + CY^2$$

be binary quadratic forms with integer coefficients. Say that P and Q are *equivalent*, written $P \sim Q$, if there is an integral change of variables

$$\begin{bmatrix} x \\ y \end{bmatrix} = \begin{bmatrix} \alpha & -\delta \\ \beta & \gamma \end{bmatrix} \begin{bmatrix} X \\ Y \end{bmatrix}$$

with det $\begin{bmatrix} \alpha & -\delta \\ \beta & \gamma \end{bmatrix} = 1$ such that

$$P(x,y) = Q(X,Y).$$

(a) Show that $\sim$ is an equivalence relation.
(b) Show that equivalent quadratic forms have the same discriminant.
(c) Show that equivalent quadratic forms represent the same set of integers:
If $P \sim Q$, then

$$\{P(x,y) \mid x,y \in \mathbb{Z}\} = \{Q(x,y) \mid x,y \in \mathbb{Z}\}.$$

Exercise 3.30. Show that a prime number p is represented by a quadratic form P if and only if there is a quadratic form equivalent to P of the form

$$px^2 + dxy + ey^2$$

for integers d and e.

Let $P(x,y) = ax^2 + bxy + cy^2$ be a quadratic form. Then P is *positive-definite* if $P(x,y) \geqslant 0$ for all x and y and is *reduced* if either

$$c > a \text{ and } -a < b \leqslant a$$

or

$$c = a \text{ and } 0 \leqslant b \leqslant a.$$

Exercise 3.31. Prove that a positive-definite binary quadratic form is equivalent to a unique reduced quadratic form.

Exercise 3.32. The *class number* of d is the number of equivalence classes of positive-definite forms with discriminant d. Prove that the class number is finite for any d.

NOTES TO CHAPTER 3: Artin's conjecture from Section 3.1 is still open – see the monograph [58, Section 3.2, 3.3] by Everest, van der Poorten, Shparlinski and Ward for descriptions of what is known and references to the literature. Shoup's result can be found in his paper [138]. There is a discussion of the history of the Chinese Remainder Theorem in many places; see the 'History of Mathematics' Web site [113] for references. Mahler's paper [102] gives an account of the method actually used by the early Chinese mathematicians, as opposed to the modern approach which follows Gauss [67]. We thank Robin Chapman for Exercise 3.12(b). Gauss was justly proud of having proved the Quadratic Reciprocity Law and many mathematicians have seen it since as foundational in the modern theory of numbers. The history and mathematics of the Quadratic Reciprocity Law and the development of reciprocity laws for higher degrees are described in Lemmermeyer's monograph [98].

4

Recovering the Fundamental Theorem of Arithmetic

This short chapter will explain how ideal theory was developed as a means of recovering from the failure of the Fundamental Theorem of Arithmetic witnessed in Chapter 3. We begin with a few historical remarks to set that development in context and go on to give a reasonably complete account of unique factorization of ideals in the ring of algebraic integers in a quadratic field. Finally we introduce the class number and the class group.

4.1 Crisis

The attempt to understand fully the problem we set out to study in the last chapter exposed a phenomenon that represented something of a historical crisis. During the nineteenth century, mathematicians had to come to terms with the breakdown of the Fundamental Theorem of Arithmetic. In March 1847, Lamé announced a proof of Fermat's Last Theorem (described in Section 2.5.1) to the Paris Academy, assuming (wrongly) that the Fundamental Theorem of Arithmetic held in the ring $\mathbb{Z}[e^{2\pi i/n}]$ for every $n \geqslant 1$. Lamé acknowledged that Liouville originally suggested this approach to Fermat's Last Theorem, but Liouville himself addressed the meeting and suggested that there might be a problem with the assumption of unique factorization into primes.

The question raised was this: Does unique factorization into primes hold in the ring $\mathbb{Z}[e^{2\pi i/n}]$? This problem became a focal point for rapid developments. On May 24th 1847, Liouville presented a letter from Kummer to the Academy that settled the arguments. Kummer had proved in 1844 that unique factorization failed in general but that his "ideal complex numbers" in a paper of 1846 allowed a form of unique factorization to be recovered. By September 1847, Kummer had presented a paper to the Berlin Academy in which he proved that for p a *regular prime*[1] Fermat's Last Theorem holds for

---

[1] A prime p is called regular if p does not divide the numerators of any of the Bernoulli numbers $B_2, B_4, \ldots, B_{p-3}$; the Bernoulli numbers are defined on p. 203.

exponent $n = p$, essentially by Lamé's method. In this paper, Kummer also showed that 37 is not regular since 37 divides the numerator of B_{32}. Thus Kummer proved Fermat's Last Theorem for many indices and showed that Lamé's approach failed for others.

These dramatic developments did not lead to a proof of Fermat's Last Theorem but contributed to algebraic number theory in a profound way by eventually leading to the result that rings such as $\mathbb{Z}[\sqrt{-5}]$ do have a kind of Fundamental Theorem of Arithmetic – but at the level of *ideals* rather than *elements*.

4.2 An Ideal Solution

Definition 4.1. *An* ideal *in a commutative ring R is a subgroup of the additive group of R that is closed under multiplication by elements of R.*

It is easy to construct ideals in a commutative ring: Take all the multiples

$$(a) = aR = \{ar \mid r \in R\}$$

of a single element a. In rings such as $\mathbb{Z}$ and $\mathbb{Z}[\sqrt{-2}]$, all ideals are of this form, and this is true for any Euclidean ring.

Exercise 4.1. (a) Using the Euclidean Algorithm, prove that any ideal in $\mathbb{Z}$ has the form $(k) = k\mathbb{Z}$ for some $k \in \mathbb{Z}$.
(b) More generally, prove that in a Euclidean ring R any ideal has the form $(k) = kR$, the multiples of a single element k.

Such singly-generated ideals are called *principal*, and any ring in which all ideals are principal is called a *principal ideal domain*.

The statement in Exercise 4.1(b) is not true in all commutative rings. It is difficult to envisage what ideals look like in general; however, a more sophisticated version of the Fundamental Theorem of Arithmetic makes them easier to understand. This is described in Section 4.3 for quadratic fields.

Any field $\mathbb{K}$ containing the rationals contains a ring $O_{\mathbb{K}}$ of *algebraic integers*; this ring is a generalization of the usual integers in the rationals, and is defined to be the set of all zeros in the field $\mathbb{K}$ of monic polynomials with coefficients in $\mathbb{Z}$.

Exercise 4.2. Show that the ring of algebraic integers in $\mathbb{Q}(\sqrt{d})$ is $\mathbb{Z}[\sqrt{d}]$ if $d \equiv 2$ or 3 modulo 4 and is $\mathbb{Z}[(1 + \sqrt{d})/2]$ if $d \equiv 1$ modulo 4. (Hint: Start by showing that any algebraic integer in $\mathbb{Q}(\sqrt{d})$ that is not in $\mathbb{Z}$ must satisfy a quadratic equation.)

Exercise 4.3. By the previous exercise, the ring of algebraic integers in $\mathbb{Q}(\sqrt{6})$ (or $\mathbb{Q}(\sqrt{14})$) is $\mathbb{Z}[\sqrt{6}]$ (resp. $\mathbb{Z}[\sqrt{14}]$). Prove that the ring of algebraic integers in $\mathbb{Q}(\sqrt{6}, \sqrt{14})$ is strictly larger than $\mathbb{Z}[\sqrt{6}, \sqrt{14}]$.

Exercise 4.4. Adapt the methods of Theorem 3.21 to show that the group of units $O_{\mathbb{K}}^*$ inside the ring of algebraic integers $O_{\mathbb{K}}$ of the field $\mathbb{K} = \mathbb{Q}(\sqrt{d})$ when $d > 0$ is square-free comprises $\{\pm u^n \mid n \in \mathbb{Z}\}$, where $u > 1$ is some unit of $O_{\mathbb{K}}$. Such an element u is called a *fundamental unit*.

Exercise 4.5. Find fundamental units for the real quadratic fields

$$\mathbb{Q}(\sqrt{2}), \quad \mathbb{Q}(\sqrt{3}), \quad \mathbb{Q}(\sqrt{5}), \quad \text{and} \quad \mathbb{Q}(\sqrt{7}).$$

Any element of $\mathbb{Q}(\sqrt{d})$ for a square-free integer d may be written uniquely in the form $\alpha = x + y\sqrt{d}$ with x and y rational. This presents $\mathbb{Q}(\sqrt{d})$ as a two-dimensional vector space over $\mathbb{Q}$ with basis $\{1, \sqrt{d}\}$.

Definition 4.2. *The norm of $\alpha = x + y\sqrt{d}$ in $\mathbb{Q}(\sqrt{d})$ is defined to be*

$$N(\alpha) = x^2 - dy^2$$

and the trace

$$T(\alpha) = 2x.$$

Exercise 4.6. The map $\beta \mapsto \alpha\beta$ on $\mathbb{Q}(\sqrt{d})$ is a $\mathbb{Q}$-linear map on the $\mathbb{Q}$ vector space $\mathbb{Q}(\sqrt{d})$. Find the 2×2 matrix determined by this map, and show that the absolute value of its determinant is $|N(\alpha)|$ and its trace is $T(\alpha)$.

Unique factorization will be recovered in Section 4.3 by working with prime ideals in the algebraic integers. These matters represent the beginnings of an important subject called algebraic number theory. The recovery of the Fundamental Theorem of Arithmetic at the level of ideals represents a major achievement that continues to influence the development of number theory and geometry.

Theorem 2.14 on p. 55 gives a different way to recover the Fundamental Theorem of Arithmetic, used to dramatic effect in Theorem 2.13, but the development of ideal theory proved to be of much greater importance.

4.3 Fundamental Theorem of Arithmetic for Ideals

We begin with a natural definition of multiplication on ideals. Subsequently, we introduce a notion of prime ideal, then we go on to show that every non-trivial ideal factorizes as a product of prime ideals in a way that is unique.

Definition 4.3. *Let I and J denote ideals in a commutative ring. The sum and product of I and J are defined by $I + J = \{a + b \mid a \in I, b \in J\}$, while IJ is the additive subgroup generated by the set $\{ab \mid a \in I, b \in J\}$.*

Exercise 4.7. If I and J denote ideals in a commutative ring, prove that the sum and the product of I and J are also ideals.

Sums and products of more than two ideals are defined in an entirely analogous fashion and again turn out to be ideals.

It might have seemed more natural to define IJ to be the set

$$\{ab \mid a \in I, b \in J\}$$

rather than the subgroup this set generates, however the set of products by itself is not always closed under addition.

Exercise 4.8. Give an example of a commutative ring R together with two ideals I and J such that the set $\{ab \mid a \in I, b \in J\}$ is not an ideal.

If $\alpha = x + y\sqrt{d} \in \mathbb{K} = \mathbb{Q}(\sqrt{d})$, write $\alpha^* = x - y\sqrt{d}$ for the *conjugate* of α. If $d < 0$, then this is the usual complex conjugate; for $d > 0$ this terminology comes from Galois theory. For an ideal I, write I^* for the set of conjugates of $\alpha \in I$.

Exercise 4.9. Let I and J denote ideals in $O_{\mathbb{K}}$.
(a) Show that I^* is an ideal in $O_{\mathbb{K}}$.
(b) Show that $(I + J)^* = I^* + J^*$.
(c) Show that $(IJ)^* = I^* J^*$.

If $\alpha_1, \ldots, \alpha_k$ are elements of $O_{\mathbb{K}}$, write

$$(\alpha_1, \ldots, \alpha_k)$$

for the ideal

$$(\alpha_1) + \cdots + (\alpha_k) = \alpha_1 O_{\mathbb{K}} + \cdots + \alpha_k O_{\mathbb{K}}$$

generated by $\alpha_1, \ldots, \alpha_k$. Also define

$$\langle \alpha_1, \ldots, \alpha_k \rangle = \alpha_1 \mathbb{Z} + \cdots + \alpha_k \mathbb{Z}$$

for the additive subgroup of $O_{\mathbb{K}}$ generated by $\alpha_1, \ldots, \alpha_k$. It is important to distinguish these different types of generation.

In what follows, we are going to work with the full ring of algebraic integers in the field $\mathbb{Q}(\sqrt{d})$ for a square-free integer d. Following Exercise 4.2, define δ to be $(1 + \sqrt{d})/2$ if $d \equiv 1$ modulo 4 and $\sqrt{d}$ if $d \equiv 2$ or 3 modulo 4. *Thus, if $\mathbb{K} = \mathbb{Q}(\sqrt{d})$, then $O_{\mathbb{K}} = \mathbb{Z}[\delta]$.* Ideals in $O_{\mathbb{K}}$, although not always principal, can always be generated as ideals by two elements.

Theorem 4.4. *Let I denote an ideal in $O_{\mathbb{K}}$. Then there are elements α, β in I with $I = (\alpha, \beta)$.*

PROOF. Since $O_{\mathbb{K}}$ as an additive group is a subgroup of $\mathbb{Q}^2$, it follows that I can be generated as an additive group by two elements. We will first show that one of these elements can be chosen to lie in $\mathbb{Z}$. Let

$$B = \{b \in \mathbb{Z} \mid a + b\delta \in I \text{ for some } a \in \mathbb{Z}\};$$

then B is an ideal of $\mathbb{Z}$. Hence $B = g\mathbb{Z}$ for some $g \in \mathbb{Z}$ and similarly $I \cap \mathbb{Z} = h\mathbb{Z}$ for some $h \in \mathbb{Z}$. Since $g \in B$, there must be $c \in \mathbb{Z}$ with $c + g\delta \in I$.

We claim that

$$I = \langle c + g\delta, h \rangle. \tag{4.1}$$

Clearly $\langle c + g\delta, h \rangle \subseteq I$. Now assume that $a + b\delta \in I$ with $a, b \in \mathbb{Z}$. Since $b \in B$, $b = eg$ for some $e \in \mathbb{Z}$. Therefore

$$a - ec = a + b\delta - e(c + g\delta).$$

This is an element of $I \cap \mathbb{Z}$, so it can be written as fh for some $f \in \mathbb{Z}$. Then

$$a + b\delta = a - ec + e(c + g\delta) = fh + e(c + g\delta) \in \langle c + g\delta, h \rangle,$$

showing Equation (4.1).

To finish the proof of the theorem, use Equation (4.1) to write $\alpha = c + h\delta$ and $\beta = g$. Then $\alpha, \beta \in I$, so $(\alpha, \beta) \subseteq I$. Conversely, if $\gamma \in I$, then for integers m and n,

$$\gamma = m\alpha + n\beta,$$

so $I \subseteq (\alpha, \beta)$, which concludes the proof. $\square$

As a final step toward proving the Fundamental Theorem of Arithmetic for ideals in $O_{\mathbb{K}}$, we note the following lemma.

Lemma 4.5. [HURWITZ'S LEMMA] *If α, β are elements of $O_{\mathbb{K}}$ and $k \in \mathbb{Z}$ divides $N(\alpha)$, $N(\beta)$, and $T(\alpha\beta^*)$, then k divides $\alpha\beta^*$ and $\alpha^*\beta$ in $O_{\mathbb{K}}$.*

Exercise 4.10. Prove Hurwitz's Lemma. (Hint: This only uses simple properties of the norm and trace functions.)

Corollary 4.6. *Let I denote any ideal of $O_{\mathbb{K}}$. Then II^* is a principal ideal $k\mathbb{Z}$ of $\mathbb{Z}$.*

PROOF. We know that $I = (\alpha, \beta)$ for some α, β, so

$$II^* = (\alpha, \beta)(\alpha^*, \beta^*) = (\alpha\alpha^*, \alpha\beta^*, \beta\alpha^*, \beta\beta^*).$$

This means that II^* contains the integers $N(\alpha) = \alpha\alpha^*$ and $N(\beta)$, as well as $T(\alpha\beta^*) = \alpha\beta^* + \alpha^*\beta$. If k denotes the greatest common divisor of these integers then $k \in II^*$, so $(k) \subseteq II^*$. Now $k \mid N(\alpha), k \mid N(\beta)$, and $k \mid T(\alpha\beta^*)$ and hence, by Hurwitz's Lemma, $k \mid \alpha\beta^*$ and $k \mid \beta\alpha^*$, so $II^* \subseteq (k)$ as claimed. $\square$

The integer k appearing in Corollary 4.6 may be taken as positive without loss of generality since $k\mathbb{Z} = -k\mathbb{Z}$.

Definition 4.7. *If I denotes any ideal of $O_{\mathbb{K}}$, then the unique integer $k > 0$ with $II^* = k\mathbb{Z}$ is called the* **norm** *of I, written $N(I)$.*

Corollary 4.8. (1) *For* $I = (\alpha, \beta)$,

$$N(I) = \gcd(N(\alpha), N(\beta), T(\alpha\beta^*)).$$

(2) *If* $I = (\alpha)$ *is a principal ideal, then* $N(I) = N(\alpha)$.
(3) *The norm is multiplicative:* $N(IJ) = N(I)N(J)$ *for all ideals* I *and* J.
(4) $N(I) = [O_{\mathbb{K}} : I]$, *the group-theoretic index of* I *as a subgroup of* $O_{\mathbb{K}}$.

PROOF.(1) This appeared in the proof of Corollary 4.6.
(2) This follows because $(\alpha)(\alpha^*) = (\alpha\alpha^*)$.
(3) $(N(IJ)) = IJI^*J^* = (II^*)(JJ^*) = (N(I))(N(J)) = (N(I)N(J))$.   □

Exercise 4.11. Prove Corollary 4.8(4). (Hint: if $(h, c + g\delta)$ is a nonzero ideal of $O_{\mathbb{K}}$, then $N(I) = gh$.)

Corollary 4.9. *If* I, J *and* K *are ideals of* $O_{\mathbb{K}}$ *with* $I \neq \{0\}$ *and* $IJ = IK$, *then* $J = K$.

PROOF. This is obvious if $I = (\alpha)$ is principal because in that case $IJ = \alpha J$ so $J = \alpha^{-1}(IJ)$. Similarly, $K = \alpha^{-1}(IK) = \alpha^{-1}(IJ) = J$. In general, the identity $IJ = IK$ implies that

$$(II^*)J = (IJ)I^* = (IK)I^* = (II^*)K,$$

and the result follows as before.   □

This important 'cancellation' property of ideals in $O_{\mathbb{K}}$ will play a key role in the proof of the Fundamental Theorem of Arithmetic for ideals.

Definition 4.10. *If* I *and* J *are ideals in* $O_{\mathbb{K}}$, *we write* $I|J$ *(* I *divides* J *) if there is an ideal* K *in* $O_{\mathbb{K}}$ *with* $J = IK$.

Notice that $IK \subseteq I$, so if $I|J$ then $J \subseteq I$.

Lemma 4.11. *Given two ideals* I *and* J *in* $O_{\mathbb{K}}$, $I|J$ *if and only if* $J \subseteq I$.

PROOF. One direction is already proved, so assume that $J \subseteq I$. Then

$$JI^* \subseteq II^* = (N(I)),$$

so

$$K = \frac{1}{N(I)}JI^*$$

is an ideal contained in $O_{\mathbb{K}}$. It follows that

$$IK = \frac{1}{N(I)}I(JI^*) = \frac{1}{N(I)}J(II^*) = \frac{1}{N(I)}J(N(I)) = J,$$

and hence $I|J$ as claimed.   □

In what follows, we see a real duplication of ideas from Chapter 1, worked out in the context of ideals. The interchangeability of inclusion and divisibility for ideals will be used repeatedly.

Definition 4.12. *A nonzero ideal $I \neq R$ in a commutative ring R is called maximal if for any ideal J, $J | I$ implies that $J = I$. An ideal P is prime if $P | IJ$ implies that $P | I$ or $P | J$.*

Exercise 4.12. In a commutative ring R, let M and P denote ideals.
(a) Show that M is maximal if and only if the quotient ring R/M is a field.
(b) Show that P is prime if and only if R/P is an integral domain (that is, in R/P the equation $ab = 0$ forces either a or b to be 0).
(c) Deduce that every maximal ideal is prime.

Theorem 4.13. [FUNDAMENTAL THEOREM OF ARITHMETIC FOR IDEALS] *Any nonzero proper ideal in $O_{\mathbb{K}}$ can be written as a product of prime ideals, and that factorization is unique up to order.*

PROOF. If I is not maximal, it can be written as a product of two nontrivial ideals. Comparing norms shows these ideals must have norms smaller than I. Keep going: The sequence of norms is descending, so it must terminate, resulting in a finite factorization of I. By Exercise 4.12, every maximal ideal is prime, so all that remains is to demonstrate that the resulting factorization is unique. This uniqueness follows from Corollary 4.9, which allows cancellation of nonzero ideals common to two products. □

4.4 The Ideal Class Group

In this section, we are going to see how the nineteenth-century mathematicians interpreted Exercise 3.32 on p. 81 in terms of quadratic fields. The major result we will present is that ideals in $O_{\mathbb{K}}$, for a quadratic field $\mathbb{K}$, can be described using a finite list of representatives $I_1, \ldots, I_h$; any nontrivial ideal I can be written $I_i P$, where $1 \leqslant i \leqslant h$ and P is a principal ideal. Thus h, known as the *class number*, measures the extent to which $O_{\mathbb{K}}$ fails to be a principal ideal domain. This statement was proved for arbitrary algebraic number fields and proved to be influential in the way number theory developed in the twentieth century.

Given two ideals I and J in $O_{\mathbb{K}}$, define a relation $\sim$ by

$$I \sim J \text{ if and only if } I = \lambda J \text{ for some } \lambda \in \mathbb{K}^*.$$

Exercise 4.13. Show that $\sim$ is an equivalence relation.

We are going to outline a proof of the following important theorem.

Theorem 4.14. *There are only finitely many equivalence classes of ideals in $O_{\mathbb{K}}$ under $\sim$.*

One class is easy to spot – namely the one consisting of all principal ideals. Of course, $O_{\mathbb{K}}$ is a principal ideal domain if and only if there is only one class under the relation. One can define a multiplication on classes: If $[I]$ denotes the class containing I, then one can show that the multiplication defined by

$$[I][J] = [IJ] \tag{4.2}$$

is independent of the representatives chosen.

Corollary 4.15. *The set of classes under* $\sim$ *forms a finite Abelian group.*

The group in Corollary 4.15 is known as the *ideal class group* of $\mathbb{K}$ (or just the *class group*).

PROOF OF COROLLARY 4.15. In the class group, associativity of multiplication is inherited from $O_{\mathbb{K}}$. The element $[O_{\mathbb{K}}]$ acts as the identity. Finally, given any nonzero ideal I, the relation $II^* = (N(I))$ shows that the inverse of the class $[I]$ is $[I^*]$. □

Lemma 4.16. *Given a square-free integer* $d \neq 1$, *there is a constant* C_d *that depends upon d only such that for any nonzero ideal I of* $O_{\mathbb{K}}$, $\mathbb{K} = \mathbb{Q}(\sqrt{d})$, *there is a nonzero element* $\alpha \in I$ *with* $|N(\alpha)| \leqslant C_d N(I)$.

Exercise 4.14. *Prove Lemma 4.16. The basic idea is a technique similar to that used in the proof of Theorem 3.21 showing that a lattice point must exist in a region constrained by various inequalities. Since the original proof, considerable efforts have gone into decreasing the constant C_d for practical application. The best techniques use the geometry of numbers, a theory initiated by Minkowski.

PROOF OF THEOREM 4.14. First show that every class contains an ideal whose norm is bounded by C_d. Given a class $[I]$, apply Lemma 4.16 with I^* replacing I. Now $(\alpha) \subseteq I^*$, so we can write $(\alpha) = I^* J$ for some ideal J. However, this gives a relation $[I^*][J] = [(\alpha)]$ in the class group. This means that $[J]$ is the inverse of $[I^*]$. However, we remarked earlier that $[I]$ and $[I^*]$ are mutual inverses in the class group. Hence $[I] = [J]$. Now

$$|N(\alpha)| = N((\alpha)) = N(I^*)N(J).$$

Since the left-hand side is bounded by $C_d N(I^*)$, we can cancel $N(I^*)$ to obtain $N(J) \leqslant C_d$.

Now the theorem follows easily: For any given integer $k \geqslant 0$, there are only finitely many ideals of norm k; this is because any ideal must be a product of prime ideals of norm p or p^2, where p runs through the prime factors of k. There are only finitely many such prime ideals and hence there are only finitely many ideals of norm k. Now apply this to the integers $k \leqslant C_d$ to deduce that there are only finitely many ideals of norm bounded by C_d. Since each class contains an ideal whose norm is thus bounded, by the first part of the proof, it follows that there are only finitely many classes. □

Exercise 4.15. Investigate the relationship between quadratic forms and ideals in quadratic fields. In particular, show that Exercise 3.32 on p. 81 is equivalent to Theorem 4.14. (Hint: If I denotes an ideal with basis $\{\alpha, \beta\}$, show that for $x, y \in \mathbb{Z}$, $N(x\alpha + y\beta)/N(I)$ is a (binary) integral quadratic form. How does a change of basis for I relate to the form? What effect does multiplying I by a principal ideal have on the form?)

4.4.1 Prime Ideals

To better understand prime ideals, we close with an exercise that links up the various trains of thought in this chapter and shows that ideal theory better explains the various phenomena encountered in Chapter 3.

Exercise 4.16. Factorize the ideal (6) into prime ideals in $\mathbb{Z}[\sqrt{-5}]$, expressing each prime factor in the form $(a, b + c\sqrt{-5})$.

Exercise 4.17. Let $O_{\mathbb{K}}$ denote the ring of algebraic integers in the quadratic field $\mathbb{K} = \mathbb{Q}(\sqrt{d})$ for a square-free integer d.
(a) If P is a prime ideal in $O_{\mathbb{K}}$, show that $P|(p)$ for some integer prime $p \in \mathbb{Z}$.
(b) Show that there are only three possibilities for the factorization of the ideal (p) in $O_{\mathbb{K}}$:

> $(p) = P_1 P_2$ where P_1 and P_2 are prime ideals in $O_{\mathbb{K}}$ (p *splits*);
> $(p) = P$, where P is a prime ideal in $O_{\mathbb{K}}$ (p is *inert*);
> $(p) = P^2$, where P is a prime ideal in $O_{\mathbb{K}}$ (p is *ramified*).

This should be compared with the possible primes in $\mathbb{Z}[i]$ described in Theorem 2.8(3). The following exercise gives a complete description of splitting types in terms of the Legendre symbol.

Exercise 4.18. Let $O_{\mathbb{K}}$ denote the ring of algebraic integers in the quadratic field $\mathbb{K} = \mathbb{Q}(\sqrt{d})$ for a square-free integer d. Let $D = d$ if $d \equiv 1$ modulo 4 and let $D = 4d$ otherwise. Show that an odd prime p is inert, ramified, or split as the Legendre symbol $\left(\frac{D}{p}\right)$ is -1, 0, or $+1$, respectively. What are the possibilities when $p = 2$?

We should say something about the terminology. Splitting and inertia are fairly obvious, the latter signifying that the prime p remains prime in this bigger ring, just as primes $p \equiv 3$ modulo 4 remain primes in $\mathbb{Z}[i]$. The term "ramify" means literally to branch, and we see here something of an overlap with the theory of functions. A function such as $y = \sqrt{x}$ really consists of two possible branches. This notion was borrowed deliberately to name the phenomenon seen in number theory, where a prime in $\mathbb{Z}$ becomes a power of a prime in a larger ring. We end this chapter with a definition because it is going to appear again in Chapter 11.

Definition 4.17. *Let* $\mathbb{K} = \mathbb{Q}(\sqrt{d})$ *denote a quadratic field, where d is a square-free integer. Define D by*

$$D = \begin{cases} d & \text{if } d \equiv 1 \text{ modulo 4 and} \\ 4d & \text{otherwise.} \end{cases} \tag{4.3}$$

Then D is called the discriminant *of the quadratic field* $\mathbb{K} = \mathbb{Q}(\sqrt{d})$.

NOTES TO CHAPTER 4: Much of this chapter was based on Robin Chapman's excellent expository notes. To see the details worked out economically in the general case, consult Lang's book [96]. Lemma 4.16 is proved as Theorem 4 on p.119 of that book; Chapter V is an excellent introduction to Minkowski's geometry of numbers.

5

Elliptic Curves

One of the many powerful ideas that have been brought to bear on problems in number theory is a connection between Diophantine problems and geometry. Exercise 2.2 on p. 45 gave a hint of this phenomenon; the geometric structure in that case was a unit circle, an object with algebraic structure in that the points of the unit circle form a group. In this chapter, we introduce a family of curves with a group structure. The main aim is to develop a working understanding of the group operation and to illustrate this with many examples. In subsequent chapters we will make these ideas more rigorous.

5.1 Rational Points

Having studied the Pythagorean equation, which has infinitely many integral solutions, perhaps the existence of so few integral solutions to the equation[1]

$$y^2 = x^3 - 2$$

seems a little disappointing. However, the solution $(3, 5)$ has an amazing property. We can use this one integral solution to generate other exotic *rational* solutions to the equation. It may not seem obvious, but we can use this solution to generate the solution $\left(\frac{129}{100}, \frac{383}{1000}\right)$. Moreover, in a precise sense, this rational solution is the next simplest solution to the equation.

---

[1] The equation $y^2 = x^3 + C$ is sometimes called *Bachet's equation* after Claude Bachet (1581–1638). Bachet is most famous for translating the *Arithmetica* of Diophantus from Greek into Latin. This is the book in which Fermat wrote his famous marginal note asserting what is now called Fermat's Last Theorem. In addition, Bachet discovered the duplication formula for this curve, showing that if (x, y) is a solution, then

$$((x^4 - 8Cx)/(2y)^2, (-x^6 - 20Cx^3 + 8C^2)/(2y)^3)$$

is also a – potentially different – solution.

To see how this is done, first construct the tangent to the curve at the point $P = (3, 5)$. This has equation

$$y = \frac{27}{10}x - \frac{31}{10}.$$

If we substitute the equation for this line into the equation of the curve, then (we claim) the line will meet the curve at another point, *and this point will have rational coordinates.*[2] To see this more explicitly, note that when we substitute, we get a cubic equation for x,

$$\left(\frac{27}{10}x - \frac{31}{10}\right)^2 = x^3 - 2.$$

We claim that $x = 3$ is a double root of this equation. Clearly, it is a single root by substituting in and getting 25 on both sides of the equation; differentiating and substituting shows it is a double root because you get 27 on both sides.

To find the third point of intersection, use the sum of roots formula. For a cubic, this says that if x_1, x_2, and x_3 are the three zeros of the cubic

$$x^3 + ax^2 + bx + c,$$

then $x_1 + x_2 + x_3 = -a$ (see Exercise 5.11 on p. 105). Applying this, and letting x denote the third root, we see that

$$3 + 3 + x = \left(\frac{27}{10}\right)^2.$$

Solving this for x gives $x = \frac{129}{100}$. To find y, use the equation of the tangent to see that $y = \frac{383}{1000}$.

It is tempting to try this again. We cannot expect anything by joining our new point back to P. However, we could join the other integral solution $(3, -5)$ to the new point to see where the line meets the curve again. Technically, it is better to reflect the new point in the x-axis and try to join that to our first point (for reasons that will become apparent later). Thus we define $P_1 = (3, 5)$ and P_2 to be $\left(\frac{129}{100}, -\frac{383}{1000}\right)$. Recursively define P_n to be the reflection in the x-axis of the third point of intersection of the line joining P to P_{n-1}. The next point is

$$P_3 = \left(\frac{164323}{29241}, -\frac{66234835}{5000211}\right),$$

from which we obtain

$$P_4 = \left(\frac{2340922881}{58675600}, \frac{113259286337279}{449455096000}\right).$$

---

[2] If you are comfortable with geometrical notions, then you will accept that since P is already a double point, the third point must be rational.

It is an amazing fact that there are infinitely many rational points on this curve and (up to reflection in the x-axis) they can all be constructed in this way, starting with P.

This example already exhibits some typical behavior; for example, the denominator of the x-coordinate of P_3 is a square while the denominator of the y-coordinate is the cube of the same number:

$$P_3 = \left(\frac{164323}{171^2}, -\frac{66234835}{171^3} \right).$$

Exercise 5.1. Prove that any rational point on the curve $y^2 = x^3 - 2$ must have the shape $P = (A/B^2, C/B^3)$ for coprime integers A, B, C.

Example 5.1. To start to understand what is going on in the geometrical iteration that produces the points P_n, consider the sequence (B_n), where B_n is the square root of the denominator of the x-coordinate of P_n. The first few values are shown in Table 5.1.

Table 5.1. Growth in the values of B_n.

n	B_n
1	1
2	10
3	171
4	7660
5	12660211
6	22652313570
7	58809175344521
8	1735132266687114280
9	35717278218714405526201
10	1154553432516829071988561920250
11	3029885420353938553602816729630205
12	68999149084295048331393516376644064606406064580
13	227433398162437271513835207416379964567358017125710
14	130198223405915703707022821246523810026556372392485847 0330
15	456878909724292243427136109000405523231826883077060806932 78173039

The lengths of the numbers B_n written out in decimal digits seem to grow quadratically in n. The number of digits in B_n is approximately $\log_{10} B_n$, so this suggests a relationship between $\log B_n$ and n^2. The following beautiful result makes this precise.

Theorem 5.2. *There is a constant $h > 0$ for which*

$$\frac{1}{n^2} \log B_n \to h \; as \; n \to \infty$$

where (B_n) is the sequence in Table 5.1.

We are not going to prove this result. An easy consequence of Theorem 5.2 is the finiteness of the number of integral points in the sequence (P_n). Later, we will prove that the maximum of $\log|A_n|$ and $2\log|B_n|$ also grows as in the statement of Theorem 5.2 (see the comments after Theorem 7.13 on p. 147.) In Section 7.4.1, Theorem 7.15, we will relate the growth rates of $\log|A_n|$ and $2\log|B_n|$ to each other.

The geometrical operation taking P_n to P_{n+1} described above is a special case of a more general one: There is a binary operation on the set of points (x, y) satisfying the equation $y^2 = x^3 - 2$ that behaves like a group law. (At this point there is no indication of an identity.) Indeed, we can define such an operation on the set of points satisfying any equation of the form $y^2 = x^3 + ax^2 + bx + c$ under the nondegeneracy condition of no repeated zeros used in Siegel's Theorem (Theorem 2.13 on p. 54).

Let E denote the set of points (x, y) with $y^2 = x^3 + ax^2 + bx + c$, assume that the cubic has no repeated zeros, and define a binary operation $+$ on the curve E as follows. If P and Q are points on E, then the line through P and Q meets E in exactly one further point, say (x, y). The reflection $R = (x, -y)$ of (x, y) in the x-axis is then defined to be $P + Q$ (see Figure 5.1). The case $P = Q$ requires a notion of tangency (which can be defined for curves over any field, using order of vanishing), and then $2P$ is obtained by reflecting the unique other point of intersection of the line tangent to the curve at P in the x-axis. The tangent is well-defined by the nondegeneracy condition.

Exercise 5.2. Draw the curve $y^2 = x^2(x+1)$. Show that the tangent at $(1, 0)$ is not well-defined.

Theorem 5.3. *The set E with binary operation $+$ forms an Abelian group after adding one point "at infinity."*

A natural question is to ask what the identity of the group is, and this will be fully resolved – and the theorem proved – in the next chapter. At this stage, we have to confess that the identity element does not appear to exist – it is the point 'at infinity'. For now, we can think of this as a formal single point added to the plane with the property that it lies on any vertical line of the form $x = $ constant. We will give more justification for this claim and will return to the question when we have described a fascinating class of functions that lie behind the theory of elliptic curves – see Chapter 6. The identity element is the point at infinity, which in Figure 5.1 may be thought of as being reached by moving infinitely far up (or down) the y-axis. Additive inverses are given by reflection in the x-axis, so if $P = (x, y)$ then $-P = (x, -y)$. In Figure 5.2, the point $P = (0, 1)$ on the curve $y^2 = x^3 - 3x + 1$ is shown, with a sequence of points approaching $-P = (0, -1)$ shown being added to P; the third point of intersection is approaching 0, the point at infinity. Take care not to confuse 0, the point at infinity, with the origin $(0, 0)$.

Exercise 5.3. Draw a picture of the (x, y) plane with a unit sphere whose South pole is tangent to the plane at $(0, 0)$. Define a map from the plane

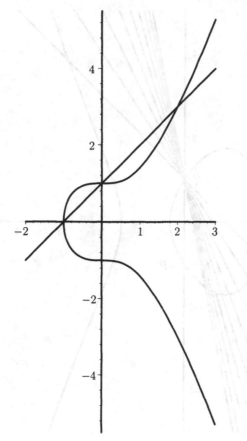

Figure 5.1. The binary operation on $y^2 = x^3 + 1$, showing $(2,3) + (0,1) + (-1,0) = 0$.

to the sphere by sending a point P on the plane to the unique point on the sphere that is collinear with P and the North pole. Show that the closure of the image of a curve $y^2 = x^3 + ax^2 + bx + c$ in the sphere contains the North pole. This single point may be thought of as giving a single "point at infinity" on the curve.

A more subtle question is how to verify the associative law for the binary operation. This is so familiar in ordinary addition that we are prone to overlook it. When it is encountered in matrix multiplication, it follows from the associative law in the underlying ring. Here a different principle is at work: Although it is still true that the law is inherited from the complex numbers, it is so via a bijection involving transcendental functions. In the twentieth century, algebraic geometers sought to understand this phenomenon in a more abstract way. The subject of Abelian varieties is a deep and powerful one

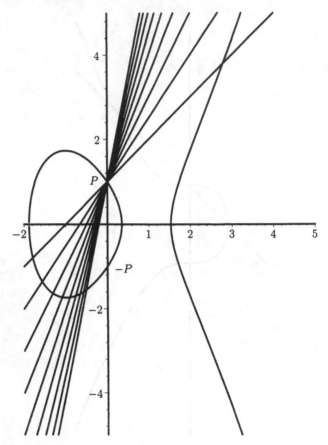

Figure 5.2. Points converging to $-P = (0, -1)$ showing the point at infinity.

about geometric objects, defined over arbitrary fields, with an Abelian group structure.

Exercise 5.4. Convince yourself that the associative law holds for an elliptic curve with the geometrical binary operation. In other words, choose a specific elliptic curve and plot it accurately. Choose three arbitrary points P, Q and R. Now demonstrate geometrically that the point you get by adding R to $P + Q$ is the same as the one you get by adding P to $Q + R$.

5.2 The Congruent Number Problem

In this section, we introduce a problem from antiquity that was recently re-interpreted using the theory of elliptic curves. A natural number-theoretic question arises with the familiar $(3, 4, 5)$ triangle in Figure 5.3. This triangle

– we may think of it as being defined by the triple of integers $(3, 4, 5)$ – has integral sides and integral area: What other triples of integers, or rationals, have this property?

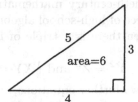

Figure 5.3. Six is a congruent number.

Example 5.4. There is a right-angled triangle with rational sides and area 5. The triple $(1\frac{1}{2}, 6\frac{2}{3}, 6\frac{5}{6})$ is Pythagorean: Expressing these as fractions over 6 and checking that

$$9^2 + 40^2 = 41^2$$

confirms the triangle with the sides given is right-angled. The area of the triangle is easily computed to be 5.

We will see later that there are arbitrarily complicated examples of this sort.

Example 5.5. The triple

$$\left(\frac{2017680}{1437599}, \frac{1437599}{168140}, \frac{2094350404801}{241717895860} \right)$$

gives a right-angled triangle with rational sides and area 6.

Examples 5.4 and 5.5 give examples of integer right-angled triangles with integral area by clearing fractions, but it is simpler to allow the sides to be rational, giving Definition 5.6.

Exercise 5.5. Find a rational right-angled triangle with area 7.

Such a triangle was known to Arab mathematicians of the twelfth century and rediscovered by Euler in the eighteenth century.

Definition 5.6. *An integer that is the area of a right-angled triangle with rational sides is called a* congruent number.

If an integer n is a congruent number and it is divisible by a square then the sides of any triangle showing that n is congruent can be scaled accordingly. Therefore we will assume without comment that it is sufficient to assume n is square-free in any discussion about whether it is a congruent number or not.

For millennia it has remained an unsolved problem to find an algorithm for checking whether a given integer is a congruent number. In recent times, Tunnell has shown how such an algorithm can be devised – see p. 243. What is remarkable about his work is the fact that although the proof uses a great deal of sophisticated twentieth-century mathematics, the way into the proof is a back-of-an-envelope piece of high-school algebra that goes as follows. If n is a congruent number, then there is a triple of rational numbers (X, Y, Z) with

$$X^2 + Y^2 = Z^2 \text{ and } \tfrac{1}{2}XY = n.$$

These two equations give two further equations,

$$(X \pm Y)^2 = X^2 \pm 2XY + Y^2 = Z^2 \pm 4n,$$

which can be written

$$\left(\frac{X \pm Y}{2}\right)^2 = \left(\frac{Z}{2}\right)^2 \pm n.$$

Multiplying the two equations (given by the choice of sign) together gives

$$\left(\frac{X^2 - Y^2}{4}\right)^2 = \left(\frac{Z}{2}\right)^4 - n^2.$$

Writing $v = (X^2 - Y^2)/4$ and $u = Z/2$, we obtain

$$v^2 = u^4 - n^2.$$

Now multiply by u^2 to obtain

$$(uv)^2 = u^6 - n^2 u^2.$$

Finally, writing $x = u^2$ and $y = uv$, we obtain a rational point (x, y) on the elliptic curve

$$y^2 = x^3 - n^2 x,$$

so a congruent number n gives rise to a rational point on an elliptic curve associated with n.

Example 5.7. If we start with the $(3, 4, 5)$ triangle, then following the steps just given, we obtain the rational point $(\frac{25}{4}, -\frac{35}{8})$ on the curve $y^2 = x^3 - 36x$ (following the convention that $X = 4$ should be the even side).

The curve $y^2 = x^3 - 36x$ has several integral points. In addition to $(0, 0)$ and $(\pm 6, 0)$, there is another, namely $(-3, 9)$. One might wonder if these also come from right-angled triangles. The answer is no, and Tunnell's Theorem (Theorem 5.8) suggests why not.

Notice that in the construction above, the x-coordinate we obtained turned out to be the square of a rational. Moreover, the denominator of x must be even. To see this, remember Theorem 2.1, which determines the Pythagorean triples. On clearing the denominators in the triple (X, Y, Z), one of X or Y must have an even numerator and Z cannot. Thus the denominator 2 in $u = Z/2$ cannot cancel.

Theorem 5.8. *Suppose n is a positive integer and (x, y) denotes a rational point on the elliptic curve $y^2 = x^3 - n^2 x$ with x equal to the square of a rational with an even denominator. Then n is a congruent number.*

PROOF. The proof uses the characterization of Pythagorean triples from Theorem 2.1. Initially we retrace some of the steps used earlier, but there is an ingenious twist at the end of the proof. Let $u = \sqrt{x} > 0$; by assumption $u \in \mathbb{Q}$. Write $v = y/u$ so

$$v^2 = y^2/u^2 = x(x^2 - n^2)/x = x^2 - n^2.$$

We therefore have a Pythagorean equation,

$$v^2 + n^2 = x^2. \tag{5.1}$$

Unfortunately, the resulting triangle does not have area n. Let t denote the denominator of v; then t is the denominator of x, by Equation (5.1). Now clear the denominators to obtain a Pythagorean triple $(t^2 v, t^2 n, t^2 x)$. Since t is even, we can write, for integers $a > b > 0$,

$$t^2 n = 2ab, \quad t^2 v = a^2 - b^2, \quad t^2 x = a^2 + b^2.$$

We claim there is a right-angled triangle with sides $2a/t, 2b/t$, and $2u$. This is easy to see:

$$\left(\frac{2a}{t}\right)^2 + \left(\frac{2b}{t}\right)^2 = 4(a^2 + b^2)/t^2 = 4t^2 x/t^2 = 4x = (2u)^2.$$

The area of this triangle is

$$\frac{1}{2}\frac{2a}{t}\frac{2b}{t} = \frac{2ab}{t^2} = n.$$

$\square$

Of course, this theorem does not solve the congruent number problem: What makes us think we know any more about the rational points on an elliptic curve than we do about congruent numbers? In fact, a great deal of research about rational points on elliptic curves took place in the twentieth century, so reducing a problem to finding rational points on elliptic curves allows many deep results to be applied. Even without invoking any of that, we already learn something quite surprising from Theorem 5.8.

Exercise 5.6. Let P be a rational point on the elliptic curve

$$y^2 = x^3 - n^2 x$$

which is neither $(0, 0)$ nor $(\pm n, 0)$. Using the algebraic doubling formula we used before, show that the x-coordinate of the resulting point is the square of a rational with an even denominator. Thus, if we can keep doing this, we obtain (potentially) infinitely many different rational right-angled triangles with area n. This was certainly not obvious from the definition of a congruent number.

The construction in Theorem 5.8 looked a little unwieldy. The next result is a neater formulation.

Theorem 5.9. *Suppose n is a positive integer and $x \in \mathbb{Q}$ has the property that $x, x + n, x - n$ are all rational squares. Put*

$$X = \sqrt{x+n} - \sqrt{x-n}, \, Y = \sqrt{x+n} + \sqrt{x-n}, \, Z = 2\sqrt{x}.$$

Then the triangle with sides X, Y, and Z is a rational right-angled triangle with area n.

Exercise 5.7. Confirm the statements in Theorem 5.9.

The shape of the equation defining the elliptic curve

$$y^2 = x(x+n)(x-n)$$

might lead you to think the conditions of Theorem 5.9 must always be satisfied for a rational point (x, y). The point $(-3, 9)$ is a counterexample however. Subsequently (see Section 7.3) we will come to understand when the conditions hold in terms of the group-theoretic structure of the curve.

Exercise 5.8. Take $n = 6$ and $P = \left(\frac{25}{4}, \frac{35}{8}\right)$. Find the rational right-angled triangle of area 6 corresponding to $2P$. Find the triangle corresponding to $4P$.

Exercise 5.9. Find a rational point P other than $(0, 0)$ or $(\pm 5, 0)$ on the curve $y^2 = x^3 - 25x$. Use P to find a rational right-angled triangle of area 5 different from Example 5.4.

The hard part of all this is to understand when rational points of the right kind exist in the first place. It is somewhat easier to show that as long as a rational point is not $(0, 0)$ or $(\pm n, 0)$, then one can go on constructing others with the right properties to guarantee the existence of many rational right-angled triangles with area n. Thus the problem comes down to finding for which n are there any nontrivial rational points. A satisfactory resolution of this problem has recently been given – see p. 243, where Tunnell's Theorem is stated. We will go on now to relate the geometric construction given before to the existence of a group structure on the curve.

Exercise 5.10. The geometric addition on elliptic curves allows us to construct new rational right-angled triangles from existing ones. In this exercise, the same construction is carried out directly on the triangle. Let (x, y, z) be a Pythagorean triple with $x < y < z$. This construction will find another Pythagorean triple (X, Y, Z) with

$$X = \frac{y^2 - x^2}{2z};$$

$$Y = \frac{2xyz}{y^2 - x^2};$$

$$Z = \frac{x^4 + y^4 + 6x^2y^2}{2z(y^2 - x^2)}.$$

Let P_x, P_y, P_z denote the vertices of the triangle, opposite the sides x, y and z respectively. Draw a circle with center P_z and radius x, and let Q be the point on this circle where a tangential line from P_x meets the circle (see Figure 5.4). Extend the line $P_x P_y$ to a point R at a distance $2z$ from P_x. Now draw a circle with center P_x through Q, and call S the point of intersection between the circle and the line $P_x P_y$. Finally, draw a line through S parallel to QR, and let T be the intersection of this line with $P_x Q$.

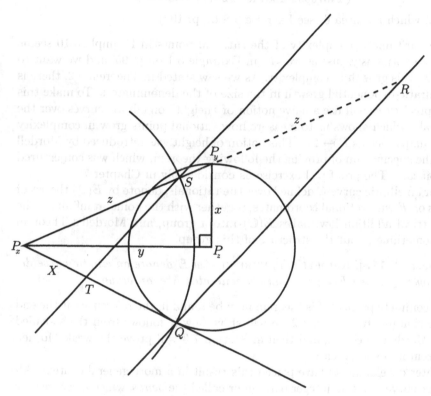

Figure 5.4. Constructing a new Pythagorean triple.

Prove that the distance from P_x to T is $X = \frac{y^2 - x^2}{2z}$, and show how to continue the construction to find the length Y.

Example 5.10. Consider the curve $y^2 = x^3 - 36x$ and the point $P = \left(\frac{25}{4}, \frac{35}{8}\right)$ on the curve. Then P is a rational point of infinite order, and we compute that

$$2P = \left(\frac{1442401}{19600}, -\frac{1726556399}{2744000}\right)$$

and

$$4P = \left(\frac{4386303618090112563849601}{23371016471594322055840 0}, -\frac{8703691090855808282759356506 26254401}{1129838581246361973721668449644 8000}\right).$$

The elliptic curve $y^2 = x^3 - 36x$ allows other rational right triangles with area 6 to be computed: Using the points above, one finds the right-angled triangles with sides

$$\left(\frac{120}{7}, \frac{7}{10}, \frac{1201}{70}\right)$$

and

$$\left(\frac{2017680}{1437599}, \frac{1437599}{168140}, \frac{2094350404801}{241717895860}\right),$$

each of which has area 6 (see Exercise 5.8 on p. 102).

The arithmetic complexity of the rational points in Example 5.10 seems to grow enormously, just as we saw in Example 5.1 on p. 95, and we want to quantify this growth in complexity. As we saw stated in Theorem 5.2, there is a quadratic-exponential growth in the size of the denominators. To make this more precise, we will use a naïve notion of "height" on elliptic curves over the rationals, which allows us to measure how rational points grow in complexity under maps such as $P \mapsto 2P$. This notion of height was introduced by Mordell with the specific aim of proving the following theorem, which was conjectured by Poincaré. The proof will exercise us considerably in Chapter 7.

For an elliptic curve E defined over the rationals, denote by $E(\mathbb{Q})$ the set of points on E with rational coordinates, together with the point 'at infinity'. The geometrical addition law makes $E(\mathbb{Q})$ into a group, and Mordell's Theorem says something about the structure of this group.

Theorem 5.11. [MORDELL'S THEOREM] *Let E denote an elliptic curve defined over $\mathbb{Q}$. Then $E(\mathbb{Q})$ is a finitely generated Abelian group.*

A complete proof of this theorem may be found in the references at the end of the chapter. In Section 7.2 we will show how it follows from the so-called weak Mordell Theorem, and then in Section 7.3 will prove the weak Mordell Theorem in a special case.

Later developments have placed this result in a more general context. Algebraic curves have an integer parameter called the *genus*, which measures the topological complexity of the underlying complex space. For an elliptic curve, the fundamental domain (this will be defined in Chapter 6 on p. 122) can be wrapped up into a torus (or doughnut) that is topologically a sphere with one handle. Roughly speaking, the genus counts the complexity in this topological sense when the underlying field of definition is the complex numbers. One of the great challenges facing mathematicians during the last century was to give a properly precise definition of genus when the base field is arbitrary. Remarkably, the genus of a curve seems to govern how many rational points it will have. Elliptic curves have genus one, giving a finitely generated group of rational points. Curves of genus greater than one have only finitely many rational points by a deep result of Faltings.

Theorem 5.11 means that $E(\mathbb{Q})$ is isomorphic to $\mathbb{Z}^r \times F$ for some $r \in \mathbb{N}$ and finite group F. The number r is called the *rank* of the curve, and it is conjectured that for any $r \in \mathbb{N}$ there is a curve defined over the rationals with rank r. The possibilities for the finite group F are more constrained – we will describe some of this in Section 5.4.

5.3 Explicit Formulas

In this section we will turn the geometric notion of addition on an elliptic curve into an algebraic formulation that allows computations to be made. We are going to work with a special form of cubic equation throughout this section. Subsequently, we will explain how the different forms of equation relate to each other. As a warm-up, we recommend the following exercise.

Exercise 5.11. Let $p(x) = x^3 + ax^2 + bx + c = (x - \lambda_1)(x - \lambda_2)(x - \lambda_3)$. Find expressions in a, b and c for $\lambda_1\lambda_2\lambda_3$, $\lambda_1\lambda_2 + \lambda_1\lambda_3 + \lambda_2\lambda_3$, and $\lambda_1 + \lambda_2 + \lambda_3$.

Given points $P_1 = (x_1, y_1)$ and $P_2 = (x_2, y_2)$ on the elliptic curve

$$y^2 = x^3 + ax + b,$$

explicit formulas may be found for x_3 and y_3, where $P_1 + P_2 = (x_3, y_3)$. Case I: If $x_1 \neq x_2$, then the line joining P_1 to P_2 has equation

$$\frac{y - y_1}{x - x_1} = \frac{y_2 - y_1}{x_2 - x_1},$$

so

$$y = \underbrace{\left(\frac{y_2 - y_1}{x_2 - x_1}\right)}_{\alpha} x + \underbrace{\left(\frac{x_2 y_1 - x_1 y_2}{x_2 - x_1}\right)}_{\beta}.$$

Substituting this into the equation

$$y^2 = x^3 + ax + b$$

for the curve gives

$$(\alpha x + \beta)^2 = x^3 + ax + b,$$

whose roots are the x-coordinates x_1, x_2, x_3 of the three points of intersection with the curve. By the sum of roots formula in Exercise 5.11, we must have

$$x_1 + x_2 + x_3 = \alpha^2,$$

so

$$x_3 = \alpha^2 - x_1 - x_2 = \left(\frac{y_2 - y_1}{x_2 - x_1}\right)^2 - x_1 - x_2.$$

Reflecting in the x-axis gives P_3, so

$$y_3 = -\alpha x_3 - \beta = -\left(\frac{y_2 - y_1}{x_2 - x_1}\right)x_3 - \left(\frac{x_2 y_1 - x_1 y_2}{x_2 - x_1}\right)$$

or

$$y_3 = \alpha(x_1 - x_3) - y_1.$$

Case II: Assume that $x_1 = x_2$ and $y_1 = y_2$. Let $y = \alpha x + \beta$ be the equation of the tangent to the curve at (x_1, y_1). By implicit differentiation of the equation $y^2 = x^3 + ax + b$, we obtain

$$\alpha = \frac{3x_1^2 + a}{2y_1},$$

and hence

$$x_3 = \left(\frac{3x_1^2 + a}{2y_1}\right)^2 - 2x_1, \quad y_3 = \left(\frac{3x_1^2 + a}{2y_1}\right)\left(\left(\frac{3x_1^2 + a}{2y_1}\right)^2 - 3x_1\right) + y_1,$$

so $y_3 = \alpha(x_1 - x_3) - y_1$.

Case III: If $x_1 = x_2$ and $y_1 = -y_2$, then $P_1 = -P_2$ so P_3 is the point at infinity.

Notice that all the formulas are *rational functions* (quotients of polynomials) with coefficients in the same field as a and b. This suggests there is a closure property as follows. Let $\mathbb{L}$ denote any field over which the curve is defined, and write $E(\mathbb{L})$ for the set of points with coefficients in $\mathbb{L}$ together with the point at infinity. Then $P_1, P_2 \in E(\mathbb{L})$ implies that $P_1 + P_2 \in E(\mathbb{L})$. Thus the group operation is well-defined on $E(\mathbb{L})$. Actually, some care needs to be taken if the characteristic of $\mathbb{L}$ is 2 or 3, starting with a different form of equation. We will discuss this further in Section 5.3.2.

5.3.1 Torsion Points

Later, we will give a more precise explanation of the identity element for the group operation. For the moment, we continue to think of the identity as the point at infinity, so an equation such as $2P = 0$ on the curve E means that a vertical line is a tangent to E at the point P. This allows us to speak of *torsion points* on an elliptic curve E: P is a point of order dividing n if $nP = 0$ in this geometrical sense. As we will see, the geometrical definition really gives a group structure to the points on the elliptic curve, and thus the usual terms from group theory such as "torsion" and "order" can be applied.

Example 5.12. Consider the curve $E : y^2 = x^3 + 1$, and let $P = (2, 3)$. Using the formulas, we find

$$2P = (0, 1), 3P = 2P + P = (-1, 0), 4P = 3P + P = (0, -1) = -2P.$$

It follows that $6P = 0$ (so P is a torsion point with respect to the group structure on the curve), and since $P, 2P, 3P \neq 0$, the point P has order 6 (see Figure 5.5).

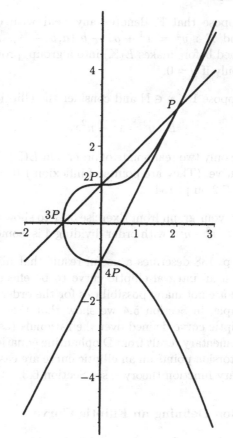

Figure 5.5. The point $P = (2, 3)$ has order 6 on $y^2 = x^3 + 1$.

Exercise 5.12. Find the order of the point $(3, 8)$ on the elliptic curve

$$y^2 = x^3 - 43x + 166.$$

Exercise 5.13. Find the order of the point $(0, 16)$ on the elliptic curve

$$y^2 = x^3 + 256.$$

Exercise 5.14. Find the order of the point $(\frac{1}{2}, \frac{1}{2})$ on the elliptic curve

$$y^2 = x^3 + \frac{1}{4}x.$$

Exercise 5.15. Find the order of the point $(-\frac{1}{3}, \frac{1}{2})$ on the elliptic curve

$$y^2 = x^3 - \frac{1}{3}x + \frac{19}{108}.$$

Exercise 5.16. Suppose that $\mathbb{K}$ denotes any field with characteristic not equal to 2 or 3, and $E : y^2 = x^3 + ax + b$ $(a, b \in \mathbb{K})$. Assuming the binary operation defined before makes $E(\mathbb{K})$ into a group, prove that $P = (x, y)$ has order 2 if and only if $y = 0$.

Exercise 5.17. Suppose $1 \leqslant n \in \mathbb{N}$ and consider the elliptic curve

$$E : y^2 = x^3 - n^2 x.$$

Prove that there are only two real points of order 3 in $E(\mathbb{R})$. Mark these points on a graph of the curve. (They are points of inflexion.) It may be interesting to look at Example 7.2 on p. 134.

Exercise 5.18. Use your graph from Exercise 5.17 to show that the subgroup of real points on $y^2 = x^3 - n^2 x$ with order dividing 4 is isomorphic to $C_2 \times C_4$.

Exercise 7.5 on p. 138 describes a useful result that allows all the rational torsion points on an integral elliptic curve to be effectively determined. When $\mathbb{K} = \mathbb{Q}$, there are not many possibilities for the orders of torsion points in $E(\mathbb{Q})$. For example, in Section 5.4 we show that there are no points of order 11 on any elliptic curve defined over the rationals (assuming a difficult but, in principle, elementary result from Diophantine equations). On the other hand, the complex torsion points on an elliptic curve are easy to describe once we have the necessary function theory – see Section 6.3.

5.3.2 The Equation Defining an Elliptic Curve

At several points we have used equations of differing shapes to define an elliptic curve. In the statement of Siegel's Theorem (Theorem 2.13) we set y^2 equal to a cubic in x with no repeated zeros. The addition formulas in the last section were computed using a special type of cubic. It is fair to ask just what is the correct definition in general. In Chapter 6 we will see that a pair of complex functions parametrize a curve of the shape $y^2 = x^3 + ax + b$ in which the right-hand side has no repeated zeros. Because of his important work in the area, this equation became known as a *Weierstrass equation* or *Weierstrass model*. We will see that the geometric definition of addition does indeed impose a group structure on the complex solutions of that equation. However, the explicit formulas define a group structure over any field of characteristic other than 2 or 3 (as does the geometrical definition of the group operation when a suitable notion of tangency is developed). We will have to ask you to take this statement on trust, or apply the Lefschetz[3] principle.

---

[3] The Lefschetz principle says, in effect, that if an algebraic formula holds in $\mathbb{C}$, then it will hold in any field where it makes sense. Although this is a valid principle, generally it is best used as a pointer toward phenomena that deserve to be better understood, rather in the way that algebraic geometers came to understand elliptic curves and their generalizations.

Suppose then that $\mathbb{F}$ denotes any field. It is possible to develop a theory of elliptic curves from one equation, regardless of the characteristic. When using the Weierstrass equation in characteristic 2 or 3, one cannot define tangents adequately (look at what happens when you differentiate). Tate used a more general equation with the following shape:

$$E : y^2 + a_1 xy + a_3 y = x^3 + a_2 x^2 + a_4 x + a_6 \qquad (5.2)$$

with $a_1, a_2, a_3, a_4, a_6 \in \mathbb{F}$ satisfying the non-degeneracy condition that every point on the curve has a unique tangent. This condition is equivalent to the non-vanishing of a complicated polynomial expression. For the special case in which Equation (5.2) takes the form $y^2 = x^3 + ax + b$, this non-degeneracy condition is equivalent to the cubic having no repeated zeros, and therefore to the condition that $4a^3 + 27b^2 \neq 0$ (see Exercise 2.14). The addition formulas can all be worked out for this general equation, in any characteristic. However, the formulas are significantly more complicated and this can hinder the development of intuition about the group law. This is why we prefer to develop the theory for the Weierstrass equation. The Equation (5.2) became known as a *generalized Weierstrass equation*, although it is becoming usual to refer to this too as a *Weierstrass equation*. The reader should beware that the modern literature on elliptic curves tends to work with the generalized equation.

The following exercise shows how gory the associative law can be when expressed in terms of the algebraic formula, even for the simplest form of equation.

Exercise 5.19. Using just the Weierstrass equation $y^2 = x^3 + ax + b$, verify the associative law for addition on an elliptic curve using the algebraic formulas from Section 5.3. Different formulas are required depending upon whether the x-coordinates are equal or not. Even doing one special case of

$$P + (Q + R) = (P + Q) + R$$

is tiresome and requires a great deal of both paper and patience.

Although we do not have the space to develop the algebraic geometry needed to properly develop a theory of elliptic curves over arbitrary fields, we recommend doing the following exercise to get a feel for elliptic curves over a finite field.

Exercise 5.20. Let E denote the elliptic curve $y^2 = x^3 - 2$. Find the order of the point $(3, 5)$ in the group $E(\mathbb{F}_7)$. What is the order of $E(\mathbb{F}_7)$? Do the same over other fields $\mathbb{F}_p$ for primes p. Can you detect any restrictions of the resulting group orders? For a precise result on this theme consult Hasse's Theorem (Theorem 11.11 on p. 240).

In several respects the group $E(\mathbb{F})$, where $\mathbb{F}$ denotes a finite field, can be studied along the lines that we studied $\mathbb{F}^*$. The two groups will often exhibit properties that can be directly related – and this phenomenon is useful in cryptography and coding theory. Earlier we proved that $\mathbb{F}^*$ is always a cyclic group. Therefore a natural question is to ask for the structure of $E(\mathbb{F})$.

Exercise 5.21. *Let $\mathbb{F}$ be a finite field. Prove that $E(\mathbb{F})$ is always a cyclic group or a direct product of two cyclic groups. Find an example where the group has two nontrivial cyclic factors.

5.4 Points of Order Eleven

The structure of the points of finite order in the group $E(\mathbb{Q})$ for an elliptic curve defined over the rationals is very constrained: A deep result of Mazur says that the torsion subgroup of $E(\mathbb{Q})$ must be isomorphic to $\mathbb{Z}/n\mathbb{Z}$ for some n, $1 \leqslant n \leqslant 12$, $n \neq 11$, or to $\mathbb{Z}/2\mathbb{Z} \oplus \mathbb{Z}/n\mathbb{Z}$, $1 \leqslant n \leqslant 4$. Proving this important result requires more material, but we can exhibit one nontrivial constraint (assuming a difficult Diophantine result and using some elementary properties of the geometry of the rational projective plane $\mathbb{P}^2(\mathbb{Q})$). If you have not encountered projective space, postpone this section until you have read Section 6.2. In what follows, we use little more than the geometric definition of addition on an elliptic curve to paint a putative rational point of order 11 into a corner where it cannot exist.

Theorem 5.13. *If E is an elliptic curve defined over $\mathbb{Q}$, then $E(\mathbb{Q})$ has no point of order 11.*

PROOF. Assume that P is a point in $E(\mathbb{Q})$ with order 11. Then no three points of $S = \{0, P, 3P, 4P\}$ could lie on a straight line because if A, B, C are collinear then $A + B + C = 0$ by the geometric definition of group addition. Since P has order 11, this last equation is impossible for three distinct points from S.

It follows that there is a nonsingular linear map on $\mathbb{P}^2(\mathbb{Q})$ sending

$$0 \to [0,1,0], P \to [1,0,0], 3P \to [0,0,1], \text{ and } 4P \to [1,1,1].$$

To see this, notice first that of the four points

$$[0,1,0], [1,0,0], [0,0,1], [1,1,1],$$

no three are collinear, by checking the various determinants. Given any four points with homogenous coordinates $\mathbf{v}_1, \mathbf{v}_2, \mathbf{v}_3, \mathbf{v}_4$, the matrix

$$M = [a\mathbf{v}_1^t | b\mathbf{v}_2^t | c\mathbf{v}_3^t]$$

will, for any $a, b, c \neq 0$, send

$$[1, 0, 0] \rightarrow \mathbf{v}_1,$$
$$[0, 1, 0] \rightarrow \mathbf{v}_2,$$
$$[0, 0, 1] \rightarrow \mathbf{v}_3, \text{ and}$$
$$[1, 1, 1] \rightarrow a\mathbf{v}_1 + b\mathbf{v}_2 + c\mathbf{v}_3.$$

The equation $a\mathbf{v}_1 + b\mathbf{v}_2 + c\mathbf{v}_3 = \mathbf{v}_4$ has a unique solution with a, b, c all nonzero by the non-collinearity assumption. Thus, by applying a change of variables in $\mathbb{P}^2(\mathbb{Q})$, we may assume that $0 = [0, 1, 0]$, $P = [1, 0, 0]$, $3P = [0, 0, 1]$, and $4P = [1, 1, 1]$.

Now let $5P = [x_1, x_2, x_3]$. Then, if ℓ_1 is the line through $5P$ and 0, and ℓ_2 is the line through $4P$ and P, $-5P \in \ell_1 \cap \ell_2$. Thus

$$r[0, 1, 0] + s[x_1, x_2, x_3] = t[1, 0, 0] + w[1, 1, 1],$$

for some $r, s, t, w \in \mathbb{Q}$. Comparing coefficients shows that

$$sx_1 = t + w; \; sx_2 + r = w; \; sx_3 = w.$$

If $s = 0$, then $P = 0$, which is impossible, so without loss of generality we may put $s = 1$. Then $r = x_3 - x_2$, and so

$$-5P = r[0, 1, 0] + s[x_1, x_2, x_3] = [x_1, x_3, x_3].$$

Similar arguments show that

$$-4P = [1, 0, 1],$$
$$-P = [x_1 - x_3, x_2, 0],$$
$$-3P = [0, x_3 - x_1 + x_2, x_3 - x_1], \text{ and}$$
$$2P = [x_1x_3 - x_1^2 + x_1x_2, x_3^2 - x_1x_3 + x_2x_3, x_3^2 - x_1x_3].$$

Since $11P = 0$, the points $5P, 4P, 2P$ are collinear. Taking the determinant of the matrix whose rows are the coefficients of these points, it follows that

$$x_3^3 - x_1^2x_2 + x_1^2x_3 + x_1x_2^2 - 2x_1x_3^2 = 0. \tag{5.3}$$

We claim that the only rational solutions to Equation (5.3) are

$$[0, 1, 0], [1, 1, 1], [1, 0, 0], [1, 0, 1], [1, 1, 0].$$

The notes at the end of the chapter provide references where this difficult result is proved. The point $5P$ must correspond to one of these possibilities. It cannot be $[0, 1, 0]$ because this is 0 and $5P \neq 0$. It cannot be $[1, 1, 1]$ because this is $4P$ and $5P = 4P$ implies $P = 0$. Similarly, it cannot be $[1, 0, 0]$ because this is P and $5P = P$ implies $4P = 0$. It cannot be $[1, 0, 1]$ because this is $-4P$ and $9P \neq 0$. It cannot be $[1, 1, 0]$ because this is $-P$ and $6P \neq 0$.

The contradiction proves that there can be no such point P. $\qquad\square$

5.5 Prime Values of Elliptic Divisibility Sequences

Elliptic curves generate a family of integer sequences that relate to several interesting parts of mathematics, including graph theory and cryptography. Suppose the elliptic curve E has a nontorsion point $P \in E(\mathbb{Q})$. Write

$$x(nP) = \frac{A_n}{B_n^2}, \tag{5.4}$$

in lowest terms, with A_n and B_n in $\mathbb{Z}$. An elliptic analog of the question about Mersenne primes asks how often B_n is prime as n varies. Because B_n grows so rapidly, this is potentially a method to find very large prime numbers.

Example 5.14. Let

$$E : y^2 = x^3 + 26, \quad P = (-1, 5).$$

The term B_{29} is a prime with 286 decimal digits.

Example 5.15. Let
$$E : y^2 = x^3 + 15, \quad P = (1, 4).$$
The term B_{41} is a prime with 510 decimal digits.

In some respects, this method for producing primes mirrors the situation with sequences such as the Mersenne and Fibonacci sequences, which are expected to produce large primes. For many years, the largest known primes have come from the Mersenne sequence. However, numerical investigation suggests that, for fixed E and P, the sequence (B_n) should only contain finitely many primes, and a non-rigorous probabilistic argument[4] suggests the number of prime terms should be uniformly bounded.

Just like the Mersenne and Fibonacci sequences, the sequence (B_n) is a *divisibility sequence*, meaning that $B_m | B_n$ whenever $m|n$. A consequence of this property, together with the rapid growth rate, is that there can only be finitely many primes in the sequence (B_n) if P is the multiple of another point, or if P is a non-integral point; moreover the terms B_n for large n cannot be prime if the index n is not itself prime. We say that a rational point is a *generator* if it is not the multiple of any other rational point.

Let E and E' be two elliptic curves defined over $\mathbb{Q}$. An *isogeny* is a nonzero homomorphism defined by rational functions on the coordinates of the points:

---

[4] Crudely, the Prime Number Theorem (Theorem 8.1) implies that the probability that a large integer N is prime is approximately $1/\log N$. The expected number of prime terms B_n with $n < x$ is (speculatively) approximately $\sum_{n<x} 1/\log B_n$. By Theorem 5.2 this sum converges as $x \to \infty$. It is known that the quantity h appearing in Theorem 5.2 is uniformly bounded below by some positive constant independent of the initial nontorsion rational point P and curve defined over the rationals E, provided the starting equation has minimal Δ.

$$\phi : E \to E'.$$

Taking $E = E'$, the multiplication-by-n map $P \mapsto nP$ for $n \in \mathbb{Z}$ is an example of an isogeny. The isogeny has an integral degree $m \geqslant 1$, which is the degree of the underlying rational functions that define it.

Exercise 5.22. *Prove that the degree of the isogeny $P \mapsto nP$ is n^2.

The curves E and E' are said to be *m-isogenous* if there is an isogeny of degree m between them. It can be proved that the multiplication-by-n map can be factorized as a composition of two isogenies, each of degree n.

Definition 5.16. *We say the point $P \in E(\mathbb{Q})$ is* magnified *if it is the image of a rational point under an isogeny of degree $m > 1$.*

The term was chosen because the height of a point increases under such a map – see Chapter 7 for more details about heights. The following result of Everest, Miller and Stephens will not be proved here.

Theorem 5.17. *If $P \in E(\mathbb{Q})$ is a magnified point, then B_n is a prime power for only finitely many n.*

Example 5.18. (1) The curve

$$y^2 = x^3 + x^2 - 4x$$

is 2-isogenous to the curve in Weierstrass form,

$$E : y^2 = x^3 + x^2 + 16x + 16.$$

The generator $(-2, 2)$ maps to the generator $P = (0, 4)$ on E. Thus the sequence of denominators for P on E contains only a finite number of prime powers.

(2) The curve

$$y^2 = x^3 - 9x + 9$$

is 3-isogenous to the curve in Weierstrass form,

$$E : y^2 = x^3 - 189x - 999.$$

The generator $(1, 1)$ maps to the generator $P = (-8, 1)$ on E. Thus the sequence of denominators for P on E contains only a finite number of prime powers.

Call the number of distinct prime divisors of an integer its *length*. The following conjecture has arisen from work of Everest and King.

Conjecture 5.19. Given a fixed bound on the length, there are only finitely many terms B_n with length below that bound.

5.5.1 The curve $u^3 + v^3 = D$

This section shows that the primality question can be answered in complete generality for curves in homogenous form.

Theorem 5.20. *Suppose E denotes a curve defined by an equation*

$$u^3 + v^3 = D \tag{5.5}$$

for some nonzero $D \in \mathbb{Q}$. Let P denote a nontorsion $\mathbb{Q}$-rational point. Write, in lowest terms,

$$P = \left(\frac{A_P}{B_P}, \frac{C_P}{B_P} \right).$$

Then the integers B_P are prime powers for only finitely many $\mathbb{Q}$-points P.

Note that the shape of the rational points is slightly different; the denominators of the x and y coordinates are not compelled to be powers. These curves, although not in the form to which we are accustomed, are still elliptic curves. The geometric addition used before works here and defines a group. As we shall see, a simple transformation puts them into the more usual form.

Example 5.21. As Ramanujan famously pointed out, the taxicab equation[5]

$$x^3 + y^3 = 1729, \tag{5.6}$$

has two distinct integral solutions. These give rise to points

$$P = (1, 12) \text{ and } Q = (9, 10)$$

on the elliptic curve defined by Equation (5.6). The only rational points on Equation (5.6) that seem to yield prime denominators are $2Q$ and $P + Q$ (and their inverses).

PROOF OF THEOREM 5.20. There is a transformation between the homogenous model given by Equation (5.5) and the Weierstrass model,

$$y^2 = x^3 - 2^4 3^3 D^2.$$

The transformations are given by

$$x = \frac{2^2 3 D}{u + v}, \qquad y = \frac{2^2 3^2 D(u - v)}{u + v},$$

$$u = \frac{2^2 3^2 D + y}{6x}, \qquad v = \frac{2^2 3^2 D - y}{6x}.$$

---

[5] Srinivasa Ramanujan was a largely self-taught mathematical genius. According to C. P. Snow, on one of G. H. Hardy's visits to Ramanujan in the hospital in Putney, Hardy said "I thought the number of my taxicab was 1729. It seemed to me rather a dull number." To which Ramanujan replied, "No, Hardy! It is a very interesting number. It is the smallest number expressible as the sum of two cubes in two different ways."

Writing $x = X/Z^2$ and $y = Y/Z^3$, where $\gcd(X, Z) = \gcd(Y, Z) = 1$, it follows that

$$u = \frac{2^2 3^2 D Z^3 + Y}{6XZ}.$$

If X divides the numerator of u, then X divides $2^6 3^3 D^2$. By Siegel's Theorem (Theorem 2.13), this can only happen finitely often. Since Z is coprime to the numerator, apart from a finite number of points, the denominator of u always has two nontrivial coprime factors. $\qquad\square$

Exercise 5.23. Prove that any integer solutions to the equation $u^3 + v^3 = D$ have $\max\{|u|, |v|\} \leqslant 2\sqrt{\frac{|D|}{3}}$.

5.5.2 Higher Rank Considerations

Let E denote an elliptic curve, defined over $\mathbb{Q}$. We say rational points P and Q are *independent* if no integer linear combination $mP + nQ$ can represent the point at infinity unless $m = n = 0$.

Theorem 5.22. *Let E denote an elliptic curve, defined over $\mathbb{Q}$, and suppose that P and Q denote independent rational points both of which are magnified under the same isogeny. Write*

$$x(nP + mQ) = \frac{A_{n,m}}{B_{n,m}^2}. \tag{5.7}$$

Then there are only finitely many pairs (m, n) for which $B_{n,m}$ is prime.

This theorem will not be proved here. Examples of the phenomenon of simultaneous magnification under the same isogeny are not easy to find: The following example uses the generalized Weierstrass form (5.2).

Example 5.23. The elliptic curve

$$y^2 + xy = x^3 + x^2 - 156x + 2070$$

has independent generators $P = (3, 39)$ and $Q = (13, 43)$ that are magnified under the same 2-isogeny.

Remark 5.24. Probabilistic arguments together with results from some numerical experiments suggest that, for certain curves in Weierstrass form (5.2), if P and Q denote independent nontorsion rational points, then the denominator of $nP + mQ$ can be the square of a prime infinitely often. Indeed, there seem to be asymptotically $c \log X$ such primes with $|m|, |n| < X$. Of course, none of the numerical examples that are considered in these arguments use magnified points.

5.5.3 Elliptic Analogs of Zsigmondy's Theorem

Zsigmondy's Theorem (Theorem 1.16 on p. 28) has an elliptic analog.

Theorem 5.25. [SILVERMAN] *Let E denote an elliptic curve defined over $\mathbb{Q}$, in generalized Weierstrass form, and let $P = (x(P), y(P))$ denote a nontorsion rational point on E. Let $x(nP) = \frac{A_n}{B_n^2}$ in lowest terms. Then the elliptic divisibility sequence (B_n) satisfies a Zsigmondy theorem: For all sufficiently large n, B_n has a primitive divisor.*

In view of the fact that sequences such as (B_n) seem likely to contain only finitely many prime terms, Theorem 5.25 takes on a more interesting status, as a means of producing large primes from elliptic divisibility sequences.

Analogs of the precise bound in Theorem 1.15 hold for certain elliptic divisibility sequences. The next result is an explicit bound for the first appearance of a primitive divisor in a congruent number curve.

Example 5.26. Let E denote the curve

$$E : y^2 = x^3 - 25x$$

and let $P = (-4, 6)$. Then B_n has a primitive divisor for every $n > 1$.

The factorizations of B_n for this example, $2 \leqslant n \leqslant 8$, with the primitive divisors in bold, are shown in Table 5.2.

Table 5.2. Primitive divisors of (B_n).

n	B_n	Factorization
2	12	$2^2 \cdot \mathbf{3}$
3	2257	$\mathbf{37} \cdot \mathbf{61}$
4	1494696	$2^3 \cdot 3 \cdot 7^2 \cdot \mathbf{31} \cdot \mathbf{41}$
5	8914433905	$5 \cdot \mathbf{13} \cdot \mathbf{17} \cdot \mathbf{761} \cdot \mathbf{10601}$
6	178761481355556	$2^2 \cdot 3^2 \cdot 11 \cdot 37 \cdot 61 \cdot \mathbf{71} \cdot \mathbf{587} \cdot \mathbf{4799}$
7	62419747600438859233	$\mathbf{197} \cdot \mathbf{421} \cdot \mathbf{215153} \cdot \mathbf{3498052153}$
8	535422986282160209291248	$2^4 \cdot 3 \cdot 7^2 \cdot 31 \cdot 41 \cdot \mathbf{113279} \cdot \mathbf{3344161} \cdot \mathbf{4728001}$

There is a difference in the proof for the odd and even terms. For a sequence (B_n), define the *even Zsigmondy bound* of (B_n) to be the greatest even integer n for which B_n does not have a primitive divisor, and similarly define the *odd Zsigmondy bound* of (B_n) to be the greatest odd integer for which B_n does not have a primitive divisor.

Theorem 5.27. *Let E denote the elliptic curve*

$$E : y^2 = x^3 - T^2 x,$$

where $T \geqslant 1$ is a square-free integer. Let $P \in E(\mathbb{Q})$ denote a nontorsion point and write B_n^2 for the denominator of $x(nP)$. Then the even Zsigmondy bound of the sequence (B_n) is not greater than 18. If $x(P) < 0$, then the odd Zsigmondy bound of (B_n) is not greater than 3. If $x(P)$ is a square, then the odd Zsigmondy bound is not greater than 21.

In specific cases, the terms not covered by Theorem 5.27 can be checked on a computer; this is how Example 5.26 was computed. Theorem 5.27 will not be proved here, but the main idea is contained in the following exercise. The condition stated there for the absence of a primitive divisor is very similar to that found for the Mersenne numbers in Exercise 1.16(b) on p. 28.

Exercise 5.24. It can be shown that if B_n does not have a primitive divisor then

$$B_n \Big| n \prod_{p|n} B_{n/p}.$$

Assuming this, use Theorem 5.2 to deduce that n must be bounded.

5.6 Ramanujan Numbers and the Taxicab Problem

In view of Example 5.21 and the story concerning Ramanujan, integers N for which the Diophantine equation

$$N = x^3 + y^3$$

has two nontrivially distinct solutions are sometimes called *Ramanujan numbers*. Table 5.3 shows the first few of these; there are infinitely many such numbers. In the table $u^3 + v^3 = x^3 + y^3$.

Table 5.3. The first few Ramanujan numbers.

N	u	v	x	y
1729	1	12	9	10
4104	2	16	9	15
13832	18	20	2	24
20683	10	27	19	24
32832	18	30	4	32

Indeed, it turns out that for any k there are infinitely many numbers N with the property that N can be expressed as a nontrivial sum of two cubes in k essentially different ways. The smallest number $T(k)$ with this property is called the kth *taxicab number* or *Hardy–Ramanujan number*. Table 5.4 shows the known taxicab numbers with the pairs whose cubes sum to the number, and the discoverer.

Table 5.4. The first few taxicab numbers.

k	$T(k)$	Pairs	Discoverer
1	2	1, 1	
2	1729	1, 12 9, 10	de Bessy (1657)
3	87539319	167, 436 228, 423 255, 414	Leech (1957)
4	6963472309248	2421, 19083 5436, 18948 10200, 18072 13322, 16630	Rosenstiel et al. (1991)
5	48988659276962496	38787, 365757 107839, 362753 205292, 342952 221424, 336588 231518, 331954	Wilson (1997)

It is suspected that

$$T(6) = 24153319581254312065344.$$

NOTES TO CHAPTER 5: The footnote about Bachet's equation on p. 93 is taken from the book of Silverman and Tate [143]. A very thorough treatment of all aspects of elliptic curves is given in Silverman's books [139], [142], and aspects of elliptic curves close to the topics in number theory we study are in Koblitz's book [89]. These books are highly recommended to any reader interested in learning more about elliptic curves. The construction in Exercise 5.10 on p. 102 was shown to us by Bartholdi, and we thank him for permission to include it here. The congruent number problem and its connection to elliptic curves are described in detail in Koblitz's book [89]. Mordell's Theorem appears first in his paper [110]; the paper of Poincaré mentioned is [116]. An attractive historical account of Mordell's theorem may be found in the paper of Cassells [26]. Faltings' Theorem on higher-genus curves appears in his papers [61] and [62]. An account of some of the background needed for this proof appears in the conference proceedings [34] edited by Cornell and Silverman. There are expositions of Faltings' proof by Deligne [41] and Szpiro [149]. The claim about the integral solutions to Equation (5.3) may be found in several places, including a paper [14] by Billing and Mahler; the presentation in Section 5.4 comes from a course taught by Silverberg at Ohio State University. Mazur's Theorem appeared first in his paper [105]; a treatment may also be found in Silverman's book [139]. Elliptic divisibility sequences are discussed in the monograph [58, Chapter 10] by Everest, van der Poorten, Shparlinski and Ward. The incidence of primes in these sequences has been studied by Chudnovsky and Chudnovsky [30] (Example 5.14 is taken from that paper), Einsiedler, Everest and Ward [48] and Rogers [131]. Theorem 5.17 appears in the paper [56] of Everest, Miller and Stephens; Example 5.18

comes from Cremona's Web site [37]. More on Conjecture 5.19 may be found in a paper of Everest and King [54]. More on Remark 5.24 may be found in a paper of Everest, Rogers and Ward [57] or Rogers' thesis [131]. Exercise 5.23 is taken from the book [143, p. 149] by Silverman and Tate. Theorem 5.25 is proved in Silverman's paper [140]; Example 5.26 and Theorem 5.27 are taken from a paper of Everest, McLaren and Ward [55]. References for the taxicab numbers in Table 5.4 may be found in Sloane's on-line encyclopedia of integer sequences [144]; there is an elementary account of the connection between $T(2)$ and elliptic curves in an accessible paper by Silverman [141], and the calculation of $T(5)$ is described in an article by Wilson [163].

Elliptic Functions

Elliptic curves can be viewed from many different mathematical perspectives. In the last chapter, they were seen as primarily geometrical objects; in this chapter, we start by emphasizing their relationship with some classical transcendental functions from complex analysis. To motivate the material in this chapter, recall that the trigonometric functions sine and cosine parametrize the points on the circle S^1. The rational points on the circle in turn parametrize Pythagorean triples. This gives a triangle of ideas involving the circle: in one corner are the classical transcendental functions, in another a compact group, and in the third a connection to a Diophantine problem. In the last chapter, we saw two corners of an analogous triangle involving elliptic curves. Rational points on elliptic curves give solutions to various Diophantine problems. Our next goal is to fill out the third corner of the elliptic triangle by finding transcendental functions that parametrize the points on elliptic curves. An important by-product of our work will be the justification that the operation defined by geometry in Chapter 5 really satisfies the axioms for a group. (See Theorem 6.5 and the comments just after.)

6.1 Elliptic Functions

Let $L \subseteq \mathbb{C}$ denote a lattice in the complex plane. This means L is the set of integer linear combinations of two complex numbers w_1 and w_2 that are linearly independent over $\mathbb{R}$. Write $\langle \omega_1, \omega_2 \rangle$ for the lattice $\omega_1 \mathbb{Z} + \omega_2 \mathbb{Z} \subseteq \mathbb{C}$. More generally, a lattice in $\mathbb{R}^n$ is any subgroup isomorphic to $\mathbb{Z}^n$; a lattice in $\mathbb{C}$ coincides with this definition by viewing $\mathbb{C}$ as $\mathbb{R}^2$.

One of the ways lattices of different dimensions arise naturally is in the study of periodic functions. The best-known example is the exponential function

$$\mathsf{e} : \mathbb{R} \to S^1 = \{z \in \mathbb{C} \mid |z| = 1\}$$
$$x \mapsto e^{ix}$$

This is a periodic function because it satisfies $e(x + 2\pi) = e(x)$ for all $x \in \mathbb{R}$, so e is periodic with respect to the one-dimensional lattice $2\pi\mathbb{Z} \subseteq \mathbb{R}$.

We are interested in complex functions f with the doubly-periodic property that

$$f(z + \omega_1) = f(z + \omega_2) = f(z),$$

that is, functions that are *periodic with respect to L* or *L-periodic*.

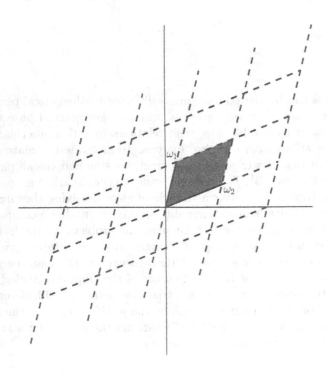

Figure 6.1. The lattice L spanned by ω_1 and ω_2 in $\mathbb{C}$.

The lattice L is represented as a discrete subset of $\mathbb{C}$ in Figure 6.1: The points of L are the points where the dashed lines intersect. The shaded region

$$\Pi = \{r_1\omega_1 + r_2\omega_2 \mid 0 \leqslant r_1, r_2 < 1\}$$

is a *fundamental domain* for the quotient $\mathbb{C}/L$ in the sense that each coset of L has exactly one representative in Π. The L-periodic function analogous to the exponential function that we will study is called the Weierstrass $\wp$-function corresponding to L. For any $z \notin L$, this is defined to be

$$\wp_L(z) = \frac{1}{z^2} + \sum_{0 \neq \ell \in L} \left\{ \frac{1}{(z - \ell)^2} - \frac{1}{\ell^2} \right\}. \tag{6.1}$$

The elements of L have to be enumerated in some way in order to define the sum. For the moment, suppose some enumeration $L\backslash\{0\} = \{\ell_1, \ell_2, \dots\}$ has been fixed and define $\sum_{0 \neq \ell \in L} f(\ell)$ to be $\sum_{n=1}^{\infty} f(\ell_n)$. We will first prove that the series in Equation (6.1) converges absolutely. It follows that the order in which the enumeration takes place does not affect the value of the sum.

Lemma 6.1. *The series*

$$\wp_L(z) = \frac{1}{z^2} + \sum_{0 \neq \ell \in L} \left\{ \frac{1}{(z-\ell)^2} - \frac{1}{\ell^2} \right\}$$

is absolutely convergent for any $z \notin L$. The series defines a meromorphic function whose only singularities are double poles at each lattice point in L.

PROOF. Let z be any point not in L. Write

$$\frac{1}{(z-\ell)^2} - \frac{1}{\ell^2} = \frac{1}{\ell^3} \cdot \frac{2z - z^2/\ell}{(z/\ell - 1)^2}.$$

Since $|z/\ell - 1|$ is bounded below by a positive constant, there is a constant C_1 depending on z such that

$$\left| \frac{1}{(z-\ell)^2} - \frac{1}{\ell^2} \right| \leq \frac{C_1}{|\ell|^3}.$$

Therefore, it is enough to prove that the series $\sum_{0 \neq \ell \in L} |\ell|^{-3}$ converges. To see this, notice first that there is a constant $C > 0$ with the property that

$$|m\omega_1 + n\omega_2| \geq \frac{1}{C} \max\{|m|, |n|\}.$$

Exercise 6.1. Prove that there are $8k$ integer pairs (m, n) with $\max\{|m|, |n|\}$ equal to k. (See Figure 6.2, which suggests an inductive proof.)

It follows that

$$\sum_{0 \neq \ell \in L} |\ell|^{-3} = \sum_{(m,n) \neq (0,0)} \frac{1}{|m\omega_1 + n\omega_2|^3}$$

$$\leq C^3 \cdot \sum_{(m,n) \neq (0,0)} \frac{1}{\max\{|m|, |n|\}^3}$$

$$= C^3 \cdot \sum_{k=1}^{\infty} \frac{8k}{k^3} = 8C^3 \sum_{k=1}^{\infty} \frac{1}{k^2},$$

which converges. We have shown that the series defining $\wp_L(z)$ converges absolutely for $z \in \mathbb{C}\backslash L$.

Finally, it is clear that the only pole of $\wp_L$ in Π is a double pole at 0 since

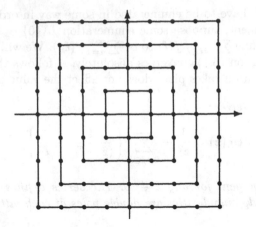

Figure 6.2. There are $8k$ integer pairs (m, n) with $\max\{|m|, |n|\} = k$.

$$\wp_L(z) - \frac{1}{z^2} = \sum_{0 \neq \ell \in L} \left\{ \frac{1}{(z - \ell)^2} - \frac{1}{\ell^2} \right\}$$

converges absolutely in Π. Similarly, for any $\ell' \in L$,

$$\wp_L(z) - \frac{1}{(z - \ell')^2} = \sum_{\substack{\ell' \neq \ell \in L; \\ \ell \neq 0}} \left\{ \frac{1}{(z - \ell)^2} - \frac{1}{\ell^2} \right\} + \frac{1}{z^2}$$

converges absolutely in $\Pi + \ell'$ for the same reason, showing that the only pole of $\wp_L$ in $\Pi + \ell'$ is a double pole at ℓ'. □

The absolute convergence of $\wp_L(z)$ means that Equation (6.1) can be differentiated term by term (see Exercise 6.2 below) to give

$$\wp_L'(z) = -2 \sum_{\ell \in L} \frac{1}{(z - \ell)^3}, \tag{6.2}$$

which also converges absolutely. It is clear that $\wp_L'(z)$ is periodic with respect to L since if $\ell_0 \in L$

$$\wp_L'(z + \ell_0) = -2 \sum_{\ell \in L} \frac{1}{(z + \ell_0 - \ell)^3} = -2 \sum_{\ell \in L} \frac{1}{(z - \ell)^3}$$

is just a rearrangement of the terms.

Our ultimate goal is to prove that $\wp_L$ is periodic with respect to L. Periodicity of $\wp_L'$ does not itself imply this, of course, but a simple argument does allow us to deduce it.

Exercise 6.2. (a) Let $f(z) = \sum_{n=0}^{\infty} c_n z^n$ be a complex power series with radius of convergence $R > 0$. Prove that f is differentiable (see Definition 8.18 on p. 170) on the set $\{z \in \mathbb{C} \mid |z| < R\}$ and that $f'(z) = \sum_{n=1}^{\infty} n c_n z^{n-1}$ on this set.

(b) Show how to use this to justify the expression Equation (6.2) by using the absolute convergence of the series defining $\wp_L$ to show that it may be expanded as a power series.

What we have done up to now might seem clumsy: Given a series whose terms are clearly differentiable, the most natural way to show it is differentiable is surely to differentiate term by term. This is a reasonable criticism, however it involves a more subtle notion of convergence called *uniform convergence* (see Section 8.5). Term-by-term differentiability is easily provable for power series, whose terms are simply monomials, but can be much trickier when the terms are more complicated functions. This alternative approach to the analyticity of $\wp_L$ is given in Exercise 8.20 on p. 173, using the concept of uniform convergence, once we have had time to introduce the concept properly.

Lemma 6.2. *The Weierstrass $\wp$-function $\wp_L$ is periodic with respect to L.*

PROOF. We want to prove that

$$\wp_L(z + \omega_1) = \wp_L(z + \omega_2) = \wp_L(z)$$

for all $z \notin L$. First, notice that by Equation (6.1) and Equation (6.2),

$$\wp_L(-z) = \wp_L(z) \text{ and } \wp_L'(-z) = -\wp_L'(z).$$

That is, $\wp_L(z)$ is an *even* function and $\wp_L'(z)$ is an *odd* function. Now fix i to be 1 or 2 and let

$$f(z) = \wp_L(z + \omega_i) - \wp_L(z).$$

Then f is differentiable for all $z \notin L$. Since $\wp_L'(z)$ is periodic with respect to L, we deduce that $f'(z) = 0$ for all $z \in \mathbb{C} \backslash L$, so f is constant on this open connected set.

To determine the constant value of f let $z = -\omega_i/2$. Then

$$f(-\omega_i/2) = \wp_L(\omega_i/2) - \wp_L(-\omega_i/2),$$

which shows that $f(-\omega_i/2) = 0$ since $\wp_L$ is an even function. It follows that f must be zero everywhere, showing that $\wp_L$ is periodic with respect to L. □

Definition 6.3. *An* elliptic function *is a meromorphic function $\mathbb{C} \to \mathbb{C}$ that is periodic with respect to a lattice L. If $L = \mathbb{Z}\omega_1 + \mathbb{Z}\omega_2$, then ω_1 and ω_2 are known as* periods. *With respect to a chosen basis $\{\omega_1, \omega_2\}$, the domain*

$$\Pi = \{r_1\omega_1 + r_2\omega_2 \mid 0 \leqslant r_1, r_2 < 1\}$$

for the lattice L is the fundamental domain.

Lemma 6.4. *An elliptic function with no poles in its fundamental domain is constant. Let $\Pi_\beta = \beta + \Pi$ be the fundamental domain translated by $\beta \in \mathbb{C}$, and let f denote an elliptic function with no zeros or poles on the boundary of Π_β. If the zeros of f in Π_β have orders m_i and the poles have orders n_j, then $\sum m_i = \sum n_j$.*

PROOF. The first statement is clear: Any such function would be bounded on Π, and therefore on all of $\mathbb{C}$, by periodicity, so it is a bounded entire function and therefore must be constant by Liouville's Theorem.

For the second statement, first notice that

$$\int_{\Pi_\beta} f(z)\, dz = 0$$

since f has the same values on opposite sides of Π_β, while dz changes sign. The result now follows by applying this to the elliptic function $g(z) = f'(z)/f(z)$. Near a zero z_0 of order m for f, g has a simple pole with residue m (that is, $g(z)$ behaves like $\frac{m}{z-z_0}$ near z_0). Near a pole z_0 of order n for f, g has a simple pole with residue $-n$ (that is, $g(z)$ behaves like $-\frac{n}{z-z_0}$).

Cauchy's Residue Theorem gives the result.   □

6.2 Parametrizing an Elliptic Curve

Lemma 6.4 will be used to prove the main result of this section: The values of $\wp_L(z)$ and $\wp_L'(z)$, for z lying in the fundamental domain, parametrize a complex elliptic curve. Before stating this important result, we return to the question raised at the end of Section 5.1: What is the identity element for the binary operation on an elliptic curve?

In order to answer this, we need to come clean about elliptic curves. The discussion in Section 5.1 concerned the set of solutions to an equation $y^2 = x^3 + ax^2 + bx + c$ in $\mathbb{R}^2$; these are just an *affine part* of the *real points* of the curve. A complex elliptic curve is really the set of *complex points* in *projective space* satisfying the projectivized version of the equation. The vague notion of adding a point 'at infinity' can be made precise by studying elliptic curves in this more natural setting of *projective space*.

Two-dimensional projective space $\mathbb{P}^2(\mathbb{C})$ is defined to be the set of equivalence classes

$$\mathbb{P}^2(\mathbb{C}) = \{(z_0, z_1, z_2) \in \mathbb{C}^3 \mid (z_0, z_1, z_2) \neq (0, 0, 0)\}/\sim,$$

where $(z_0, z_1, z_2) \sim (z_0', z_1', z_2')$ if there is a constant $\lambda \neq 0$ with

$$(z_0, z_1, z_2) = (\lambda z_0', \lambda z_1', \lambda z_2').$$

An element of $\mathbb{P}^2(\mathbb{C})$ is then an equivalence class, and we write

$$[z_0, z_1, z_2] = \{(z_0', z_1', z_2') \mid (z_0', z_1', z_2') \sim (z_0, z_1, z_2)\}$$

for the equivalence class containing (z_0, z_1, z_2).

The complex elliptic curve $E(\mathbb{C})$ associated with the equation

$$E : y^2 = x^3 + ax^2 + bx + c$$

is the subset of $\mathbb{P}^2(\mathbb{C})$ defined by

$$E(\mathbb{C}) = \{[z_0, z_1, z_2] \mid z_1^2 z_2 = z_0^3 + az_0^2 z_2 + bz_0 z_2^2 + cz_2^3\}.$$

Notice that this curve contains two parts. If $z_2 \neq 0$, then we can assume without loss of generality that $z_2 = 1$, so all the points $[z_0, z_1, 1]$ with

$$z_1^2 = z_0^3 + az_0^2 + bz_0 + c$$

lie on E. This is the *complex affine* part of the curve. There is exactly one point with $z_2 = 0$ (if $z_2 = 0$, then $z_0 = 0$ so z_1 must be nonzero), namely $[0, 1, 0]$. This point is the "point at infinity" on the curve.

We will write

$$E : y^2 = x^3 + ax^2 + bx + c$$

for the complex projective curve, suppressing the third variable (because it only contributes one point to the curve). We will always assume that the right-hand side has no repeated zeros. (See Exercise 2.14 for a simple formulation of this condition in the case $a = 0$.)

It will be useful to talk about the $\mathbb{K}$-points of an elliptic curve for other fields $\mathbb{K}$. The curve $E : y^2 = x^3 + ax^2 + bx + c$ is said to be *defined over* a field $\mathbb{L}$ if the coefficients a, b, c come from $\mathbb{L}$. For any field $\mathbb{K}$ containing $\mathbb{L}$, the $\mathbb{K}$-points of the curve, $E(\mathbb{K})$, are the points in E whose projective coordinates can be chosen in $\mathbb{K}$. Thus $E(\mathbb{C})$ is the complex projective curve. The following is a major result and most of this section will be devoted to the proof.

Theorem 6.5. *Let $L \subseteq \mathbb{C}$ denote a lattice with fundamental domain Π.*

(1) *There are constants $a = a(L)$ and $b = b(L)$ with $4a^3 + 27b^2 \neq 0$ such that, for all $z \in \mathbb{C} \backslash L$,*

$$\tfrac{1}{4} \wp_L'(z)^2 = \wp_L(z)^3 + a\wp_L(z) + b.$$

(2) *For $z \in \mathbb{C}/L$, the map $\pi : \Pi \to \mathbb{P}^2(\mathbb{Q})$ defined by $\pi(0) = [0, 1, 0]$ and*

$$\pi(z) = [\wp_L(z), \tfrac{1}{2}\wp_L'(z), 1], \quad z \neq 0,$$

defines a bijection between Π and the set of complex projective points on the elliptic curve $E : y^2 = x^3 + ax + b$.

(3) *Suppose $z_1, z_2, z_3 \in \Pi$ have images $\pi(z_i) = P_i, i = 1, 2, 3$. Then*

$$z_1 + z_2 + z_3 = 0$$

in Π if and only if P_1, P_2, and P_3 lie on a straight line.

The last part of the theorem is the long-awaited justification that the operation defined on the points of an elliptic curve in Chapter 5 is a group operation. Under the bijection

$$\pi : z \to [\wp_L(z), \tfrac{1}{2}\wp_L'(z), 1], \quad z \neq 0,$$

the fact that the point $0 = z \in \mathbb{C}/L$ corresponds to the point at infinity relates the geometrical idea of infinity on the projective curve to the analytic idea that $\wp_L(z) \longrightarrow \infty$ as $z \longrightarrow 0$. This is important if we work with the projective curve because the set of projective points forms a group with the point at infinity as the identity. Notice that this arises simply by transporting the group structure of $\mathbb{C}/L$ to the curve E. Theorem 6.5(3) says that the familiar addition in $\mathbb{C}$ is related, via the transcendental functions $\wp_L$ and $\wp_L'$, to the geometric addition on the projective curve. This transport of structure from the additive group $\mathbb{C}$ to the curve proves that the geometric binary operation on the projective curve really does satisfy the group axioms. Now the 'Lefschetz principle' (see the footnote on p. 108) shows that this result over $\mathbb{C}$ extends to verify the group law for elliptic curves over arbitrary fields in characteristic not equal to 2 or 3.

Exercise 6.3. Show that

$$\wp_L'(\omega_1/2) = \wp_L'(\omega_2/2) = \wp_L'((\omega_1 + \omega_2)/2) = 0.$$

Show that there are no other solutions of $\wp_L'(z) = 0$ with $z \in \Pi$.

Exercise 6.3 identifies the 2-torsion points on the elliptic curve with reference to the lattice L. The complex torsion on an elliptic curve can easily be described. We will take a brief interlude to apply Theorem 6.5 to the study of the complex torsion points on an elliptic curve. The proof of Theorem 6.5 will follow in Section 6.4.

6.3 Complex Torsion

Theorem 6.5 allows the torsion points on an elliptic curve to be understood in a way that is analogous to our understanding of torsion points on the circle: Since $e : \mathbb{R} \to \mathbb{S}^1$ has kernel $2\pi\mathbb{Z}$, it induces an isomorphism

$$e : \mathbb{R}/2\pi\mathbb{Z} \longrightarrow \mathbb{S}^1.$$

The distinct points of order dividing n in the additive group $\mathbb{R}/2\pi\mathbb{Z}$ are those of the form $\frac{2\pi j}{n} + 2\pi\mathbb{Z}$ for $j = 0, 1, \ldots, n-1$. We deduce that the points of order dividing n in $\mathbb{S}^1$ are those of the form $e(2\pi j/n) = e^{2\pi i j/n}$ for $j = 0, 1, \ldots, n-1$.

It is not difficult to find the points of order dividing n on $\mathbb{S}^1$. Theorem 6.5 repeats the trick for the problem of finding all points of order dividing n for the group operation on a complex elliptic curve. Given $1 \leqslant n \in \mathbb{N}$, the points

$$z = (r_1\omega_1 + r_2\omega_2)/n \text{ for } 0 \leqslant r_1, r_2 \leqslant n$$

all have $nz \equiv 0$ modulo L, and these are the n^2 points with order dividing n in the group $\mathbb{C}/\mathbb{L}$. These are torsion points on the complex curve. Deciding which of these points correspond to *rational* torsion points on the curve is a different and difficult question.

Exercise 6.4. Let $E_n(\mathbb{C})$ for $n \in \mathbb{N}$ denote the subgroup of points on a complex elliptic curve E whose order divides n. Show that

$$E_n(\mathbb{C}) \cong \mathbb{Z}/n\mathbb{Z} \oplus \mathbb{Z}/n\mathbb{Z}.$$

6.4 Partial Proof of Theorem 6.5

We are not going to prove all of Theorem 6.5; in particular we will not prove that the quantity $4a^3 + 27b^2$ is not zero. A complete account may be found in the references. What we will show is how the important equation in Theorem 6.5(1) arises.

PROOF OF THEOREM 6.5(1). Assume first that z has $|z| < |\ell|$ for all nonzero $\ell \in L$. Then the Taylor expansion about $z = 0$ gives

$$\frac{1}{(z-\ell)^2} - \frac{1}{\ell^2} = \frac{1}{\ell^2}(1 - z/\ell)^{-2} - \frac{1}{\ell^2} = \frac{2z}{\ell^3} + \frac{3z^2}{\ell^4} + \frac{4z^3}{\ell^5} + \cdots.$$

By absolute convergence of the series defining $\wp_L(z)$, we can rearrange the terms in

$$\wp_L(z) = \frac{1}{z^2} + \sum_{0 \neq \ell \in L} \left(\frac{2z}{\ell^3} + \frac{3z^2}{\ell^4} + \frac{4z^3}{\ell^5} + \cdots \right)$$

to get

$$\wp_L(z) = \frac{1}{z^2} + 2z\sum{}'\ell^{-3} + 3z^2\sum{}'\ell^{-4} + 4z^3\sum{}'\ell^{-5} + \cdots.$$

The $\sum'$ indicates that the sum is over the nonzero lattice points $\ell \in L$ only. For any $n \in \mathbb{N}$, the terms of the form $\ell^{-(2n+1)}$ as ℓ runs through the nonzero terms of L cancel out in pairs: $(-\ell)^{-2n-1} = -\ell^{-2n-1}$. It follows that $\sum'\ell^{-2n-1} = 0$, so the Laurent expansion of $\wp_L(z)$ about $z = 0$ looks like

$$\wp_L(z) = \frac{1}{z^2} + 3z^2 G_4(L) + 5z^4 G_6(L) + \cdots, \tag{6.3}$$

where

$$G_{2n}(L) = \sum{}'\ell^{-2n}, \ 1 \leqslant n \in \mathbb{N}.$$

This expression agrees with the classical result that even meromorphic functions only have even powers in their Laurent expansion at 0.

Consider the function

$$g(z) = \wp'_L(z)^2 - 4\wp_L(z)^3 + 60G_4(L)\wp_L(z) + 140G_6(L).$$

This function is analytic on Π, moreover g is periodic with respect to L because it is an algebraic expression in periodic functions. Finally, it can be checked that the Laurent expansion of $g(z)$ contains only positive powers of z. By Lemma 6.4, g must be a constant. Setting $z = 0$ shows that this constant value must be zero, so g is the zero function and hence the equation stated in the theorem holds (after dividing by 4).

Notice that $a = -15G_4(L)$ and $b = -35G_6(L)$. $\square$

Notice that Theorem 6.5(1) is a statement about all $z \in \mathbb{C} \backslash L$. In the proof, we have assumed that $|z| < |\ell|$ for all nonzero lattice points. This means in particular that the proof is valid for all points in the region $-\left(\frac{\omega_1 + \omega_2}{2}\right) + \Pi$; it follows for all $z \in \mathbb{C} \backslash L$ by periodicity.

Exercise 6.5. (a) Let $L = \langle 1, i \rangle$. Show that the corresponding curve E_L has equation $y^2 = x^3 + ax$ for some $a \in \mathbb{R}$.
(b) Let $L = \langle 1, \omega \rangle$, where ω denotes a cube root of unity. Show that the corresponding elliptic curve E_L has equation $y^2 = x^3 + b$ for some $b \in \mathbb{R}$.

PROOF OF THEOREM 6.5(2). We show that the map is a bijection, beginning with surjectivity. Suppose $\alpha \in \mathbb{C}$ is given. The function $\wp_L(z) - \alpha$ has two poles (actually one double pole) in Π so, by Lemma 6.4, it must have two zeros. To prove injectivity (which appears to be threatened by the existence of the two zeros) note that the two zeros are negatives of each other. This is because, for $z \notin L$, $\wp_L(-z) = \wp_L(z)$. However, $\wp'_L(-z) = -\wp'_L(z)$. Thus, the images of z and $-z$ will (usually) be distinct points on the curve, the only counterexamples arising when $\wp'_L(z) = 0$. By Exercise 6.3, this happens for only three values of z, namely $w_1/2, w_2/2$, and $(w_1 + w_2)/2$, but this is exactly when z and $-z$ define the same element of $\mathbb{C}/L$. $\square$

Finally, we show how an argument using complex analysis gives the third part of Theorem 6.5.

PROOF OF THEOREM 6.5(3). Let the equation of the line containing the points P_1 and P_2 be $y = mz + b$. Consider the function

$$f(z) = \wp'_L(z) - m\wp_L(z) - b.$$

This has three poles in Π (actually one triple pole) so, by Lemma 6.4, it has three zeros. Two of these are z_1 and z_2; let z_3 denote the third. Then P_1, P_2, and P_3 lie on the line $y = mz + b$ and (3) is seen by integrating the function $h(z) = zf'(z)/f(z)$ over a displaced parallelogram $\Pi_\beta = \beta + \Pi$, where β is chosen so that h has no singularities on the boundary Γ_β of Π_β shown in Figure 6.3.

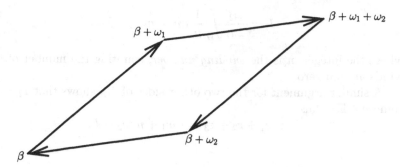

Figure 6.3. Integrating along the four sides of Γ_β.

The main part of the proof is to show that $z_1 + z_2 + z_3 \in L$. By Cauchy's Residue Theorem,

$$\frac{1}{2\pi i} \int_{\Gamma_\beta} h(z) \, dz = z_1 + z_2 + z_3 \qquad (6.4)$$

because h has a simple pole at each z_i with residue z_i. Now break the integral in Equation (6.4) into two parts corresponding to pairs of opposite sides in Γ_β:

$$\frac{1}{2\pi i} \int_{\Gamma_\beta} h(z) \, dz = \frac{1}{2\pi i} \left(\int_\beta^{\beta+\omega_1} h(z) \, dz + \int_{\beta+\omega_1+\omega_2}^{\beta+\omega_2} h(z) \, dz \right)$$

$$+ \frac{1}{2\pi i} \left(\int_{\beta+\omega_1}^{\beta+\omega_1+\omega_2} h(z) \, dz + \int_{\beta+\omega_2}^{\beta} h(z) \, dz \right)$$

$$= I_1 + I_2.$$

Substitute $z = w + \omega_2$ in the second integral of I_1, and use the periodicity of f to obtain

$$I_1 = \frac{1}{2\pi i} \left(\int_\beta^{\beta+\omega_1} \frac{z f'(z)}{f(z)} \, dz - \int_\beta^{\beta+\omega_1} \frac{(z+w_2) f'(z)}{f(z)} \, dz \right)$$

$$= \frac{w_2}{2\pi i} \int_\beta^{\beta+\omega_1} \frac{f'(z)}{f(z)} \, dz.$$

Now make the substitution $u = f(z)$ to deduce that

$$I_1 = \frac{w_2}{2\pi i} \int_\Omega \frac{1}{u} \, du,$$

where Ω is the image of the line joining β to $\beta+\omega_1$ in the variable u. Periodicity with respect to L means that Ω is a closed curve, so we finally obtain

$$I_1 = \frac{\omega_2}{2\pi i} \int_\Omega \frac{1}{u}\, du = m\omega_2 \in \mathbb{Z}\omega_2,$$

where the integer m is the *winding number*, counting the number of times Ω winds around zero.

A similar argument for the two other sides of Π_β shows that $I_2 = nw_1$ for some $n \in \mathbb{Z}$. Thus

$$z_1 + z_2 + z_3 = nw_1 + mw_2 \in L.$$

$\square$

Exercise 6.6. (a) Prove that, for any lattice $L \subseteq \mathbb{C}$,

$$G_8(L) = \tfrac{3}{7} G_4^2(L).$$

(b) More generally, prove that all the G_i $(i \geqslant 8)$ can be expressed as polynomials in G_4 and G_6 with rational coefficients.

Exercise 6.7. (a) Given any nonzero $c \in \mathbb{C}$, consider the map $L \to cL = L'$. Let E_L and $E_{L'}$ denote the corresponding elliptic curves. Prove that the map defines a group isomorphism between $E_L(\mathbb{C})$ and $E_{L'}(\mathbb{C})$.
(b) Prove that the map in (a) has the following effect upon the coordinates of the corresponding curves. If $y^2 = x^3 + ax + b$ is the equation defining E_L and $y^2 = x^3 + a'x + b'$ is the equation defining $E_{L'}$, show that the effect of the map in (a) is to take (x, y) to $(c^{-2}x, c^{-3}y)$. (Hint: Recall the definition of a and b from Theorem 6.5(1).)

Exercise 6.8. (a) Show that, for any lattice L and $c \in \mathbb{C}^*$,

$$G_4(cL) = c^{-4} G_4(L) \text{ and } G_6(cL) = c^{-6} G_6(L).$$

(b) Prove that any elliptic curve $y^2 = x^3 + ax + b$ with $ab = 0$ is parametrized by the Weierstrass $\wp$-function for some lattice L.

NOTES TO CHAPTER 6: The Lefschetz principle is discussed in Silverman [139, Section VI.6]. Theorem 6.5 is also proved in [139] along with a converse result: Given a and b with $4a^3 + 27b^2 \neq 0$, there exists a lattice L such that $\wp_L(z)$ and $\frac{1}{2}\wp'_L(z)$ parametrize the elliptic curve with equation $y^2 = x^3 + ax + b$. For an explanation of the remarkable phenomenon described in Exercise 6.6(b), consult Koblitz [89]. A classical treatment of elliptic functions from the analytic viewpoint is contained in Whittaker and Watson [160]; there are sophisticated accounts of elliptic functions and their role in number theory in the books of Apostol [5], Chandrasekharan [29], Lang [95] and Weil [159].

7

Heights

In this chapter we introduce a way to measure the arithmetic complexity of points on elliptic curves. This measurement of the *height* turns out to be an essential ingredient in understanding the structure of the rational points on an elliptic curve. Our understanding of heights will be a key ingredient in the proof of Mordell's Theorem in Section 7.2.

7.1 Heights on Elliptic Curves

Given a rational affine point $P = (\frac{M}{N}, *)$, where M and N are coprime integers, define the *naïve height* of P to be

$$H(P) = \begin{cases} \max\{|M|, |N|\} & \text{if } \frac{M}{N} \neq 0, \\ 1 & \text{if } M = 0. \end{cases}$$

Write $x(P)$ and $y(P)$ for the coordinates of an affine point $P = (x(P), y(P))$. Define the *logarithmic height* to be $h(P) = \log H(P)$.

The definition of the complex projective plane $\mathbb{P}^2(\mathbb{C})$ on p. 126 extends to higher dimensions: For any field $\mathbb{K}$, projective N-space over $\mathbb{K}$ is defined by

$$\mathbb{P}^N(\mathbb{K}) = \{(x_0, \ldots, x_N) \mid (x_0, \ldots, x_N) \neq (0, \ldots, 0)\}/\sim,$$

where $(x_0, \ldots, x_N) \sim (x_0', \ldots, x_N')$ if there is a constant $\lambda \in \mathbb{K}^*$ with

$$(x_0, \ldots, x_N) = \lambda(x_0', \ldots, x_N').$$

As before, we write $[x_0, \ldots, x_N]$ for the equivalence class (or point in projective space) containing the affine point $(x_0, \ldots, x_N)$.

The naïve height extends to projective space $\mathbb{P}^N(\mathbb{Q})$. Given a point $[\mathbf{y}]$ in $\mathbb{P}^N(\mathbb{Q})$, choose $\mathbf{x} = (x_0, \ldots, x_N) \in \mathbb{Z}^{N+1}$ in such a way that $[\mathbf{y}] = [\mathbf{x}]$ and

$$\gcd(x_0, \ldots, x_N) = 1.$$

Then the projective height

$$H : \mathbb{P}^N(\mathbb{Q}) \to \mathbb{R}$$

is defined by

$$H([\mathbf{x}]) = \max_{i=0,\dots,N}\{|x_i|\}.$$

Notice that this is compatible with the naïve height in the following sense: If $P = (x, y)$ is a point on the affine piece of $E(\mathbb{Q})$, then $H(P) = H([x, 1])$, where $[x, 1] \in \mathbb{P}^1(\mathbb{Q})$.

The logarithmic quantity $h(P)$ is a simple example of a Mahler measure:

$$\log \max\{|M|, |N|\} = m(Nx - M)$$

(see p. 150).

Examples 5.1 and 5.10 suggest that the number of decimal digits in the numerator and the denominator roughly quadruples each time a point is doubled. This is a manifestation of a general phenomenon, the *duplication formula*.

Theorem 7.1. [DUPLICATION FORMULA] *Let E denote an elliptic curve defined over the rationals, and let P be a point in $E(\mathbb{Q})$. Then*

$$h(2P) = 4h(P) + \mathrm{O}(1), \tag{7.1}$$

where the implied constant in O depends on E but not on the point P.

This will be proved on p. 137 after some more machinery has been developed.

In multiplicative notation, the duplication formula may be written

$$H(P)^4 \ll H(2P) \ll H(P)^4.$$

Example 7.2. Consider the curve $E : y^2 = x^3 - n^2 x$ with $1 \leqslant n \in \mathbb{N}$. Let P be a rational point on E. A calculation gives

$$x(2P) = \left(\frac{x^2 + n^2}{2y}\right)^2 = \frac{(x^2 + n^2)^2}{4(x^3 - n^2 x)},$$

so if $x(P) = \frac{M}{N}$ in lowest terms, then

$$x(2P) = \frac{(M^2 + n^2 N^2)^2}{4MN(M^2 - n^2 N^2)}. \tag{7.2}$$

It may be checked that any cancellation in Equation (7.2) is bounded: Explicitly, if d divides both numerator and denominator, then $d | 16 n^6$. Examining the cases $|M| \geqslant |N|$ and $|M| < |N|$ separately shows that

$$\max\{|M^2 + n^2 N^2|^2, |N^2(M^2 - n^2 N^2)|\}$$

is commensurate[1] with

$$\max\{|M|^4, |N|^4\} = \max\{|M|, |N|\}^4,$$

and the duplication formula Equation (7.1) follows.

Exercise 7.1. Verify Theorem 7.1 for the curve $y^2 = x^3 + C$, $C \neq 0$.

The duplication formula is a special case of a general principle about polynomial maps on projective space, and so we prove Theorem 7.1 in greater generality.

A polynomial f in N variables is called *homogenous* if there is a constant $d \in \mathbb{N}$ (the *degree* of f) with

$$f(\lambda x_0, \ldots, \lambda x_N) = \lambda^d f(x_0, \ldots, x_N).$$

Exercise 7.2. Let $f_0, \ldots, f_M$ be polynomials in $N+1$ variables. Show that the map $\mathbf{x} \mapsto (f_0(\mathbf{x}), \ldots, f_M(\mathbf{x}))$ between $\mathbb{K}^{N+1}$ and $\mathbb{K}^{M+1}$ induces a well-defined map $\mathbb{P}^N(\mathbb{K}) \to \mathbb{P}^M(\mathbb{K})$ if and only if the polynomials $f_0, \ldots, f_M$ are all homogenous of the same degree and the only common zero of the polynomials is the point $(0, \ldots, 0)$.

Definition 7.3. *A map*

$$f : \mathbb{P}^N(\mathbb{Q}) \longrightarrow \mathbb{P}^M(\mathbb{Q})$$

is called a morphism of degree d *if*

$$f([\mathbf{x}]) = f([x_0, \ldots, x_N]) = [f_0([\mathbf{x}]), \ldots, f_M([\mathbf{x}])],$$

where the f_j, $0 \leqslant j \leqslant M$ are homogenous polynomials of degree d with the property that the only common zero is 0.

Lemma 7.4. *Let $f : \mathbb{P}^N(\mathbb{Q}) \to \mathbb{P}^M(\mathbb{Q})$ be a morphism of degree d. Then*

$$H([\mathbf{x}])^d \ll H(f([\mathbf{x}])) \ll H([\mathbf{x}])^d.$$

PROOF. Write $f([\mathbf{x}]) = [f_0(\mathbf{x}), \ldots, f_M(\mathbf{x})]$, where $[\mathbf{x}] = [x_0, \ldots, x_N] \in \mathbb{P}^N(\mathbb{Q})$. By clearing denominators, we may assume that each x_j is an integer. Since each f_i is homogenous of degree d, they may be written

$$f_i(\mathbf{x}) = \sum c_{\mathbf{e}} x_0^{e_0} \cdots x_N^{e_N},$$

with $c_{\mathbf{e}} \in \mathbb{Q}$, $e_i \in \mathbb{N}$, $e_0 + \cdots + e_N = d$, and only finitely many $c_{\mathbf{e}}$ nonzero. It follows that there is a constant C such that

---

[1] In the sense that the ratio is bounded above and below by positive constants independent of N and M.

$$|f_i(\mathbf{x})| \leqslant C \cdot (\max\{|x_j|\})^d,$$

for each i and all j, so there is a similar bound for $\max\{|f_i(\mathbf{x})|\}$. To find the height, notice that the only possible denominators that need to be cleared come from the coefficients of the polynomials f_i, which is a bounded quantity in total. It follows that there is an upper bound for the height of the form

$$C \cdot H(\mathbf{x})^d.$$

To get the lower bound, we use Hilbert's Nullstellensatz: There exists $e \in \mathbb{N}$ and polynomials $g_{ij} \in \mathbb{Q}[\mathbf{x}]$ such that

$$x_0^e = g_{00}(\mathbf{x})f_0(\mathbf{x}) + \cdots + g_{0N}(\mathbf{x})f_N(\mathbf{x})$$
$$\vdots$$
$$x_N^e = g_{N0}(\mathbf{x})f_0(\mathbf{x}) + \cdots + g_{NN}(\mathbf{x})f_N(\mathbf{x}).$$

The g_{ij}s can be taken to be homogenous polynomials of degree $(e - d)$ so

$$|g_{ij}(\mathbf{x})| \ll (\max\{|x_k|\})^{e-d}.$$

On the other hand,

$$x_j^e = g_{j0}(\mathbf{x})f_0(\mathbf{x}) + \cdots + g_{jN}(\mathbf{x})f_N(\mathbf{x})$$

for $j = 0, \ldots, N$, so

$$(\max\{|x_j|\})^{e-d}\max\{|f_0|, \ldots, |f_N|\} \gg (\max\{|x_j|\})^e.$$

It follows that
$$\max\{|f_0|, \ldots, |f_N|\} \gg (\max\{|x_j|\})^d,$$

and since the only possible denominators are those arising from the coefficients of the f_i, the lower bound is proved.   □

Example 7.5. To see that $e > d$ really occurs in the Nullstellensatz, define

$$f : \mathbb{P}^1(\mathbb{Q}) \to \mathbb{P}^1(\mathbb{Q})$$

by
$$f : [x_0, x_1] \mapsto [x_0^2, (x_0 + x_1)^2] = [f_0(x_0, x_1), f_1(x_0, x_1)].$$

Then f is a morphism of degree 2. Now $x_0^2 = 1 \cdot f_0$, but there are no rational polynomials A, B for which $x_1^2 = A \cdot f_0 + B \cdot f_1$. However,

$$x_0^3 = x_0 \cdot f_0$$
$$x_1^3 = (2x_0 + 3x_1) \cdot f_0 + (-2x_0 + x_1) \cdot f_1.$$

Exercise 7.3. (a) Using the explicit formulas from Example 7.2, prove that the map defined by $[x(P), 1] \mapsto [x(2P), 1]$ is a morphism of degree 4 for the curve $y^2 = x^3 - n^2 x$.

(b) Do the same for the curve $y^2 = x^3 + c$.

PROOF OF THEOREM 7.1. By Lemma 7.4, all we need to show is that the map

$$[x, 1] \mapsto [x(2P), 1]$$

on $\mathbb{P}^1(\mathbb{Q})$ is a morphism of degree 4. Assume that the curve is

$$E : y^2 = x^3 + ax + b$$

and $P = (x, y)$. Then

$$
\begin{aligned}
x(2P) &= \left(\frac{3x^2 + a}{2y} \right)^2 - 2x \\
&= \frac{(3x^2 + a)^2}{4y^2} - 2x \\
&= \frac{9x^4 + 6x^2 a + a^2}{4(x^3 + ax + b)} - 2x \\
&= \frac{x^4 - 2x^2 a - 8xb + a^2}{4(x^3 + ax + b)}.
\end{aligned}
$$

Write $x = \frac{x_0}{x_1} \in \mathbb{Q}$ (in lowest terms as usual). Then, writing $x(2P) = \frac{f_0(x_0, x_1)}{f_1(x_0, x_1)}$ and dropping a factor of 4,

$$f_0(x_0, x_1) = x_0^4 - 2x_0^2 x_1^2 a - 8x_0 x_1^3 b + a^2 x_1^4, \text{ and}$$
$$f_1(x_0, x_1) = x_0^3 x_1 + ax_0 x_1^3 + bx_1^4.$$

To show that these define a morphism of degree 4, it only remains to show that the unique common zero of f_0 and f_1 is $(0, 0)$. If $f_0 = f_1 = 0$ and $x_1 = 0$, then $x_0 = 0$. Assume $x_1 \neq 0$. Then we may assume that $x_1 = 1$ and $x_0 = x$. We now need to show that

$$f(x) = x^4 - 2x^2 a - 8xb + a^2, \text{ and}$$
$$g(x) = x^3 + ax + b$$

cannot have a common zero. One way to see this is using resultants (see Exercise 7.4); we will use the Euclidean Algorithm (see Example 2.3(2)) to find the greatest common divisor of f and g. Assume first that $a \neq 0$ and recall we are assuming that $4a^3 + 27b^2 \neq 0$. The Euclidean Algorithm gives

$$x^4 - 2x^2 a - 8xb + a^2 = (x^3 + ax + b) \, x - 3ax^2 - 9bx + a^2;$$

$$x^3 + ax + b = (-3ax^2 - 9bx + a^2) \left(-\frac{1}{3a} x + \frac{b}{a^2} \right) + \left(\frac{9b^2}{a^2} + \frac{4}{3} a \right) x;$$

$$-3ax^2 - 9bx + a^2 = \left(\left(\frac{4}{3} a + 9 \frac{b^2}{a^2} \right) x \right) \left(-\frac{9a^3}{4a^3 + 27b^2} x - \frac{27ba^2}{(4a^3 + 27b^2)} \right) + a^2,$$

which shows that the greatest common divisor of f and g is a nonzero constant.
    If $a = 0$ then $b \neq 0$ since $4a^3 + 27b^2 \neq 0$, so the Euclidean Algorithm gives

$$x^4 - 8xb = (x^3 + b)x - 9xb;$$
$$(x^3 + b) = (-9xb)\left(-\tfrac{1}{9b}x\right) + b,$$

which, again, shows the greatest common divisor of f and g is a nonzero
constant.   □

Exercise 7.4. Show that the resultant of the polynomials

$$f(x) = x^4 - 2x^2 a - 8xb + a^2 \text{ and } g(x) = x^3 + ax + b$$

is $(4a^2 + 27b^2)^2$. This shows that the condition $4a^3 + 27b^2 \neq 0$ implies that f
and g have no common zero.

Exercise 7.5. Let $E : y^2 = x^3 + ax + b$ with $a, b \in \mathbb{Z}$ denote an elliptic
curve. Using arguments from p-adic analysis, it can be shown that any nonzero
torsion point $Q \in E(\mathbb{Q})$ must have $x(Q)$ and $y(Q)$ integral. Assuming this,
prove that $y(Q) = 0$ or $y(Q)^2$ divides $4a^3 + 27b^2$ for any rational torsion point.

Exercise 7.6. Recall from Exercise 5.12 that the point $P = (3, 8)$ has order 7
on the elliptic curve

$$y^2 = x^3 - 43x + 166.$$

Using Exercise 7.5, show that there are no rational torsion points other than
those in the subgroup generated by P.

7.2 Mordell's Theorem

In this section, we will see how Mordell's Theorem follows from the *weak
Mordell Theorem*. In the next section, we will give a proof of the weak Mordell
Theorem for the congruent number curve and discuss how the proof can be
extended to cover a wider class of curves. The proof in full generality requires
more algebraic number theory than we have at our disposal. Complete proofs
may be found in the references discussed at the end of the chapter.

Theorem 7.6. [WEAK MORDELL THEOREM] *Let E denote an elliptic curve
defined over $\mathbb{Q}$. Then $E(\mathbb{Q})/2E(\mathbb{Q})$ is a finite Abelian group.*

Lemma 7.7. *Let $E : y^2 = x^3 + ax + b$ be an elliptic curve defined over the
rationals.*

(1) *If $P_0 \neq 0$ is a point in $E(\mathbb{Q})$, then there is a constant $c_1 = c_1(E, P_0) > 0$
    such that*

$$h(P + P_0) < 2h(P) + c_1. \tag{7.3}$$

(2) *Given $h_0 > 0$, there are only finitely many points $P \in E(\mathbb{Q})$ with*

$$h(P) < h_0.$$

PROOF. (2) is clear since only finitely many rationals $\frac{m}{n}$ (in lowest terms) have $\log \max\{|m|, |n|\} < h_0$.

To prove (1), write $P = (x, y)$ and $P_0 = (x_0, y_0)$. From the equation

$$y^2 = x^3 + ax + b,$$

write (in lowest terms)

$$x = \frac{r}{t^2}, \; y = \frac{s}{t^3}, \; x_0 = \frac{r_0}{t_0^2}, \; y_0 = \frac{s_0}{t_0^3},$$

with r, s, t, r_0, s_0, t_0 all integers. Then

$$x(P + P_0) = \left(\frac{y_0 - y}{x_0 - x}\right)^2 - x - x_0$$

$$= \frac{y_0^2 - 2y_0 y + y^2}{(x_0 - x)^2} - \frac{(x_0 + x)}{(x_0 - x)^2}(x_0^2 - 2x_0 x + x^2)$$

$$= \frac{x_0^3 + ax_0 + b + x^3 + ax + b - 2y_0 y}{(x_0 - x)^2}$$

$$- \frac{(x_0^3 - 2x_0^2 x + x_0 x^2 + x x_0^2 - 2x_0 x^2 + x^3)}{(x_0 - x)^2}$$

$$= \frac{a(x_0 + x) + 2b - 2y_0 y - (-x_0^2 x - x^2 x_0)}{(x_0 - x)^2}$$

$$= \frac{a(x_0 + x) + 2b - 2y_0 y + x_0 x(x_0 + x)}{(x_0 - x)^2}$$

$$= \frac{(a + x_0 x)(x_0 + x) + 2b - 2y_0 y}{(x_0 - x)^2}.$$

Substituting r, s, t then gives

$$x(P + P_0) = \frac{\left(a + \frac{r_0 r}{t_0^2 t^2}\right)\left(\frac{r_0}{t_0^2} + \frac{r}{t^2}\right) + 2b - 2\frac{s s_0}{t_0^3 t^3}}{\left(\frac{r_0}{t_0^2} - \frac{r}{t^2}\right)^2}$$

$$= \frac{(at_0^2 t^2 + r_0 r)(r_0 t^2 + r t_0^2) + 2b t^4 t_0^4 - 2 s s_0 t t_0}{(r_0 t^2 - r t_0^2)^2}.$$

The effect of clearing denominators in the rationals a and b appearing as coefficients in the elliptic curve can be absorbed into the constant c_1. It is therefore sufficient to check that the numerator and denominator satisfy the inequality in Equation (7.3).

First

$$|\text{numerator}| < \underbrace{(c_2|t|^2 + c_3|r|)(c_4|t|^2 + c_5|r|)}_{<c_8(\max\{|r|,|t|^2\})^2} + c_6|t|^4 + c_7|st|. \qquad (7.4)$$

The first two terms have

$$c_8(\max\{|r|,|t|^2\})^2 + c_6|t|^4 \leqslant c_9 H(P)^2.$$

Looking at the third term, we need to show that

$$|st| \leqslant c_{10} H(P)^2.$$

Since $y^2 = x^3 + ax + b$, we have

$$(st)^2 = r^3 t^2 + art^6 + bt^8.$$

There are two cases to consider.
I: $|r| \geqslant |t|^2$. In this case

$$|st|^2 \leqslant c_{11}|r|^4 + c_{12}|r|^4 + c_{13}|r|^4 = c_{14} H(P)^4.$$

II: $|r| < |t|^2$. In this case

$$|st|^2 < c_{15}|t|^8 + c_{16}|t|^8 + c_{17}|t|^8 = c_{18} H(P)^4.$$

In both cases, $|st| < c_{10} H(P)^2$, as required. Therefore

$$|\text{numerator}| < c_{19} H(P)^2,$$

or in logarithmic form

$$\log|\text{numerator}| < c_{20} + 2h(P).$$

The denominator is simpler:

$$\begin{aligned}
|\text{denominator}| &= |r_0 t^2 - r t_0^2|^2 \\
&< \left(c_{21}|t|^2 + c_{22}|r|\right)^2 \\
&< c_{23} H(P)^2,
\end{aligned}$$

so

$$\log|\text{denominator}| < c_{24} + 2h(P).$$

$\square$

PROOF OF THEOREM 5.11 ASSUMING THEOREM 7.6. Let $\mathcal{Q} = \{Q_1, \ldots, Q_s\}$ denote a fixed set of coset representatives for $2E(\mathbb{Q})$ in $E(\mathbb{Q})$. By Theorem 7.6, it is enough to prove the following: There is a constant $R = R(E, \mathcal{Q})$ with the property that every point $P \in E(\mathbb{Q})$ can be expressed as an integral combination of the Q_i, $i = 1, \ldots, s$ and those $Q \in E(\mathbb{Q})$ with $h(Q) < R$ (finite in number by Lemma 7.7(2)).

Let P be a rational point on E, and write (for some $i_0, i_1, \cdots \in \{1, \ldots, s\}$)

$$P = P_0 = Q_{i_0} + 2P_1$$
$$P_1 = Q_{i_1} + 2P_2$$
$$\vdots$$
$$P_n = Q_{i_n} + 2P_{n+1}.$$

The duplication formula Equation (7.1) says that

$$4h(P_{n+1}) - c_1 < h(2P_{n+1}) \qquad (7.5)$$

for some $c_1 = c_1(E) > 0$. On the other hand, Lemma 7.7(1) shows that

$$h(2P_{n+1}) = h(P_n - Q_{i_n}) < 2h(P_n) + c_2 \qquad (7.6)$$

for some $c_2 = c_2(E, \mathcal{Q})$. Combining Equation (7.5) and Equation (7.6) gives

$$h(P_{n+1}) < \frac{1}{2}h(P_n) + c_3$$

for some $c_3 = c_3(E, \mathcal{Q})$. Iterating this gives

$$h(P_{n+1}) < \frac{1}{2}\left(\frac{1}{2}h(P_{n-1}) + c_3\right) + c_3$$

$$= \frac{1}{2^2}h(P_{n-1}) + c_3\left(1 + \frac{1}{2}\right)$$

$$\vdots$$

$$< \frac{1}{2^{n+1}}h(P_0) + c_3\left(1 + \frac{1}{2} + \frac{1}{2^2} + \cdots + \frac{1}{2^n}\right).$$

As $n \to \infty$, for fixed P_0,

$$\frac{1}{2^{n+1}}h(P_0) \to 0 \quad \text{and} \quad c_3\left(1 + \frac{1}{2} + \frac{1}{2^2} + \cdots + \frac{1}{2^n}\right) \longrightarrow 2c_3. \qquad (7.7)$$

Now $P = Q_{i_0} + 2P_1$, $P_1 = Q_{i_1} + 2P_2$ and so on gives

$$P = Q_{i_0} + 2(Q_{i_1} + 2P_2)$$
$$= Q_{i_0} + 2Q_{i_1} + 2^2 P_2$$
$$= Q_{i_0} + 2Q_{i_1} + 2^2 Q_{i_2} + 2^3 P_3$$
$$\vdots$$

so P is being written as an integer combination of elements of $\mathcal{Q}$ and a point whose height is uniformly bounded as $n \to \infty$ by Equation (7.7).

Take $R = (2 + \frac{1}{10})c_3$, a constant depending on E and $\mathcal{Q}$. We have shown that any rational point P can be written as an integral combination of the points of $\mathcal{Q}$ and a point with height bounded by R. $\square$

7.3 The Weak Mordell Theorem: Congruent Number Curve

We will now give a proof of Theorem 7.6 for the congruent number curve

$$y^2 = x^3 - n^2x,$$

where $n > 0$ denotes an integer. The proof uses a homomorphism from $E(\mathbb{Q})$ to a quotient group of the group of nonzero rationals and we begin my introducing this group. Let $\mathfrak{Q}$ denote $\mathbb{Q}^*/\mathbb{Q}^{*2}$, which is the quotient of the group of nonzero rationals by the subgroup of all nonzero squares. The representatives for this group can be taken to be all nonzero integers r which are not divisible by the square of a prime. We will write $\bar{r}$ for the coset containing r. Notice that the identity of the group is $\bar{1}$ and the element $\overline{-1}$ is an element of order 2 in $\mathfrak{Q}$. In this section, the point $(0,0)$ will play a distinguished role and will be denoted $T = (0,0)$.

Lemma 7.8. *Define a map* $\phi_1 : E(\mathbb{Q}) \to \mathfrak{Q}$ *by*

$$\phi_1(0) = \bar{1}$$
$$\phi_1(T) = \overline{-1}$$
$$\phi_1((x,y)) = \bar{x} \ \textit{otherwise.}$$

Then ϕ_1 *is a group homomorphism.*

This is a remarkable claim. If you try to prove it simply using the addition formula it can be difficult to dig out, and might even start to look impossible. We will use a simple trick already encountered to make it come out quite smoothly. The reason ϕ in the definition carries the suffix 1 is because we will define two other similar maps shortly.

PROOF OF LEMMA 7.8. Let P_1 and P_2 denote rational points with

$$P_1 + P_2 = P_3.$$

We wish to deduce that $\phi_1(P_3) = \phi_1(P_1)\phi_1(P_2)$. There are a number of special cases to be considered before we can deal with the general situation. The only nontrivial special case which requires any work arises when one of P_1 or P_2 is the 2-torsion point $T = (0,0)$. Say we add $P = (x,y)$ to T, where $x \neq 0$. The image of the sum under ϕ_1 is

$$\overline{(y/x)^2 - x} = \overline{(y^2 - x^3)/x^2} = \overline{-n^2x} = \overline{-x} = \phi_1(P)\phi_1(T),$$

hence the result is true in this special case. An almost identical proof gives the case where $P_3 = T = (0,0)$.

Recall Section 5.3, where we converted the geometric addition on an elliptic curve into explicit formulas. The group law on an elliptic curve tells us that the points P_1, P_2 and $-P_3$ lie on the same straight line. Writing $P_1 + P_2 = P_3$ with $P_i = (x_i, y_i)$, we need to show that $x_1 x_2 x_3$ is a rational square. From the above, we may assume each of x_1, x_2, and x_3 are nonzero rational numbers. Let the line containing the points P_1, P_2 and $-P_3$ be written $y = \alpha x + \beta$, for rationals α and β. Our assumptions guarantee that $\beta \neq 0$. Substitute the equation of the line into the equation of the curve to get

$$x^3 - n^2 x - (\alpha x + \beta)^2 = 0.$$

The roots of this equation are the three rational numbers x_1, x_2 and x_3 because it is this equation which defines them. Hence we can factorize the left-hand side as

$$(x - x_1)(x - x_2)(x - x_3).$$

Now if we compare the two equations (see Exercise 5.11 on p. 105) we see that $x_1 x_2 x_3$ is equal to β^2, the square of a rational. In other words, up to a rational square $x_1 x_2$ and x_3 are equal; hence

$$\phi_1(P_1 + P_2) = \phi_1(P_1)\phi_1(P_2).$$

$\square$

Exercise 7.7. Verify that Lemma 7.8 is true for an elliptic curve of the form

$$y^2 = x^3 + ax^2 + bx$$

with the same definition of ϕ_1.

We have already indicated that $E(\mathbb{Q})$ can be an infinite group. The second lemma says that even if that is true, the image of this group under ϕ_1 is a finite group.

Lemma 7.9. *The image of $E(\mathbb{Q})$ under ϕ_1 is a finite subgroup of $\mathfrak{Q}$.*

PROOF. Suppose $\bar{r}$ lies in the image of ϕ_1. Without loss of generality, assume r is a square-free integer. We claim that $r \mid n$. To prove this, suppose p is a prime with $p \mid r$, then we will show $p \mid n$. The statement $\phi_1((x, y)) = \bar{r}$ amounts to two equations

$$x^2 - n^2 = rs^2$$
$$x = rt^2$$

for rationals s and t. Now clear denominators by writing $t = a/b$ for coprime integers a and b. Eliminating x, we obtain an equation

$$r^2 a^4 - n^2 b^4 = rc^2$$

for some integer c. If $p|r$ but $p \nmid n$ then $p|b$ and therefore $p^4|n^2b^4$. Thus p^2 must divide the left-hand side and it follows that p^2 must divide the right-hand side. Since r is square-free, it follows that $p|c$ so $p^2|c^2$ and hence p^3 divides the right-hand side. This forces p^3 to divide r^2a^4 so $p|a$ (since r is square-free). Thus p divides a and b which contradicts the assumption that they are coprime.   □

Exercise 7.8. For the elliptic curve E defined by $y^2 = x^3 - 36x$, the torsion-free part of $E(\mathbb{Q})$ is generated by the rational point $(-3, 9)$ (you may assume this). Find the image of $E(\mathbb{Q})$ under the map ϕ_1.

Exercise 7.9. Suppose E denotes an elliptic curve and p and q denote rational numbers. The map $x \mapsto x - p, y \mapsto y - q$ takes rational points on this curve to rational points on a new elliptic curve. Assume that the point at infinity on the first curve maps to the point at infinity on the second (this can be verified by taking limits as before). Show that the resulting map is a group isomorphism. In the language of Section 5.5, the map is an isogeny of degree one.

Exercise 7.10. Define a map $\phi_2 : E(\mathbb{Q}) \to \mathfrak{Q}$ by

$$\phi_2(O) = \bar{1};$$
$$\phi_2((n, 0)) = \overline{-1};$$
$$\phi_2((x, y)) = \overline{x - n} \text{ otherwise.}$$

Prove that ϕ_2 is a group homomorphism. (Hint: Compose this map with a suitable translation map and use Exercise 7.9.)

In a similar vein to Exercise 7.10, we can define a map $\phi_3 : E(\mathbb{Q}) \to \mathfrak{Q}$ by

$$\phi_3((x, y)) = \overline{x + n}$$

whenever $x \neq -n$.

Exercise 7.11. Show that both of the maps ϕ_2 and ϕ_3 have finite image in $\mathfrak{Q}$.

Our goal is in sight now. Combine the three maps into one by defining

$$\phi : E(\mathbb{Q}) \to \mathfrak{Q}^3$$

to be

$$\phi(P) = (\phi_1(P), \phi_2(P), \phi_3(P)).$$

Earlier on we showed that the doubling map on a rational point on the congruent number curve $E : y^2 = x^3 - n^2x$ produced an x-coordinate which is the square of a rational, provided the starting point does not have order 2. This suggests that we might find $2E(\mathbb{Q})$ inside the kernel of ϕ. More is true.

Lemma 7.10. *The kernel of ϕ is precisely $2E(\mathbb{Q})$. In other words the rational point $P = (x, y)$ is the double of a rational point if and only if x and $x \pm n$ are all rational squares. Explicitly, write*

$$x = r_1^2;$$
$$x + n = r_2^2;$$
$$x - n = r_3^2 \text{ for } r_i \in \mathbb{Q}.$$

Then $P = 2Q$ where $Q = (X, Y)$ and X and Y are given by the formulas

$$X = x + r_1 r_2 + r_1 r_3 + r_2 r_3,$$
$$Y = (r_1 + r_2 + r_3)(X - x) - y,$$

provided the signs of the r_i are chosen so that $r_1 r_2 r_3 = y$.

Example 7.11. Let $n = 6$. The point $Q = (-3, 9)$ doubles to the point $\left(\frac{25}{4}, -\frac{35}{8}\right)$ on the curve $y^2 = x^3 - 36x$. This is verified by taking $r_1 = \frac{5}{2}, r_2 = -\frac{7}{2}, r_3 = \frac{1}{2}$. As expected,

$$X = \frac{25}{4} - \frac{5}{2} \cdot \frac{7}{2} + \frac{5}{2} \cdot \frac{1}{2} - \frac{1}{2} \cdot \frac{7}{2} = -3,$$

and similarly $Y = 9$.

The proof of Lemma 7.10 is purely computational and we leave the verification as an exercise. The burden of explanation rests on the question of why it should be true in the first place. In one sense it is not wrong to say it comes down to Mordell's genius. The notes at the end of the chapter include a useful reference which suggests how Mordell might have come upon this remarkable phenomenon.

Exercise 7.12. Suppose E is an elliptic curve defined by the equation

$$E : y^2 = x^3 + ax^2 + bx + c$$

where a, b and c are rational. Assuming the roots of the cubic are all rational, adapt the proof above to deduce the weak Mordell Theorem for E.

In the general case, the technicalities of the proof are no greater from the point of view of elliptic curves. What is required is a deeper knowledge of algebraic number fields.

During this section, we have seen how homomorphisms between elliptic curves, or homomorphisms from elliptic curves to other groups, played an important role. Although we will not develop this any further, it is worth being aware of the importance of the map which reduces modulo p, for a prime p. This map takes an elliptic curve defined over $\mathbb{Q}$ to one defined over $\mathbb{F}_p$. Since all the group operations are defined by rational functions, we should not be surprised that the map is a group homomorphism (though this does of course require that the reduced curve is really an elliptic curve.) More remarkably, the notion of "infinity" as the identity of the group is quite robust. The following exercise gives an opportunity to encounter this phenomenon.

Exercise 7.13. Suppose E is an elliptic curve defined by the equation

$$E : y^2 = x^3 + ax + b, \quad a, b \in \mathbb{Z}.$$

Let p denote a prime number coprime to $4a^3 + 27b^2$ and let $E_1(\mathbb{Q})$ denote the set of rational points (x, y) on the curve with the property that the denominators of x and y are divisible by p together with the point at infinity. Prove that $E_1(\mathbb{Q})$ is a subgroup of $E(\mathbb{Q})$. (Hint: Resist the temptation to do this using the functions defining addition. What is the kernel of the reduction map?)

7.3.1 The Generation Game

We have seen some examples of elliptic curves with rank 1; for example the curve given by the equation $y^2 = x^3 - 2$, with the generator $(3, 5)$, also the congruent number curve for $n = 6$ which is generated by $(-3, 9)$. It is natural to ask how the rank can be proved to be 1 and how these can be proved to be the generators. Although many special cases have been worked out, in general there is no algorithm known for determining the rank of an elliptic curve nor for finding a set of generators. In the notes at the end of the chapter, several of the references provide details about how special cases can be approached, as well as links to massive tables of curves whose ranks have been computed, along with systems of generators. We recommend as a worthwhile exercise, doing some computations with some of these curves using a computer algebra package.

7.4 The Parallelogram Law and the Canonical Height

The duplication formula Equation (7.1) says that for any $P \in E(\mathbb{Q})$

$$h(2P) = 4h(P) + O(1),$$

or, equivalently, there is a constant $c = c(E)$ such that

$$\left| h(P) - \tfrac{1}{4}h(2P) \right| < c. \tag{7.8}$$

The next result exploits this to produce a height function with better functorial properties, the *canonical height*. The approach below is due to Tate; the canonical height was discovered independently by Neron.

Theorem 7.12. *For any rational point P on an elliptic curve E defined over the rationals,*

$$\lim_{n \to \infty} \frac{h(2^n P)}{4^n} = \widehat{h}(P) \tag{7.9}$$

exists. The limit $\widehat{h}(P)$ is called the canonical height of P.

PROOF. Let $a_N = \frac{1}{4^N} h(2^N P)$. If $N > M \geqslant 1$, then

$$a_M - a_N = \frac{1}{4^M} h(2^M P) - \frac{1}{4^N} h(2^N P)$$

$$= \frac{1}{4^M} h(2^M P) - \frac{1}{4^{M+1}} h(2^{M+1} P)$$

$$+ \frac{1}{4^{M+1}} h(2^{M+1} P) - \frac{1}{4^{M+2}} h(2^{M+2} P)$$

$$+ \cdots$$

$$+ \frac{1}{4^{N-1}} h(2^{N-1} P) - \frac{1}{4^N} h(2^N P)$$

which may be grouped into

$$a_M - a_N = \frac{1}{4^M} \left(h(2^M P) - \frac{1}{4} h(2 \cdot 2^M P) \right)$$

$$+ \frac{1}{4^{M+1}} \left(h(2^{M+1} P) - \frac{1}{4} h(2 \cdot 2^{M+1} P) \right)$$

$$+ \cdots$$

$$+ \frac{1}{4^{N-1}} \left(h(2^{N-1} P) - \frac{1}{4} h(2 \cdot 2^{N-1} P) \right).$$

By the duplication formula (Theorem 7.1), this gives

$$|a_M - a_N| < \frac{1}{4^M} c \left(1 + \frac{1}{4} + \frac{1}{4^2} + \cdots \right) = \frac{1}{4^M} c \left(\frac{4}{3} \right) \to 0 \text{ as } M \to \infty,$$

showing that (a_N) is a Cauchy sequence. $\square$

If the order of P is a power of 2, then $\widehat{h}(P) = 0$. In fact, any torsion point P has $\widehat{h}(P) = 0$, and moreover $\widehat{h}(P) = 0$ implies that P is a torsion point by Theorem 7.13(4).

Theorem 7.13. *Let E be an elliptic curve defined over the rationals.*

(1) *For every point $P \in E(\mathbb{Q})$,*

$$\widehat{h}(P) = h(P) + O(1)$$

uniformly.

(2) *For all $P, Q \in E(\mathbb{Q})$,*

$$\widehat{h}(P + Q) + \widehat{h}(P - Q) = 2\widehat{h}(P) + 2\widehat{h}(Q). \tag{7.10}$$

(3) *For every $m \in \mathbb{Z}$ and $P \in E(\mathbb{Q})$,*

$$\widehat{h}(mP) = m^2 \widehat{h}(P).$$

(4) *For $P \in E(\mathbb{Q})$,*

$$\widehat{h}(P) = 0 \text{ if and only if } P \text{ is a torsion point.}$$

This is proved below. Equation (7.10) is called the *parallelogram law*. It follows from (1) and (3) in Theorem 7.13 that

$$h(mP) = m^2 h(P) + \mathrm{O}(1),$$

which is a weaker version of Theorem 5.2. A more useful generalization of this formula is the parallelogram law for the naïve height. This will be stated now, then Theorem 7.13 will be proved. The parallelogram law for the naïve height will be proved in Section 7.5.

Lemma 7.14. *For all $P, Q \in E(\mathbb{Q})$,*

$$h(P + Q) + h(P - Q) = 2h(P) + 2h(Q) + \mathrm{O}(1) \qquad (7.11)$$

uniformly.

PROOF OF THEOREM 7.13.
(1) By iterating the relation

$$h(P) = \frac{1}{4}\left(h(2P) + \mathrm{O}(1)\right),$$

we have

$$h(P) = \frac{1}{4}\left(\left(\frac{h(2^2 P)}{4} + \frac{1}{4}\mathrm{O}(1)\right) + \mathrm{O}(1)\right)$$

$$= \frac{h(2^2 P)}{4^2} + \mathrm{O}(1)\left(\frac{1}{4} + \frac{1}{4^2}\right)$$

$$\vdots$$

$$= \frac{h(2^N P)}{4^N} + \underbrace{\mathrm{O}(1)\left(\frac{1}{4} + \frac{1}{4^2} + \cdots + \frac{1}{4^N}\right)}_{\mathrm{O}(1)}.$$

Letting $N \to \infty$ gives

$$h(P) = \widehat{h}(P) + \mathrm{O}(1).$$

(2) Applying a similar limiting procedure to the naïve parallelogram law Equation (7.11) gives

$$\widehat{h}(P+Q) + \widehat{h}(P-Q) - 2\widehat{h}(P) - 2\widehat{h}(Q)$$

$$= \lim_{N\to\infty} \left(\frac{1}{4^N} h(2^N(P+Q)) + \frac{1}{4^N} h(2^N(P-Q)) \right.$$

$$\left. - \frac{2}{4^N} h(2^N P) - \frac{2}{4^N} h(2^N Q) \right)$$

$$= \lim_{N\to\infty} \left(\frac{1}{4^N} O(1) \right) = 0.$$

(3) This is proved by induction on $m \geqslant 1$. The case $m \leqslant -1$ follows since

$$h(-P) = h(P) \Rightarrow \widehat{h}(P) = \widehat{h}(-P).$$

For $m = 0$, $h(0) = 0 = \widehat{h}(0)$. Assume therefore that

$$\widehat{h}(mP) = m^2\widehat{h}(P),$$

and substitute mP for P and P for Q in the parallelogram law Equation (7.10):

$$\widehat{h}(mP+P) = 2\widehat{h}(mP) + 2\widehat{h}(P) - \widehat{h}((m-1)P)$$
$$= 2m^2\widehat{h}(P) + 2\widehat{h}(P) - (m-1)^2\widehat{h}(P)$$
$$= (m+1)^2\widehat{h}(P).$$

(4) If P is a torsion point, then $mP = 0$ for some $m \neq 0$, so by (3) $\widehat{h}(P) = 0$. Conversely, suppose that $\widehat{h}(P) = 0$ for some $P \in E(\mathbb{Q})$. Then

$$\widehat{h}(mP) = m^2\widehat{h}(P) = 0 \text{ for all } m,$$

so $h(mP)$ must be uniformly bounded for all m by (1). By Lemma 7.7(2), this means that the set $\{mP\}_{m\in\mathbb{Z}}$ must be finite, so P is a torsion point. $\qquad\square$

7.4.1 A Strong Form of Siegel's Theorem

The result that follows we call the *Strong Siegel Theorem*; it was proved by Silverman and we will not prove it here. It relates the growth rates of the numerators and the denominators of the multiples nP of a nontorsion rational point.

Theorem 7.15. [STRONG SIEGEL THEOREM] *Let E denote an elliptic curve defined over $\mathbb{Q}$ and suppose $P \in E(\mathbb{Q})$ denotes a nontorsion point. Let (P_n) be any sequence of rational points with $\widehat{h}(P_n) \to \infty$ as $n \to \infty$, and write*

$$P_n = \left(\frac{A_n}{B_n^2}, \frac{C_n}{B_n^3} \right).$$

Then

$$\frac{\log|A_n|}{2\log|B_n|} \longrightarrow 1 \quad as \quad n \longrightarrow \infty.$$

This can be interpreted as saying that the numerators and denominators of P_n have roughly the same number of decimal digits for large n. Theorem 5.2 follows from this, together with Theorem 7.13. A particular case of a rational sequence (P_n) with $\widehat{h}(P_n) \to \infty$ is given by taking $P_n = nP$ for a rational nontorsion point P on an elliptic curve defined over the rationals.

7.5 Mahler Measure and the Naïve Parallelogram Law

In proving Lemma 7.14, some simple estimates on polynomials will be needed, and one way to phrase them is to use the *Mahler measure*, which is of independent interest. There are several natural ways to measure the size of a polynomial in such a way that an integer polynomial with zeros of large arithmetic complexity will have large measure.

Definition 7.16. *For any nonzero polynomial*

$$F(x) = a_d x^d + a_{d-1} x^{d-1} + \cdots + a_0 = a_d \prod_{i=1}^{d} (x - \alpha_i)$$

in $\mathbb{C}[x]$, define three measures as follows.

(1) *The* Mahler measure *of F is $M(F) = |a_d| \cdot \prod_{i=1}^{d} \max\{1, |\alpha_i|\}$.*
(2) *The* height *of F is $H(F) = \max_{0 \leqslant 1 \leqslant d}\{|a_i|\}$.*
(3) *The* length *of F is $L(F) = \sum_{i=0}^{d} |a_i|$.*

In (1), an empty product is assumed to be 1, so the Mahler measure of the nonzero constant polynomial $F(x) = a_0$ is $|a_0|$. Write $m(F) = \log M(F)$ for the *logarithmic Mahler measure* of F.

Mahler showed that

$$|a_i| \leqslant \binom{d}{i} M(F) \text{ for all } i = 0, \ldots, d$$

and also showed that all three measures are commensurate in the sense that

$$H(F) \ll M(F) \ll H(F)$$

and

$$L(F) \ll M(F) \ll L(F), \tag{7.12}$$

with the implied constants depending only on the degree d.

The absolute value of the discriminant of F is defined to be

$$|\Delta(F)| = |a_d|^{2d-2} \prod_{i \neq j} |\alpha_i - \alpha_j|.$$

Mahler also showed that

$$|\Delta(F)| \leqslant d^d M(F)^{2d-2}. \tag{7.13}$$

Exercise 7.14. (a) Prove that

$$-d\log 2 + \ell(F) \leqslant m(F) \leqslant \ell(F),$$

where we write $\ell = \log L$. This is equivalent to an exact description of the implied constants in Equation (7.12):

$$2^{-d}L(F) \leqslant M(F) \leqslant L(F).$$

(b) Prove a weaker form of the inequality (7.13) as follows. Assume that

$$F(x) = x^d + a_{d-1}x^{d-1} + \cdots + a_0 = \prod_{1 \leqslant i \leqslant d} (x - \alpha_i)$$

is monic, so the absolute value of the discriminant is

$$|\Delta(F)| = \prod_{i \neq j} |\alpha_i - \alpha_j|.$$

Prove that

$$|\Delta(F)| \leqslant 2^{d(d-1)} M(F)^{2d-2}.$$

Exercise 7.15. Fix a polynomial

$$F(x) = a_d x^d + a_{d-1}x^{d-1} + \cdots + a_0 = a_d \prod_{i=1}^{d} (x - \alpha_i)$$

in $\mathbb{Z}[x]$. Call F *hyperbolic* if $|\alpha_i| \neq 1$ for all $i = 1, \ldots, d$ and *ergodic* if $\alpha_i^k = 1$ for some $k \geqslant 0$, and any i implies that $k = 0$.
(a) Prove that

$$m(F) = \int_0^1 \log |F(e^{2\pi i s})| \, ds \qquad (7.14)$$

when F is hyperbolic.
(b) Prove Equation (7.14) without assuming that F is hyperbolic.
(c) Prove that $\Delta_n(F) = \prod_{i=1}^{d} |\alpha_i^n - 1|$ is an integer for all n. For F hyperbolic, prove that

$$\lim_{n \to \infty} \Delta_n(F)^{1/n} = \lim_{n \to \infty} \frac{\Delta_{n+1}(F)}{\Delta_n(F)} = M(F).$$

(d) Prove that an ergodic polynomial of degree $d \leqslant 3$ is hyperbolic.
(e) Find a polynomial that is ergodic but not hyperbolic.
(f)*For F ergodic but not hyperbolic, prove that

$$\lim_{n \to \infty} \Delta_n(F)^{1/n} = M(F)$$

but that $\frac{\Delta_{n+1}(F)}{\Delta_n(F)}$ does not converge as $n \to \infty$.

Exercise 7.16. [KRONECKER'S LEMMA] Prove that a polynomial $F \in \mathbb{Z}[x]$ has $m(F) = 0$ if and only if every zero λ of F satisfies $\lambda^k = 1$ for some $k \geqslant 1$.

Exercise 7.17. Considerable interest has been shown in the set of values of the Mahler measure of integer polynomials.
(a) Compute $m(F)$ to 3 decimal places when

$$F(x) = x^{10} + x^9 - x^7 - x^6 - x^5 - x^4 - x^3 + x + 1. \qquad (7.15)$$

(b)*Explore the mathematical literature on *Lehmer's Problem*: Is there an integer polynomial G with $m(G) > 0$ and with $m(G) < m(F)$? More generally, given arbitrary $\epsilon > 0$, is there an integer polynomial H with $m(H) > 0$ and $m(H) < \epsilon$? Extensive calculations have been made of values of the Mahler measure for monic polynomials, and no nonzero value smaller than $m(F)$ has been found.

PROOF OF LEMMA 7.14. Let $E : y^2 = x^3 + ax + b$ be the elliptic curve. Let P and Q be points in $E(\mathbb{Q})$, and write $x(P) = x_1, x(Q) = x_2, x(P + Q) = x_3$, and $x(P - Q) = x_4$.

The values of x_3 and x_4 depend on the y coordinates of P and Q, which complicates the proof considerably. We will work in the coordinates

$$x_1 x_2, \quad x_1 + x_2, \quad x_3 x_4, \quad \text{and} \quad x_3 + x_4,$$

because these only depend on the x coordinates. Now

$$x_3 + x_4 = \frac{2(x_1 + x_2)(a + x_1 x_2) + 4b}{(x_1 - x_2)^2},$$

$$x_3 x_4 = \frac{(x_1 x_2 - a)^2 - 4b(x_1 + x_2)}{(x_1 - x_2)^2},$$

and we may write

$$(x_1 - x_2)^2 = (x_1 + x_2)^2 - 4x_1 x_2,$$

giving $x_3 + x_4$ and $x_3 x_4$ in terms of $x_1 x_2, x_1 + x_2$.

We claim that for any $x_1, x_2 \in \mathbb{Q}$,

$$h([x_1 + x_2, x_1 x_2, 1]) = h([x_1, 1]) + h([x_2, 1]) + O(1). \qquad (7.16)$$

To see this, write $x_1 = \frac{s}{t}, x_2 = \frac{u}{v}$ in lowest terms, and define

$$F_1(x) = tx - s, \quad F_2(x) = vx - u.$$

Then

$$m(F_1 F_2) = m(F_1) m(F_2). \qquad (7.17)$$

Now by Equation (7.12) and Exercise 7.14,

$$m\left(\sum_{i=0}^{d} a_i x^i\right) = h\left(\sum_{i=0}^{d} a_i x^i\right) + O(d),$$

where $h(\sum_{i=0}^{d} a_i x^i) = \log \max\{|a_i|\}$. Applying this to Equation (7.17) gives

$$h(F_1 F_2) = h(F_1) + h(F_2) + O(1). \tag{7.18}$$

Now

$$h(F_1) = h([x_1, 1]), \quad h(F_2) = h([x_2, 1]). \tag{7.19}$$

On the other hand,

$$F_1(x)F_2(x) = (tx - s)(vx - u) = tvx^2 - x(sv + tu) + su$$

so

$$h(F_1 F_2) = \max\{|tv|, |sv + tu|, |su|\}.$$

Now

$$x_1 + x_2 = \frac{sv + tu}{tv}, \quad \text{and}$$

$$x_1 x_2 = \frac{su}{tv},$$

so

$$h([x_1 + x_2, x_1 x_2, 1]) = h\left(\left[\frac{sv + tu}{tv}, \frac{su}{tv}, 1\right]\right) = h([sv + tu, su, tv]).$$

Now $sv + tu, su$, and tv cannot have a common factor by Gauss' Lemma, so $h(F_1 F_2) = h([x_1 + x_2, x_1 x_2, 1])$, and Equations (7.18) and (7.19) give Equation (7.16).

Change variables and work with $x_1 + x_2$ and $x_1 x_2$:

$$x_3 + x_4 = \frac{2(x_1 + x_2)(a + x_1 x_2) + 4b}{(x_1 + x_2)^2 - 4x_1 x_2} \quad \text{and}$$

$$x_3 x_4 = \frac{(x_1 x_2 - a)^2 - 4b(x_1 + x_2)}{(x_1 + x_2)^2 - 4x_1 x_2}.$$

Lemma 7.17. *Assume that $4a^3 + 27b^2 \neq 0$. Then the map $\mathbb{P}^2(\mathbb{Q}) \to \mathbb{P}^2(\mathbb{Q})$ defined by*

$$[u, v, t] = [2u(at + v) + 4bt^2, (v - at)^2 - 4btu, u^2 - 4tv]$$

is a morphism of degree 2.

The formulas in Lemma 7.17 come from setting $u = x_1 + x_2$ and $v = x_1 x_2$ and using t to make the expressions homogenous.

PROOF OF LEMMA 7.17. Suppose the three polynomials vanish:

$$2u(at + v) + 4bt^2 = 0, \tag{7.20}$$
$$(v - at)^2 - 4btu = 0, \text{ and} \tag{7.21}$$
$$u^2 - 4tv = 0. \tag{7.22}$$

If $t = 0$, then $u = 0$ by Equation (7.22), so $v = 0$ by Equation (7.21).
Suppose therefore that $t \neq 0$, and divide by t^2 in each equation. Write

$$x = \frac{u}{2t}, \text{ so } x^2 = \frac{v}{t}$$

by Equation (7.22). Equations (7.20) and (7.21) give

$$(x^2 - a)^2 - 8bx = 0 \text{ and}$$
$$x(a + x^2) + b = 0,$$

or

$$x^4 - 2ax^2 - 8bx + a^2 = 0 \text{ and}$$
$$x^3 + ax + b = 0.$$

These polynomials arose in the proof of the duplication formula on p. 138, where it was shown that they have no common zero. □

Now apply Lemma 7.17 to the vectors

$$[x_1 + x_2, x_1 x_2, 1] \text{ and } [x_3 + x_4, x_3 x_4, 1].$$

Since the map from the first to the second is a morphism of degree 2,

$$h([x_3 + x_4, x_3 x_4, 1]) = 2h([x_1 + x_2, x_1 x_2, 1]) + O(1)$$

by Lemma 7.4. Equation (7.16) shows that

$$h([x_3 + x_4, x_3 x_4, 1]) = h([x_3, 1]) + h([x_4, 1]) + O(1)$$

and

$$h([x_1 + x_2, x_1 x_2, 1]) = h([x_1, 1]) + h([x_2, 1]) + O(1),$$

so

$$h([x_3, 1]) + h([x_4, 1]) = 2h([x_1, 1]) + 2h([x_2, 1]) + O(1),$$

and therefore

$$h(P + Q) + h(P - Q) = 2h(P) + 2h(Q) + O(1).$$

□

Notes to Chapter 7: The polynomial in Equation (7.15) is taken from Lehmer's paper [97]; a starting point for Exercises 7.15 and 7.17(b) is [59] and the references therein. Hilbert's Nullstellensatz may be found in any book on algebraic geometry; an accessible account is in Reid's notes [122]. The statement about integrality of torsion points in Exercise 7.5 is due to Lutz [101] and Nagell [112]. Accessible proofs are in Cassels [27, Chapter 12], Husemöller [79, Chapter 5, Section 6] and Silverman [139, Chapter VIII, Section 7]. The characterization of all torsion points on the curves $y^2 = x^3 + ax$ (for a integral and not divisible by a fourth power) and $y^2 = x^3 + b$ (for b integral and not divisible by a sixth power) is given in Cassels [27, Exercise to Chapter 12]. Theorem 7.6 is proved in Lang [94], and Silverman [139]; a sketch proof from an advanced point of view is in the article by Milne [108, Theorem 20.10]; see also Lemmermeyer's excellent Web notes on elliptic curves. Cassels' article [26] is an excellent piece of background reading, in which he gives a plausible explanation as to how Mordell might have discovered what became known as the weak Mordell Theorem. Cremona's Web site [37] gives the rank and a set of generators for thousands of elliptic curves. The strong form of Siegel's Theorem (Theorem 7.15) can be found in Silverman [139, Chapter IX, Section 3]. The Mahler measure in Section 7.5 was introduced in two papers of Mahler [103], [104]. There are extensive references to the many places where the Mahler measure arises in [59], especially connections between heights and dynamical systems. Computational material related particularly to the Mahler measure from Section 7.5 appears in a book by Borwein [18].

8

The Riemann Zeta Function

We saw in Chapter 1 that estimates for sums of arithmetic functions are an essential step in understanding arithmetic problems. One of the themes we wish to pursue is the following strange phenomenon: Arithmetic properties of integers, especially primes, can be deduced from analytic properties of functions. A serious instance is afforded by the Prime Number Theorem itself (see p. 3 for the $\sim$ and $o(x)$ notation).

Theorem 8.1. [PRIME NUMBER THEOREM] *Asymptotically, the number of primes is given by*

$$\pi(x) = |\{p \in \mathbb{P} \mid p \leqslant x\}| \sim \frac{x}{\log x}.$$

The Prime Number Theorem is of major importance in number theory. The notes at the end of the chapter give references where proofs can be found, including elementary approaches (in this context "elementary" means "without recourse to complex analysis", and not "easy"). Around the beginning of the nineteenth century, Legendre published a conjecture equivalent to the Prime Number Theorem. Gauss also studied the values of $\pi(x)$ at a similar time and conjectured[1] that

$$\pi(x) \sim \mathrm{Li}(x) = \int_2^x \frac{1}{\log t} dt. \tag{8.1}$$

The Prime Number Theorem was first proved in 1896 by two mathematicians independently – Hadamard and de la Vallée Poussin. Their proofs used the Riemann zeta function and they were able to give an estimate for the error

---

[1] For all small values of x, $\pi(x) < \mathrm{Li}(x)$, and several prominent mathematicians conjectured that the inequality always holds. In 1914, Littlewood proved that the inequality reverses infinitely often. Amazingly, the smallest value of x where the inequality first reverses is still not known, although it is known to be below 10^{371}. This is a compelling instance of a situation where even enormous amounts of numerical evidence can be completely deceptive.

term in the formula, based upon an estimate for a zero-free region of the zeta function. We will have more to say about the zeros of the Riemann zeta function in Section 9.2.1.

In this chapter, we will start by giving a far-reaching refinement of the integral test that quickly gives sharper estimates for some arithmetic functions. We then develop the algebra of arithmetic functions with respect to a natural notion of multiplication, Dirichlet convolution. Finally, we apply these results to show how the Riemann zeta function may be extended to the whole complex plane with a simple pole at 1.

8.1 Euler's Summation Formula

The integral test used on p. 10 compares the sum $\sum_{n=1}^{N} f(n)$ with the integral $\int_{1}^{N} f(x)\, dx$. Euler's Summation Formula is a refinement of this tool that allows us to derive sharper asymptotic formulas. Recall that

$$\{t\} = t - \lfloor t \rfloor \tag{8.2}$$

denotes the fractional part of a real number t, where $\lfloor t \rfloor$ is the greatest integer smaller than or equal to t.

Theorem 8.2. *Let $a < b$ be real numbers, and suppose that f is a complex-valued function defined on $[a,b]$ with a continuous derivative on (a,b). Then*

$$\sum_{a<n\leqslant b} f(n) = \int_{a}^{b} f(t)\, dt + \int_{a}^{b} \{t\}f'(t)\, dt - f(b)\{b\} + f(a)\{a\}. \tag{8.3}$$

PROOF. We give the proof in the case $a, b \in \mathbb{N}$ for simplicity.
Suppose that $a < n - 1 < n \leqslant b$. Now

$$\int_{n-1}^{n} \lfloor t \rfloor f'(t)\, dt = (n-1)[f(n) - f(n-1)] = nf(n) - (n-1)f(n-1) - f(n).$$

Sum this from $a + 1$ to b:

$$\int_{a}^{b} \lfloor t \rfloor f'(t)\, dt = \sum_{n=a+1}^{b} (n-1)[f(n) - f(n-1)]$$

$$= bf(b) - af(a) - \sum_{n=a+1}^{b} f(n).$$

Rearrange to give

$$\sum_{n=a+1}^{b} f(n) = bf(b) - af(a) - \int_{a}^{b} \lfloor t \rfloor f'(t)\, dt. \tag{8.4}$$

On the other hand, integrating by parts,

$$\int_a^b f(t)\,dt = \left[tf(t) \right]_a^b - \int_a^b tf'(t)\,dt. \tag{8.5}$$

Equations (8.4) and (8.5) together give

$$\sum_{n=a+1}^b f(n) = \int_a^b f(t)\,dt + \int_a^b \{t\}f'(t)\,dt,$$

completing the proof.   □

Applying the Euler Summation Formula to the harmonic series $\sum_{k=2}^N \frac{1}{k}$ already gives a nontrivial result. Here $a = 1$, $b = N$, and $f(t) = 1/t$. This gives

$$\sum_{k=2}^N \frac{1}{n} = \int_1^N \frac{1}{t}\,dt - \int_1^N \frac{\{t\}}{t^2}\,dt = \log N - \int_1^N \frac{\{t\}}{t^2}\,dt. \tag{8.6}$$

Clearly

$$\int_1^N \frac{\{t\}}{t^2}\,dt = \int_1^\infty \frac{\{t\}}{t^2}\,dt - \int_N^\infty \frac{\{t\}}{t^2}\,dt$$

and the last term is less than $\int_N^\infty \frac{1}{t^2}\,dt = \frac{1}{N}$. Adding 1 to each side of Equation (8.6) gives the following result, which should be compared with Exercise 1.2 to appreciate the power of the Euler Summation Formula.

Theorem 8.3.

$$\sum_{n=1}^N \frac{1}{n} = \log N + \gamma + O\!\left(\frac{1}{N}\right), \text{ where}$$

$$\gamma = 1 - \int_1^\infty \frac{\{t\}}{t^2}\,dt$$

is the Euler–Mascheroni constant.

Definition 8.4. *For $1 \leqslant n \in \mathbb{N}$, let $d(n)$ denote the* number of divisors of n.

For example, $d(n) = 2$ if and only if n is a prime. It follows that information about d reflects something about the distribution of the primes themselves.

Exercise 8.1. Prove that $d(n)$ is odd if and only if n is a square.

Theorem 8.5.

$$\sum_{n=1}^N d(n) = N \log N + (2\gamma - 1)N + O(\sqrt{N}).$$

PROOF. The Euler Summation Formula in the usual form with integer boundaries gives a much larger remainder term of the form $O(N)$, which swamps the $(2\gamma-1)N$ term. For the sharper result with a $O(\sqrt{N})$ error term, we apply the Euler Summation Formula in the more general form given in Theorem 8.2.

Notice that

$$|\{(m,q): mq \leqslant x\}| = 2|\{(m,q): mq \leqslant x, m < q\}| + O(\sqrt{x}). \qquad (8.7)$$

It follows that

$$\sum_{n\leqslant x} d(n) = \sum_{m\leqslant x} \sum_{q\leqslant x/m} 1$$

$$= 2 \sum_{\substack{m,q; \\ mq\leqslant x, m<q}} 1 + O(\sqrt{x})$$

$$= 2 \sum_{m<\sqrt{x}} \left(\left[\frac{x}{m}\right] - m\right) + O(\sqrt{x})$$

$$= 2x \sum_{m<\sqrt{x}} \frac{1}{m} - 2 \sum_{m<\sqrt{x}} m + O(\sqrt{x}).$$

Now we can apply the Euler Summation Formula to each sum to obtain

$$\sum_{n\leqslant x} d(n) = 2x(\log\sqrt{x} + \gamma + O(x^{-1/2})) - 2\left(\frac{x}{2} + O(\sqrt{x})\right) + O(\sqrt{x})$$

$$= x\log x + (2\gamma - 1)x + O(\sqrt{x})$$

as required.   □

The sharper form of the Euler Summation Formula used in the proof of Theorem 8.5 does not always give more precise results. For example, Theorem 8.2 applied to the harmonic series gives

$$\sum_{1\leqslant n\leqslant x} \frac{1}{n} = \log x + \gamma + O\left(\frac{1}{x}\right) + \frac{\{x\}}{x};$$

the last summand is $O(\frac{1}{x})$ and so goes into the remainder term. In this case, the general form does not give us more information, and for many summands this will be the case.

Exercise 8.2. Prove Equation (8.7). (Hint: If $mq \leqslant x$ and $m < q$, then $m < \sqrt{x}$. The number of q in this set for fixed m is $[x/m] - m$; draw a picture to see why.)

Another application of the Euler Summation Formula gives Stirling's Formula.

Theorem 8.6. [STIRLING'S FORMULA]

$$\log N! = N \log N - N + O(\log N).$$

PROOF. Clearly $\log N! = \sum_{n=2}^{N} \log n$. Put $f(t) = \log t$, and then by the Euler Summation Formula

$$\log N! = \int_{1}^{N} \log t \; dt + \int_{1}^{N} \frac{\{t\}}{t} \, dt = N \log N - N + O(\log N).$$

$\square$

Stirling's Formula and its refinements are extremely important in many parts of mathematics. A refinement of the Euler Summation Formula using derivatives of higher order gives the more precise estimate

$$\sqrt{2\pi} n^{n+1/2} e^{-n} < n! < \sqrt{2\pi} n^{n+1/2} e^{-n+1/12n}. \tag{8.8}$$

Exercise 8.3. *Prove the inequality (8.8).

Exercise 8.4. Use the Euler Summation Formula to prove the following asymptotic formulas, where A and B are constants.

(a) $\displaystyle\sum_{n=1}^{N} \frac{\log n}{n} = \frac{1}{2}(\log N)^2 + A + O\left(\frac{\log N}{N}\right)$.

(b) $\displaystyle\sum_{n=2}^{N} \frac{1}{n \log n} = \log \log N + B + O\left(\frac{1}{N \log N}\right)$.

Exercise 8.5. Prove that

$$\sum_{n=1}^{N} \frac{d(n)}{n} = \frac{1}{2}(\log N)^2 + 2\gamma \log N + O(1),$$

where γ is the Euler–Mascheroni constant.

8.2 Multiplicative Arithmetic Functions

Recall from Definition 3.4 that an arithmetic function f is *multiplicative* if

$$f(m)f(n) = f(m)f(n) \quad \text{when} \quad \gcd(m,n) = 1.$$

It turns out that many functions of interest in number theory have this property.

Definition 8.7. *The Möbius function μ is defined by $\mu(1) = 1$ and*

$$\mu(n) = \begin{cases} (-1)^k & \text{if } n \text{ is a product of } k \text{ distinct primes, and} \\ 0 & \text{otherwise.} \end{cases}$$

Exercise 8.6. By Fermat's Little Theorem (Theorem 1.12), for any prime p and integer a, $a^p - a \equiv 0$ modulo p. The Möbius function gives a natural generalization to a composite modulus.
(a) For any $a, n \in \mathbb{N}$, prove that

$$\sum_{d|n} \mu(n/d)a^d \equiv 0 \pmod{n}.$$

(b)*For any $a, k, n \in \mathbb{N}$, prove that

$$\sum_{d|n} \mu(n/d)a^{d^k} \equiv 0 \pmod{n}.$$

Remarkably, the Prime Number Theorem is also intimately connected with growth properties of the Möbius function, and in fact

$$\text{Prime Number Theorem} \iff \sum_{n \leqslant x} \mu(n) = o(x). \tag{8.9}$$

That is, the Prime Number Theorem follows from and implies the fact that $\frac{1}{x} \sum_{n \leqslant x} \mu(n) \to 0$. Similarly, the important Riemann Hypothesis (Conjecture 9.7 in the next chapter) is equivalent to a conjectured result about partial sums of the Möbius function.

Exercise 8.7. Prove Tchebychef's weak form of the Prime Number Theorem by the following steps.
(a) Let N be an integer and p a prime. Show that the largest power of p dividing $N!$ is exactly

$$\left\lfloor \frac{N}{p} \right\rfloor + \left\lfloor \frac{N}{p^2} \right\rfloor + \left\lfloor \frac{N}{p^3} \right\rfloor + \cdots.$$

(b) Use (a) to show that

$$\log(N!) = \sum_{p \leqslant N} \left(\left\lfloor \frac{N}{p} \right\rfloor + \left\lfloor \frac{N}{p^2} \right\rfloor + \left\lfloor \frac{N}{p^3} \right\rfloor + \cdots \right) \log p.$$

(c) Use Stirling's Formula (Theorem 8.6) to show that

$$N \sum_{p \leqslant N} \frac{\log p}{p} = N \log N + O(N).$$

(d) Deduce that there exist constants $A, B > 0$ with the property that

$$A\frac{N}{\log N} < \pi(N) < B\frac{N}{\log N} \quad \text{for all } N \geqslant 2.$$

Exercise 8.8. Let $0 < a < b$ be fixed real numbers and assume the Prime Number Theorem.

(a) Prove that $\frac{\pi(ax)}{\pi(bx)} \to \frac{a}{b}$ as $x \to \infty$.

(b) Deduce that there is some X such that for any $x > X$ there is a prime p with $ax < p < bx$.

(c) Deduce that there is a rational $\frac{p}{q}$, $a < \frac{p}{q} < b$, with p and q both prime.

The Möbius function appears in a striking prime formula due to Gandhi.

Exercise 8.9. *Let P_n denote the product of the first n primes. Prove that the next prime p_{n+1} is the unique integer m with the property that

$$1 < 2^m \left(\sum_{d|P_n} \frac{\mu(d)}{2^d - 1} - \frac{1}{2} \right) < 2.$$

Theorem 8.8. *The Möbius function is multiplicative. Moreover,*

$$\sum_{d|n} \mu(d) = \begin{cases} 1 & \text{if } n = 1, \\ 0 & \text{otherwise.} \end{cases}$$

PROOF. Let m and n be integers with $\gcd(m, n) = 1$, and factorize m and n as products of prime powers. The primes involved must all be distinct. If, in the factorization there is an exponent of 2 or more, then we are done since both sides of the equation $\mu(mn) = \mu(m)\mu(n)$ are zero. If m (and n) is a product of k (resp. ℓ) distinct primes, then mn is a product of $k + \ell$ primes, all of which are distinct since m and n are coprime. It follows that $\mu(m) = (-1)^k$ and $\mu(n) = (-1)^\ell$, and so

$$\mu(mn) = (-1)^{k+\ell} = \mu(m)\mu(n).$$

For the next claim, it is sufficient to check the prime power case since again both sides of the equation of Theorem 8.8 are multiplicative (by the argument used in the proof of Theorem 3.9). If $n = p^r$ with $r \geqslant 1$, the left-hand side is

$$\mu(1) + \mu(p) = 1 - 1 = 0,$$

which completes the proof. □

Theorem 8.9. *For all integers $n \geqslant 1$, $\phi(n) = \sum_{d|n} \mu(d)\frac{n}{d}$.*

PROOF. By Corollary 3.6, $\phi(n) = n \prod_{p_i|n} \left(1 - \frac{1}{p_i}\right)$, so

$$\frac{\phi(n)}{n} = 1 - \sum_{p_i|n} \frac{1}{p_i} + \sum_{i \neq j} \frac{1}{p_i p_j} - \cdots = \sum_{d|n} \frac{\mu(d)}{d}.$$

□

8.3 Dirichlet Convolution

The proof of Theorem 8.9 is a special instance of a general technique.

Definition 8.10. *The* convolution *of arithmetic functions* f *and* g *is the function* f $*$ g *defined by*

$$(f * g)(n) = \sum_{d|n} f(d) g\left(\frac{n}{d}\right).$$

Theorem 8.11. *Convolution is commutative and associative. In other words,*

$$f * g = g * f \quad and \quad (f * g) * h = f * (g * h)$$

for any arithmetic functions f, g, *and* h.

PROOF. The sum in

$$\sum_{d|n} f(d) g\left(\frac{n}{d}\right)$$

runs over all pairs $d, e \in \mathbb{N}$ with $de = n$, so it is equal to

$$\sum_{de=n} f(d) g(e),$$

and the latter expression is symmetric in f and g.

To see that convolution is associative, check the property for $n = p$ a prime by hand. The proof in the general case goes in much the same way as the proof of commutativity:

$$\left((f * g) * h\right)(n) = \left(f * (g * h)\right)(n) = \sum_{cde=n} f(c) g(d) h(e),$$

from which associativity is clear.   □

Lemma 8.12. *Define the arithmetic function* I *by* $I(1) = 1$ *and* $I(n) = 0$ *for all* $n > 1$. *Then, for any arithmetic function* f,

$$f * I = I * f = f.$$

PROOF.
$$(f * I)(n) = \sum_{d|n} f(d) I\left(\frac{n}{d}\right) = f(n) I(1) = f(n)$$

since all the other summands are zero by the definition of I.   □

Theorem 8.13. *If* f *is an arithmetic function with* $f(1) \neq 0$, *then there is a unique arithmetic function* g *such that* $f * g = I$. *This function is denoted* f^{-1}.

PROOF. The equation $(f * g)(1) = f(1)g(1)$ determines $g(1)$. Then define g recursively as follows. Assuming that $g(1), \ldots, g(n-1)$ have been defined uniquely, the equation

$$(f * g)(n) = f(1)g(n) + \sum_{1 < d \mid n} f(d)g\left(\frac{n}{d}\right)$$

allows us to calculate $g(n)$ uniquely.   □

Example 8.14. Let $u(n) = 1$ for all n. Then, by Theorem 8.8,

$$u^{-1} = \mu. \tag{8.10}$$

Exercise 8.10. Let f be a multiplicative arithmetic function with $f(1) \neq 0$.
(a) Prove that $f^{-1}(n) = \mu(n)f(n)$ for all square-free n.
(b) Prove that $f^{-1}(p^2) = f(p)^2 - f(p^2)$ for all primes p.

Exercise 8.11. Let f be an arithmetic function, and consider the (formal) relationship

$$\prod_{n=1}^{\infty} (1 - x^n)^{f(n)/n} = \sum_{n=0}^{\infty} R(n)x^n. \tag{8.11}$$

(a) Prove that $R(n) = -\frac{1}{n}\sum_{a=1}^{n}(f * u)(a) \cdot R(n-a)$ for all $n \geqslant 1$.
(b) Assume that $R(0) = 1$. Prove that f is uniquely determined by Equation (8.11).
(c) For $f(n) = n^\alpha$, prove that $R(n) = -\frac{1}{n}\sum_{a=1}^{n}(n^\alpha + 1) \cdot R(n-a)$ for all $n \geqslant 1$.

Exercise 8.12. If f is multiplicative, prove that f is completely multiplicative if and only if $f^{-1}(p^a) = 0$ for all primes p and $a \geqslant 2$.

Exercise 8.13. Define an arithmetic function $\nu(n)$ to be 1 when $n = 0$ and the number of distinct prime factors of n for $n \geqslant 1$. Let $f = \mu * \nu$. Prove that $f(n) \in \{0, 1\}$ for all $n \in \mathbb{N}$.

Exercise 8.14. (a) Prove that the collection of all arithmetic functions f with $f(1) \neq 0$ forms an Abelian group under Dirichlet convolution.
(b) Prove that the multiplicative arithmetic functions form a subgroup.
(c) Show by example that the completely multiplicative functions do *not* form a subgroup.

Theorem 8.15. [MÖBIUS INVERSION FORMULA] *Given arithmetic functions f and g, $f(n) = \sum_{d \mid n} g(d)$ if and only if $g(n) = \sum_{d \mid n} f(d)\mu\left(\frac{n}{d}\right)$.*

PROOF. Assume that $f(n) = \sum_{d \mid n} g(d)$, and let $u(n) = 1$ for all n as in Example 8.14. Then $f = g * u$. Convolve both sides of $f = g * u$ with μ and use Equation (8.10) to see that

$$f * \mu = g * u * \mu = g * I = g,$$

so

$$g(n) = \sum_{d|n} f(d)\mu\left(\frac{n}{d}\right).$$

For the converse, convolve $g = f * \mu$ with u. □

Thus Theorem 3.9 and Theorem 8.9 are equivalent: We can move from one to the other by convolving with the Möbius function or its convolution inverse.

Exercise 8.15. Suppose σ denotes a real number for which

$$F(\sigma) = \sum_{n=1}^{\infty} \frac{f(n)}{n^{\sigma}} \quad \text{and} \quad G(\sigma) = \sum_{n=1}^{\infty} \frac{g(n)}{n^{\sigma}}$$

are absolutely convergent series. Prove that

$$F(\sigma) \cdot G(\sigma) = \sum_{n=1}^{\infty} \frac{(f * g)(n)}{n^{\sigma}}.$$

Example 8.16. If $f * g = I$, then $F(\sigma)G(\sigma) = 1$, so

$$\frac{1}{\zeta(\sigma)} = \sum_{n=1}^{\infty} \frac{\mu(n)}{n^{\sigma}}.$$

Series such as F, G, and $F \cdot G$ are called *Dirichlet series*. We next study the Riemann zeta function in the context of Dirichlet series.

Exercise 8.16. For all $s \in \mathbb{C}$ with $\Re(s) > 2$, show that

$$\frac{\zeta(s-1)}{\zeta(s)} = \sum_{n=1}^{\infty} \frac{\phi(n)}{n^s}.$$

The traditional notation for the variable s in Definition 1.4 of the Riemann zeta function is

$$s = \sigma + it \qquad \text{with } \sigma, t \in \mathbb{R}.$$

For s with real part $\sigma = \Re(s) > 1$, we claim that the series $\sum_{n=1}^{\infty} \frac{1}{n^s}$ converges absolutely. To prove this, notice that

$$n^{-s} = n^{-\sigma - it} = n^{-\sigma} e^{-it \log n}$$

has modulus $n^{-\sigma}$ and that $\sum_{n=1}^{\infty} \frac{1}{n^{\sigma}}$ is a convergent series by the integral test.

8.3.1 Application of Möbius Inversion to Zsigmondy's Theorem

Before showing how Theorem 8.15 can be used to prove Zsigmondy's Theorem (Theorem 1.15), a preliminary observation needs to be made. The polynomial $x^n - 1$ already has some natural factorization according to the divisors of n. If $d \geqslant 1$ denotes any integer, let ϕ_d denote the monic polynomial whose zeros are the *primitive*[2] dth roots of unity. The polynomial ϕ_d is known as the dth *cyclotomic* polynomial.

A simple application of Galois theory says that

$$\phi_d(x) \in \mathbb{Z}[x] \text{ for every } d \geqslant 1.$$

If you are not familiar with Galois theory we ask you take this on trust. A natural factorization of $x^n - 1$ into integral polynomials follows at once, by dividing the nth roots of unity into the dth primitive roots of unity for d running over the divisors of n,

$$x^n - 1 = \prod_{d|n} \phi_d(x). \tag{8.12}$$

Exercise 8.17. Compute the polynomials ϕ_d for $1 \leqslant d \leqslant 15$.

The factorization given by Equation (8.12) into integral polynomials yields a partial factorization of $M_n = 2^n - 1$ into integers,

$$2^n - 1 = \prod_{d|n} \phi_d(2). \tag{8.13}$$

The first thing to notice about Equation (8.13) is that, by definition, any primitive divisor of M_n must divide $\phi_n(2)$. The proof of Theorem 1.15 proceeds by showing that any factor of $\phi_n(2)$ which is common to M_d for some $d < n$ must itself already divide n. Therefore, as soon as $\phi_n(2)$ exceeds n, M_n is guaranteed to have a primitive divisor.

We claim that for every $n > 6$, $\phi_n(2) > n$. To prove this, note first that for all $n > 1$,

$$\phi_n(2) > \frac{2^{\phi(n)}}{e}. \tag{8.14}$$

To prove this inequality, apply Möbius inversion to Equation (8.13) to see that the logarithm of the left-hand side of the inequality (8.14) is

---

[2] A complex number z is a primitive dth root of unity if $z^d = 1$, but $z^e \neq 1$ for any e with $0 < e < d$. Thus the primitive dth roots of unity are precisely the complex numbers

$$e^{2\pi ik/d} \text{ with } \gcd(k, d) = 1.$$

Thus there are exactly $\phi(d)$ distinct primitive dth roots of unity, where ϕ denotes the Euler function defined on p. 61.

$$\sum_{d|n} \mu(n/d) \log(2^d - 1).$$

This is bounded below by $\phi(n) \log 2 - 1$ using Theorem 8.9 and an easy estimate for the logarithm.

The right-hand term of the inequality (8.14) can be estimated using Corollary 3.6 and the bound (1.4). We deduce that if $\phi_n(2) \leqslant n$ then

$$\log n < 2 \log \log n + C, \tag{8.15}$$

for a constant $C > 0$ which can be made explicit. Clearly, this inequality bounds n and so we have completed the proof that for large enough n, the logarithm of the right-hand side of the inequality (8.14) is greater than $\log n$. This completes the proof of the original claim, albeit in an inexplicit way.

Exercise 8.18. Find an explicit value for the constant C in the inequality (8.15) and use this to find an explicit bound for n.

Applying the multiplicative form of Theorem 8.15 to Equation (8.12) yields

$$\phi_n(x) = \prod_{d|n} (x^d - 1)^{\mu(n/d)}, \tag{8.16}$$

so in particular

$$\phi_n(2) = \prod_{d|n} (2^d - 1)^{\mu(n/d)}. \tag{8.17}$$

Thus, Equation (8.17) gives

$$\mathrm{ord}_p(\phi_n(2)) = \sum_{d|n} \mu(n/d) \, \mathrm{ord}_p(M_d). \tag{8.18}$$

Now suppose that p is a prime with $p|M_n$ and $p|M_d$ for some d, $1 < d < n$ (so p is not a primitive divisor of M_n). Let d_0 be the smallest value of d for which $p|M_d$. Now the sequence (M_n) has the strong divisibility property of Exercise 1.15(c) on p. 27, namely,

$$\gcd(M_n, M_m) = M_{\gcd(n,m)} \text{ for all } m, n.$$

Thus we may assume that $d_0|n$. By Exercise 1.15(b),

$$\mathrm{ord}_p(M_{nd_0}) = \mathrm{ord}_p(M_{d_0}) + \mathrm{ord}_p(n), \tag{8.19}$$

and $\mathrm{ord}_p(M_d) = 0$ unless d is a multiple of d_0. Use Equation (8.19) to write Equation (8.18) as

$$\mathrm{ord}_p(\phi_n(2)) = \sum_{d|(n/d_0)} \mu(n/dd_0) \left(\mathrm{ord}_p(M_{d_0}) + \mathrm{ord}_p(d) \right)$$

$$= \mathrm{ord}_p(M_{d_0}) \sum_{d|(n/d_0)} \mu(n/dd_0) + \sum_{d|(n/d_0)} \mu(n/dd_0) \, \mathrm{ord}_p(d).$$

Since $n > d_0$, the first term vanishes because

$$\sum_{d|n} \mu(d) = 0 \text{ for } n > 1$$

by Theorem 8.8. The second term is bounded above by $\text{ord}_p(n)$. As shown above, $\phi_n(2) > n$ for all $n > 6$, which concludes the proof of Theorem 1.15 on p. 27.

The next exercise uses similar methods to solve a special case of a deep general result due to Bilu, Hanrot and Voutier. The result was first shown by Carmichael in 1913.

Exercise 8.19. [CARMICHAEL] Let A and B be nonzero integers. The *Lucas sequence* associated with the pair (A, B) is the sequence (u_n) defined by

$$u_n = \frac{\alpha^n - \beta^n}{\alpha - \beta},$$

where α and β are the roots of the equation

$$x^2 - Ax + B = 0.$$

Assume that α and β are real, and prove the following theorem: The term u_n of the associated Lucas sequence has a primitive divisor for $n \neq 1, 2, 6$, except when $A = B = -1$ and $n = 12$. (Hint: Let ζ be a primitive nth root of unity, and write

$$Q_n = \prod_{\substack{1 \leqslant k \leqslant n, \\ \gcd(k,n)=1}} (\alpha - \zeta^k \beta).$$

Show that u_n will have a primitive divisor if Q_n is not too small relative to the size of n, and then show that the smallest values of Q_n arise for $A = 1, B = -1$ and $A = 3, B = 2$.)

8.4 Euler Products

Recall from Chapter 1 that the Riemann zeta function has a decomposition as an Euler product. By Theorem 1.5,

$$\zeta(\sigma) = \prod_p \left(1 - \frac{1}{p^\sigma}\right)^{-1},$$

where, as before, the product over p means a product over all the primes, and $\sigma > 1$. We will now show that this holds for all complex s with $\Re(s) > 1$. Since it is no more difficult, we prove a more general theorem.

Theorem 8.17. *If* f *is a multiplicative arithmetic function, and* $\sum_{n=1}^{\infty} f(n)$ *converges absolutely, then*

$$\sum_{n=1}^{\infty} f(n) = \prod_{p} \left(f(1) + f(p) + f(p^2) + \cdots \right).$$

If, in addition, f *is completely multiplicative, then*

$$\sum_{n=1}^{\infty} f(n) = \prod_{p} \frac{1}{1 - f(p)}.$$

PROOF. Let

$$P(x) = \prod_{p \leqslant x} \left(f(1) + f(p) + f(p^2) + \cdots \right).$$

Each factor is absolutely convergent by hypothesis, and there are a finite number of factors, so

$$P(x) = \sum_{n \in A(x)} f(n),$$

where

$$A(x) = \{ n \in \mathbb{N} \mid \text{all prime factors of } n \text{ are less than or equal to } x \}.$$

Now consider the difference

$$\left| \sum_{n=1}^{\infty} f(n) - \sum_{n \in A(x)} f(n) \right| \leqslant \sum_{n \notin A(x)} |f(n)| \leqslant \sum_{n > x} |f(n)|.$$

The last sum tends to zero as x tends to infinity because it is the tail of a convergent series. The identity for completely multiplicative functions follows from the general Euler product expansion because in this case each factor of the infinite product is a convergent geometric series.   □

The nonvanishing condition of Remark 1.6 on p. 15 is automatically satisfied in the setting of Theorem 8.17 for all completely multiplicative functions f since the limit of a nontrivial convergent geometric series cannot be zero.

Much of the study of the Riemann zeta function involves complex analysis.

Definition 8.18. *Let S be an open subset of* $\mathbb{C}$. *A function* f $: S \to \mathbb{C}$ *is called* complex differentiable *or* holomorphic *on S if the limit*

$$\lim_{h \to 0} \frac{f(z + h) - f(z)}{h}$$

exists and is finite for all $z \in S$. *If for all* $z \in S$, f *equals its own Taylor series in a small neighborhood of* z,

$$f(z + h) = \sum_{n=0}^{\infty} \frac{f^{(n)}(z)}{n!} h^n \text{ for small } h,$$

then f *is called* analytic *on S.*

Recall that all functions holomorphic on S are analytic on S and vice versa, in contrast with the case of real functions of a real variable, where "analytic" is a strictly stronger condition than "differentiable infinitely often."

Our next goal is to show that the function $s \mapsto \zeta(s)$ is analytic on the half-plane $\Re(s) > 1$. It is tempting to argue as follows. Each of the individual functions $s \mapsto n^{-s}$ is analytic, so the convergent sum should be analytic with derivative $\sum_{n=1}^{\infty} \frac{-\log n}{n^s}$. Unfortunately, an infinite sum of analytic functions might not be analytic – indeed it might not even be continuous.

Example 8.19. Let

$$f_n(x) = \frac{x^2}{(1+x^2)^n}$$

for $n \geqslant 1$, and let $f(x) = \sum_{n=1}^{\infty} f_n(x)$. We can sum the f_n because they form a geometric progression,

$$\sum_{n=0}^{N} f_n(x) = \frac{x^2}{1 - \dfrac{1}{1+x^2}} - \frac{x^2}{\left(1 - \dfrac{1}{1+x^2}\right)(1+x^2)^{N+1}}.$$

Now when $x \neq 0$ we can let N tend to infinity, the second term tends to zero, and the whole sum converges to $f(x) = 1 + x^2$. However, $f_n(0) = 0$ for all $n \geqslant 1$, so $f(0) = \sum_{n=1}^{\infty} f_n(0) = 0$ also. Thus the limit function f is not even continuous, although all the summands f_n are analytic on a neighborhood of 0. The same phenomenon for complex z is seen in the region $\{z \mid |\arg(z)| < \pi/4\}$.

One useful criterion to make sure that nothing goes wrong in manipulating series of functions is *uniform convergence*.

8.5 Uniform Convergence

Definition 8.20. *Let $S \subseteq \mathbb{C}$ be a nonempty set. A sequence (F_n) converges pointwise to F on S if, for every $s \in S$, $F_n(s) \to F(s)$. The sequence (F_n) converges uniformly to F on S if for all $\epsilon > 0$ there exists $N = N(\epsilon)$ such that for all $n > N$*

$$|F(s) - F_n(s)| < \epsilon \text{ for all } s \in S.$$

The uniformity in this definition is that the number N is not allowed to depend on s. Many useful properties of the terms of a sequence of functions are inherited by the limit function if the convergence is uniform.

Theorem 8.21. *Suppose that the sequence of functions (F_n) converges to F uniformly on S. If each F_n is continuous on S, then F is continuous on S.*

PROOF. Fix $s_0 \in S$ and $\epsilon > 0$. Choose N such that, for all $s \in S$,

$$|F(s) - F_N(s)| < \frac{\epsilon}{3}. \qquad (8.20)$$

This is possible because the sequence of functions (F_n) is converging uniformly on S. Next, choose $\delta > 0$ such that, for all $s \in S$ with $|s - s_0| < \delta$,

$$|F_N(s) - F_N(s_0)| < \frac{\epsilon}{3}. \qquad (8.21)$$

Now we have set the stage: For all $s \in S$ with $|s - s_0| < \delta$, we have

$$\begin{aligned}
|F(s) - F(s_0)| &= |F(s) - F_N(s) + F_N(s) - F_N(s_0) + F_N(s_0) - F(s_0)| \\
&\leqslant |F(s) - F_N(s)| + |F_N(s) - F_N(s_0)| + |F_N(s_0) - F(s_0)| \\
&< \frac{\epsilon}{3} + \frac{\epsilon}{3} + \frac{\epsilon}{3} = \epsilon.
\end{aligned}$$

We have used the inequality (8.20) twice, in order to estimate the first and third terms, and the inequality (8.21) for the second term. This proves that F is continuous at each point $s_0 \in S$.   □

Theorem 8.22. *For every $\delta > 0$, the partial sums of the Riemann zeta function converge uniformly on $S_{1+\delta} = \{s \in \mathbb{C} : \Re(s) > 1 + \delta\}$. That is,*

$$\sum_{n=1}^{N} \frac{1}{n^s} \longrightarrow \zeta(s) \text{ as } N \to \infty$$

uniformly on $S_{1+\delta}$.

In particular, by Theorem 8.21, the Riemann zeta function is continuous on

$$\bigcup_{\delta>0} S_{1+\delta} = \{s \in \mathbb{C} : \Re(s) > 1\} = S_1.$$

The convergence is not uniform on the whole of S_1.

PROOF OF THEOREM 8.22. Notice that for any $s \in S_{1+\delta}$,

$$\left| \zeta(s) - \sum_{n=1}^{N} \frac{1}{n^s} \right| = \left| \sum_{n=N+1}^{\infty} \frac{1}{n^s} \right|$$

$$< \sum_{n=N+1}^{\infty} \frac{1}{n^{1+\delta}}$$

$$< \int_{N}^{\infty} \frac{1}{x^{1+\delta}} \, dx = \frac{N^{-\delta}}{\delta}.$$

Given any $\epsilon > 0$, this is less than ϵ when N is large, independently of s, showing uniform convergence.   □

Exercise 8.20. (see Section 6.1) Let L denote a lattice in $\mathbb{C}$ with an associated Weierstrass $\wp$-function defined by

$$\wp_L(z) = \frac{1}{z^2} + \sum_{0 \neq \ell \in L} \left\{ \frac{1}{(z-\ell)^2} - \frac{1}{\ell^2} \right\} \text{ for } z \notin L. \tag{8.22}$$

Prove that

$$\frac{1}{z^2} + \sum_{\substack{0 \neq \ell \in L, \\ |\ell| \leqslant n}} \left\{ \frac{1}{(z-\ell)^2} - \frac{1}{\ell^2} \right\} \longrightarrow \wp_L(z)$$

uniformly as $n \to \infty$ on any compact subset of $\mathbb{C} \backslash L$.

8.6 The Zeta Function Is Analytic

The next consequence of uniform convergence is that it preserves the analyticity of complex functions.

Theorem 8.23. *Suppose $S \subseteq \mathbb{C}$ is open, and we have a function* $\mathsf{F} : S \to \mathbb{C}$ *and a sequence of functions* $\mathsf{F}_N : S \to \mathbb{C}$ *converging to* F *uniformly on S. If each F_N is analytic, then F is analytic.*

Example 8.24. The function $\mathsf{F}_N(s) = \sum_{n=1}^{N} \frac{1}{n^s}$ is analytic and converges uniformly to $\zeta(s)$ on every $S_{1+\delta}$, $\delta > 0$ as in Theorem 8.22, so by Theorem 8.23, the Riemann zeta function is analytic on $S_{1+\delta}$ for every $\delta > 0$ – in other words, ζ is analytic on $\{s \in \mathbb{C} \mid \Re(s) > 1\}$.

PROOF OF THEOREM 8.23. Given a fixed point $a \in S$, we have to prove that F is analytic on a neighborhood of a. We use complex analysis, in particular Cauchy's formula. Let γ be a closed simple curve, that is a finite join of smooth curves such that $a \in \mathrm{Int}(\gamma)$ and the closure $\overline{\mathrm{Int}}(\gamma) \subseteq S$. Then Cauchy's formula says that for any function f that is analytic on S, and for any $b \in \mathrm{Int}(\gamma)$,

$$\mathsf{f}(b) = \frac{1}{2\pi \mathrm{i}} \int_\gamma \frac{\mathsf{f}(z)}{z-b} \, dz. \tag{8.23}$$

Since S is open, it contains a small disk around a. We will need the following result.

Lemma 8.25. *Suppose a sequence of continuous functions* $\mathsf{G}_N : \gamma \to \mathbb{C}$ *converges uniformly on γ to a function* $\mathsf{G} : \gamma \to \mathbb{C}$. *Then G is continuous and*

$$\lim_{N \to \infty} \int_\gamma \mathsf{G}_N(s) \, ds = \int_\gamma \mathsf{G}(s) \, ds.$$

PROOF. The continuity of G follows from Theorem 8.21, so in particular G is integrable. Now

$$\left| \int_\gamma G(s)\, ds - \int_\gamma G_N(s)\, ds \right| = \left| \int_\gamma [G(s) - G_N(s)]\, ds \right|$$

$$\leqslant \int_\gamma |G(s) - G_N(s)|\, ds$$

$$\leqslant \text{length}(\gamma) \max_{s \in \gamma} |G(s) - G_N(s)|.$$

The last quantity tends to zero by the definition of uniform convergence. This completes the proof of the lemma. □

Exercise 8.21. (a) Prove *Morera's Theorem:* If f is a continuous function on a domain $D \subseteq \mathbb{C}$ with the property that $\int_\gamma f(z)\, dz = 0$ for every closed contour in D, then f is analytic on D.
(b) Use this to give a different proof of Theorem 8.23.

Now a magic wand can be waved to complete the proof of Theorem 8.23. By hypothesis, the functions F_N are analytic on S, so for all $b \in \text{Int}(\gamma) \subseteq S$, by Cauchy's formula Equation (8.23),

$$F_N(b) = \frac{1}{2\pi i} \int_\gamma \frac{F_N(s)}{s - b}\, ds.$$

Define $G_N(s) = \frac{F_N(s)}{s-b}$. Then G_N converges to $G(s) = \frac{F(s)}{s-b}$ uniformly on γ since

$$|G(s) - G_N(s)| = \left| \frac{F(s) - F_N(s)}{s - b} \right|$$

$$\leqslant C \max_{s \in \gamma} |F(s) - F_N(s)|,$$

where $C = \max_{s \in \gamma} \frac{1}{|s-b|}$. This proves that for all $b \in \text{Int}(\gamma)$

$$F(b) = \lim_{N \to \infty} F_N(b) = \lim_{N \to \infty} \frac{1}{2\pi i} \int_\gamma G_N(s)\, ds = \frac{1}{2\pi i} \int_\gamma \frac{F(s)}{s - b}\, ds. \qquad (8.24)$$

Here, we have applied Lemma 8.25 to interchange a uniform limit and the integral in the last step. Finally, recall that any function F satisfying Cauchy's formula (Equation (8.24)) on $\text{Int}(\gamma)$ is analytic there; this may be seen as follows. For all $b \in \text{Int}(\gamma)$,

$$\frac{F(b + h) - F(b)}{h} = \frac{1}{2\pi i h} \int_\gamma F(s) \left(\frac{1}{s - b - h} - \frac{1}{s - b} \right) ds$$

$$= \frac{1}{2\pi i} \int_\gamma \frac{F(s)}{(s - b)(s - b - h)}\, ds,$$

since the hs cancel. In the last integral, the limit $h \to 0$ may be taken, which gives the derivative. Strictly speaking, we should establish uniform convergence (as with the G_N above) to be able to apply Lemma 8.25 again, but this is straightforward. □

Corollary 8.26. *For all s with $\Re(s) > 1$,*

$$\frac{d}{ds}\,\zeta(s) = -\sum_{n=1}^{\infty} \frac{\log n}{n^s}.$$

We will see later that many of the deeper properties of the Riemann zeta function and their consequences for number theory take place for complex values $s = \sigma + it$ with $\sigma < 1$, and all we have done so far (including the definition of the Riemann zeta function) does not apply to such values.

8.7 Analytic Continuation of the Zeta Function

A very important idea from complex analysis is that of *analytic continuation.* Given a function f defined by a convergent power series on a disk D of positive radius, an analytic function defined on any domain containing D that coincides with f on D is called a continuation of f.

Exercise 8.22. Prove the Uniqueness Theorem: Suppose that G is a domain in $\mathbb{C}$, and f and g are differentiable functions on G with $f(z) = g(z)$ for all $z \in S$, where $S \subseteq G$ has a limit point in G. Then $f(z) = g(z)$ for all $z \in G$.

Example 8.27. Consider the function defined by the power series

$$g(s) = 1 + s + s^2 + \cdots,$$

which converges for $|s| < 1$. Then g can be continued to a function that is analytic on the whole of $\mathbb{C}$ except for a simple pole at $s = 1$. To see this, notice that for $|s| < 1$, $g(s) = \frac{-1}{s-1}$. The latter expression is defined on $\mathbb{C}$ apart from a simple pole at $s = 1$ with residue -1. Of course, g is *not* defined by the series for $|s| \geqslant 1$.

Theorem 8.28. *The Riemann zeta function has an analytic continuation to the whole of the complex plane with the exception of a simple pole with residue 1 at $s = 1$.*

We will present two different proofs of this: a standard proof that may be found in any of the books on the topic and an alternative proof no doubt known to the experts but not readily available in the literature.

In Chapter 9, we will find a functional equation for the Riemann zeta function. Using this, Theorem 8.28 may be deduced from the weaker statement that there is a continuation to $\Re(s) > 0$. The first proof is of this weaker statement.

Theorem 8.29. *The Riemann zeta function has an analytic continuation to the set $\{s \in \mathbb{C} \mid \Re(s) > 0\}$ with the exception of a simple pole at $s = 1$ with residue 1.*

FIRST (STANDARD) PROOF OF THEOREM 8.29. This involves a careful use of the Euler Summation Formula – the care is needed because the formula as stated only applies to finite intervals. Assume first that $\Re(s) > 1$; then, by the Euler Summation Formula,

$$\sum_{n=2}^{N} \frac{1}{n^s} = \int_1^N t^{-s}\, dt + \int_1^N \frac{-s\{t\}}{t^{s+1}}\, dt. \tag{8.25}$$

The first term is

$$\left[\frac{t^{1-s}}{1-s}\right]_1^N = \frac{N^{1-s}}{1-s} - \frac{1}{1-s},$$

and since we suppose $\Re(s) > 1$, we get $N^{1-s} \to 0$ as N tends to infinity. The second term in Equation (8.25) also converges as N tends to infinity since

$$\left|\int_1^\infty \frac{-s\{t\}}{t^{s+1}}\, dt\right| \leqslant \int_1^\infty \frac{|s|}{t^{\sigma+1}}\, dt < \infty.$$

This shows that $\sum_{n=1}^{\infty} \int_n^{n+1} \frac{-s\{t\}}{t^{s+1}}\, dt$ is absolutely convergent, so we are justified in writing

$$\zeta(s) = 1 + \sum_{n=2}^{\infty} \frac{1}{n^s} = 1 - \frac{1}{1-s} - s \int_1^\infty \frac{\{t\}}{t^{1+s}}\, dt. \tag{8.26}$$

Lemma 8.30. *The integral in* Equation (8.26) *represents an analytic function on the range $\Re(s) > 0$.*

PROOF. Write

$$I(s) = \sum_{n=1}^{\infty} f_n(s),$$

where $f_n(s) = \int_n^{n+1} \frac{\{t\}}{t^{s+1}}\, dt$. We will prove that for any $\delta > 0$

   (a) the series for $I(s)$ converges uniformly on $\Re(s) > \delta$ and
   (b) each f_n is analytic on the range $\Re(s) > \delta$.

Then we may apply Theorem 8.23 to complete the proof that $I(s)$ is analytic on $\Re(s) > 0$. As for (a),

$$\left|I(s) - \sum_{n=1}^{N} f_n(s)\right| = \left|\sum_{n=N+1}^{\infty} f_n(s)\right| \leqslant \sum_{n=N+1}^{\infty} |f_n(s)|$$

$$\leqslant \int_{N+1}^\infty \frac{1}{t^{\sigma+1}}\, dt = \left[\frac{t^{-\sigma}}{-\sigma}\right]_{N+1}^\infty$$

$$= \frac{(N+1)^{-\sigma}}{\sigma},$$

and in absolute value this is smaller than $\frac{1}{\delta(N+1)^\delta}$, so it tends to zero. The bound depends on δ only, not on σ or s, which proves the uniform convergence.

For (b), consider the difference quotient

$$\frac{f_n(s+h) - f_n(s)}{h} = \frac{1}{h} \int_n^{n+1} \frac{\{t\}}{t^{s+1}} \left(\frac{1}{t^h} - 1 \right) dt. \tag{8.27}$$

Use a first-order Taylor approximation for the exponential

$$t^{-h} = e^{-h \log t} = 1 - h \log t + f(h,t),$$

where

$$f(h,t) = O\big((h \log t)^2\big). \tag{8.28}$$

Substituting into Equation (8.27) gives

$$\frac{1}{h} \left(f_n(s+h) - f_n(s) \right) = \int_n^{n+1} \frac{\{t\}}{t^{s+1}} \left(-\log t + \frac{1}{h} f(h,t) \right) dt.$$

The left-hand side for small h should be close to the derivative of f_n, and we can make an intelligent guess at what this derivative will be:

$$\left| \frac{1}{h} \left(f_n(s+h) - f_n(s) \right) + \int_n^{n+1} \frac{\{t\}}{t^{s+1}} \log t \, dt \right| \leqslant \int_n^{n+1} \frac{1}{h t^{\sigma+1}} |f(h,t)| \, dt.$$

The right-hand side tends to zero as $h \to 0$ by Equation (8.28). This completes the first proof of the analytic continuation of the zeta function to $\Re(s) > 0$. $\qquad \square$

Lemma 8.30 gives the analytic continuation of ζ to the half-plane $\Re(s) > 0$ by Equation (8.26), apart from a simple pole at $s = 1$ with residue 1. $\qquad \square$

A natural question is to ask why it was necessary to split the integral from 1 to ∞ into a sum of subintegrals. The reason is that the Taylor approximation in Equation (8.28) is only valid for bounded values of $h \log t$. If we had tried to make a similar argument for the integral from 1 to ∞, t would be unbounded and the quantity $h \log t$ would be unbounded. By the splitting of the integral, we had only to consider $t \in [n, n+1]$ for a *fixed* n at each stage.

These are treacherous waters! We are catching a glimpse here of how quite reasonable questions about the Riemann zeta function turn out to involve potentially subtle analytic problems. The methods we have just used can be squeezed to give a little more. The second proof of Theorem 8.29 (given below) will give an analytic continuation to the whole plane.

Exercise 8.23. (a) Show that

$$A(s) = 1 - \frac{1}{2^s} + \frac{1}{3^s} - \frac{1}{4^s} + \cdots$$

is analytic for $\Re(s) > 0$.

(b) Show that $A(s) = \left(1 - \frac{1}{2^{s-1}}\right)\zeta(s)$, and deduce the analytic continuation of the Riemann zeta function to $\Re(s) > 0$.

(c) Repeat (a) and find a suitable analog of (b) for

$$B(s) = 1 + \frac{1}{2^s} - \frac{1}{3^s} + \frac{1}{4^s} + \frac{1}{5^s} - \frac{1}{6^s} + \cdots.$$

(d) Deduce from (c) that the only pole of ζ in $\Re(s) > 0$ is at $s = 1$.

Exercise 8.24. Prove that the Laurent expansion of ζ about $s = 1$ begins

$$\zeta(s) = \frac{1}{s-1} + \gamma + \phi(s), \tag{8.29}$$

where ϕ denotes a function that is analytic at $s = 1$ and vanishes there.

The next proof gives the full continuation to the whole complex plane (with the exception of the simple pole at $s = 1$).

SECOND PROOF OF THEOREM 8.28. Consider

$$\int_1^\infty x^{-s}\, dx = \frac{-1}{1-s} = \frac{1}{s-1}.$$

We subtract this from $\zeta(s)$ to remove the pole at $s = 1$. For $\Re(s) > 1$,

$$\zeta(s) - \frac{1}{s-1} = \sum_{n=1}^\infty \frac{1}{n^s} - \sum_{n=1}^\infty \int_n^{n+1} x^{-s}\, dx$$

$$= \sum_{n=1}^\infty \left(\frac{1}{n^s} - \int_0^1 (n+x)^{-s}\, dx\right)$$

$$= \sum_{n=1}^\infty \frac{1}{n^s}\left(1 - \int_0^1 \left(1 + \frac{x}{n}\right)^{-s} dx\right). \tag{8.30}$$

The sums involved here converge absolutely in the region $\Re(s) > 1$.

Now assume that we have continued the zeta function to the domain

$$\Re(s) > 1 - K$$

for some integer $K \geqslant 0$. We want to continue it further to $\Re(s) > -K$. To do this, put $h = x/n$ and use a Taylor approximation for

$$f_s(h) = (1+h)^{-s}$$

of order K. Recall that the Taylor polynomial of degree K for f_s at $h = 0$ is defined by[3]

---

[3] If you only want the continuation to $\Re(s) > 0$, think of $K = 1$: The Taylor polynomial for $(1 + x/n)^{-s}$ in this case is simply $1 - \frac{sx}{n}$, and the error term is $O(n^{-2})$. Substitute Equation (8.31) into Equation (8.30), and you get a series convergent for $\Re(s) > 0$.

$$T_{f,s,K}(h) = \sum_{k=0}^{K} \frac{f_s^{(k)}(0)}{k!} h^k. \tag{8.31}$$

We have to calculate higher derivatives of f_s at $h = 0$. These are given by setting $h = 0$ in the relation

$$f_s^{(k)}(h) = \frac{(-1)^k(s+k-1)(s+k-2)\cdots(s+1)s}{(h+1)^k} f_s(h),$$

which is easy to prove by induction on k. Since f_s is analytic on the neighborhood of $h = 0$, we have an estimate for the error term of the form

$$|f_s(h) - T_{f,s,K}(h)| \leqslant \frac{\left|f_s^{(K+1)}(h')\right|}{(K+1)!} |h|^{K+1} \tag{8.32}$$

for some h' with $|h'| < |h|$. For bounded values of s, this is $O(|h|^{K+1})$. Use the Taylor polynomial with $h = x/n$ in Equation (8.30). We evaluate the inner integral first:

$$\int_0^1 \left(1 + \frac{x}{n}\right)^{-s} dx = 1 + \sum_{k=1}^{K} \frac{f_s^{(k)}(0)}{(k+1)! \, n^k} + O\left(\frac{1}{n^{K+1}}\right).$$

Putting this into Equation (8.30) gives the nice identity

$$\zeta(s) - \frac{1}{s-1} = -\sum_{k=1}^{K} \frac{(-1)^k(s+k-1)\cdots(s+1)s}{(k+1)!} \zeta(s+k)$$

$$+ \sum_{n=1}^{\infty} \frac{1}{n^s} \int_0^1 T_{f,s,K}\left(\frac{x}{n}\right) - \left(1 + \frac{x}{n}\right)^{-s} dx. \tag{8.33}$$

The last sum converges by the inequality (8.32), and for all $s \neq 0, -1, -2, \ldots$ the values

$$\zeta(s+1), \zeta(s+2), \ldots, \zeta(s+K-1)$$

are all defined by hypothesis, giving the continuation of the zeta function to

$$\Re(s) > -K.$$

In the case $s = -m = 0, -1, -2, \ldots, 1 - K$, one of the arguments of ζ in the first summand of Equation (8.33) becomes 1, but this is no problem since the pole is cancelled out by the appropriate factor $(s+m)$ in the coefficient. (The right-hand side of Equation (8.33) has a removable singularity there.)

Thus, by induction, there is an analytic continuation of the zeta function to

$$\Re(s) > -1, -2, \ldots,$$

in fact to the whole of the complex plane.   □

Exercise 8.25. Recall that $\pi(x)$ denotes the number of primes less than or equal to x. Prove that if $s = \sigma + it$ with $\sigma > 1$, then

$$\log \zeta(s) = s \int_2^\infty \frac{\pi(x)}{x(x^s - 1)}\, dx.$$

To do this, convert the sum over all primes to a sum over all natural numbers using

$$\pi(n) - \pi(n-1) = \begin{cases} 1 & \text{if } n \text{ is prime,} \\ 0 & \text{otherwise.} \end{cases}$$

The next exercise is an easy version of what has been done for the Riemann zeta function $\zeta(s) = \sum_{n=1}^\infty e^{-s \log n}$.

Exercise 8.26. Define a function F by

$$F(s) = \sum_{n=0}^\infty e^{-ns}.$$

(a) Find the domain of convergence for this series.
(b) Prove that the series converges uniformly for $\Re(s) > \delta$ for any fixed $\delta > 0$.
(c) Using (b), find the derivative of F by differentiating term by term.
(d) Find a simple expression for F by viewing it as a geometric progression and summing it.
(e) Differentiate this closed form and expand the answer using the Binomial Theorem. Check that you get the series in (c) again.
(f) Obtain the analytic continuation of F to the whole complex plane. Describe the location and order of all the poles of F.
(g) Compute $F(-1)$

Exercise 8.27. Prove that $\frac{1}{\zeta(s)} = \sum_{n=1}^\infty \frac{\mu(n)}{n^s}$ for $\Re(s) > 1$.

Exercise 8.28. Prove that $\zeta^2(s) = \sum_{n=1}^\infty \frac{d(n)}{n^s}$ for $\Re(s) > 1$.

Exercise 8.29. Prove that $\frac{\zeta(s-1)}{\zeta(s)} = \sum_{n=1}^\infty \frac{\phi(n)}{n^s}$ for $\Re(s) > 2$.

Exercise 8.30. Prove that $\displaystyle\sum_{\substack{m,n \in \mathbb{N}; \\ \gcd(m,n)=1}} \frac{1}{(mn)^2} = \frac{\zeta^2(2)}{\zeta(4)}.$

NOTES TO CHAPTER 8: A treatment of the Prime Number Theorem at a level similar to ours may be found in Jameson's book [81]; this book also includes Selberg's elementary proof of the Prime Number Theorem [136] (not using analytic methods). The quantity $\theta(n) = \sum_{p \leqslant n} \log p$ from Lemma 1.8 plays a central role in the elementary proof; indeed the statement $\frac{\theta(n)}{n} \to 1$ as $n \to \infty$ is equivalent to the theorem. Erdös [52] also published an elementary proof of the Prime Number Theorem; a careful account of the controversy is provided by Goldfeld [70]. An accessible proof of the important inequality (8.8) may be found in Spivak's lovely book [145, p. 543]. Exercise 8.6(b) is a special case of a result due to Moss [111]. The implications (8.9) are in the book of Apostol [4, p. 91]. Tchebychef's proof of Exercise 8.7 appeared originally in his paper [152] of 1852. Exercise 8.9 is due to Gandhi [65]; an accessible treatment is in a paper of Golomb [71]. Exercise 8.11(a) and (c) are taken from a paper of Brent [20]; Exercise 8.11(b) comes from a paper of Jänichen [82]. The proof of Zsigmondy's Theorem in Section 8.3.1 is adapted from a more general result of Schinzel [133]. A readable account of Exercise 8.19 appears in a paper of Yabuta [165]. For further reading on Section 8.7, consult Apostol [4] or Titchmarsh [153]. Edwards' book [47] contains a translation of Riemann's original paper [128]. Fourier analysis – and its more august cousin, harmonic analysis – plays a central role in number theory. For sophisticated accounts, see Tate's thesis [150] or Weil's book [158]. A more accessible account of this advanced material may be found in the book of Ramakrishnan and Valenza [121].

9

The Functional Equation of the Riemann Zeta Function

A *functional equation* is simply an identity involving functions. The trigonometric identity $\sin^2(x) + \cos^2(x) = 1$ is one of the most familiar examples. In this chapter, a highly nontrivial functional equation satisfied by the Riemann zeta function is found, that allows the function to be extended to the whole complex plane (apart from the singularity at $s = 1$).

The proof is difficult, and it might seem that we have strayed far from the arithmetic path that started with the Fundamental Theorem of Arithmetic. However, the proof relies crucially on some observations that arose because of the way mathematicians, particularly in the nineteenth century, thought about functions. If a function $f : \mathbb{C} \to \mathbb{C}$ has distinct zeros $z_1, z_2, \ldots$, then it seems natural to "factorize" it, and hope that

$$f(z) = (z - z_1)(z - z_2) \cdots .$$

Of course, convergence issues arise, and occasionally some careful doctoring is needed to make this idea precise and useful.

9.1 The Gamma Function

The Gamma function is one of many classical special functions. It is surprising how the Gamma function helps us to understand properties of the zeta function and some other arithmetic problems.

Definition 9.1. *The* Gamma function Γ *is defined by*

$$\Gamma(s) = \int_0^\infty e^{-t} t^{s-1} \, dt \tag{9.1}$$

for any $s \in \mathbb{C}$ with $\Re(s) > 0$.

Exercise 9.1. Prove that the integral in Equation (9.1) exists for any $s \in \mathbb{C}$ with $\Re(s) > 0$.

As with the Riemann zeta function, it is important to establish the analytic properties of the Gamma function.

Exercise 9.2. Prove that $\Gamma(s)$ is an analytic function of s for $\Re(s) > 0$ by proving the following statements.

(a) $\Gamma(s) = \sum_{n=0}^{\infty} \Gamma_n(s)$, where $\Gamma_n(s) = \int_n^{n+1} e^{-t} t^{s-1}\, dt$.

(b) For any fixed $\delta > 0$, $\sum_{n=0}^{N} \Gamma_n(s) \to \Gamma(s)$ uniformly on $\{s \in \mathbb{C} \mid \Re(s) > \delta\}$ as $N \to \infty$.

(c) Γ_n is analytic for any $n \geqslant 0$.

All these steps are very similar to the argument for $\int_0^{\infty} \frac{\{t\}}{t^{s+1}}\, dt$.

Later on, another (better) proof that Γ is analytic will be given.

Lemma 9.2. *The Gamma function has the following properties.*

(1) *For all s with $\Re(s) > 0$,*

$$\Gamma(s+1) = s\Gamma(s).$$

(2) *For all integers $N \geqslant 0$,*

$$\Gamma(N+1) = N!.$$

PROOF. The first relation is found by integrating,

$$\Gamma(s+1) = \int_0^{\infty} e^{-t} t^s\, dt = \left[-e^{-t} t^s\right]_0^{\infty} + s \int_0^{\infty} e^{-t} t^{s-1}\, dt.$$

The first term vanishes at $t = 0$ because $\Re(s) > 0$.

The second statement follows from the first by induction together with the easy calculation that $\Gamma(1) = 1$. $\qquad\square$

Proposition 9.3. *The Gamma function can be analytically continued to all of $\mathbb{C}$, where it is analytic apart from simple poles at $0, -1, -2, \ldots$ and so on.*

PROOF. By Lemma 9.2(1), we may write

$$\Gamma(s) = \frac{1}{s}\Gamma(s+1).$$

The right-hand side is defined for $\Re(s) > -1$ apart from $s = 0$, where it has a simple pole with residue $\Gamma(1) = 1$. Iterating this gives

$$\Gamma(s) = \frac{1}{s(s+1)}\Gamma(s+2). \tag{9.2}$$

The right-hand side of Equation (9.2) is defined for $\Re(s) > -2$, apart from $s = 0, -1$, where there are simple poles again. In this way, we can inductively continue the Gamma function to the whole plane, where it is analytic apart from simple poles at $0, -1, -2, \ldots$. $\qquad\square$

Theorem 9.4. $\Gamma(s) \neq 0$ *for all $s \in \mathbb{C}$.*

This will be proved in Section 9.6.

9.2 The Functional Equation

Our goal throughout this chapter will be the proof of the following theorem.

Theorem 9.5. [THE FUNCTIONAL EQUATION] *Let*

$$F(s) = \pi^{-s/2} \Gamma\left(\frac{s}{2}\right) \zeta(s)$$

for $\Re(s) > 0$. *Then* F *satisfies the functional equation*

$$F(1 - s) = F(s).$$

Corollary 9.6. *The function* F *has an analytic continuation to the whole complex plane apart from poles at 1 and 0. The Riemann zeta function has an analytic continuation to the complex plane where it is analytic apart from a simple pole at $s = 1$. The zeta function vanishes at negative even integers.*

PROOF. Expand Theorem 9.5 to give

$$\pi^{-(1-s)/2} \Gamma\left(\frac{1-s}{2}\right) \zeta(1-s) = \pi^{-s/2} \Gamma\left(\frac{s}{2}\right) \zeta(s),$$

$$\zeta(1-s) = \frac{\pi^{-s+1/2} \Gamma\left(\frac{s}{2}\right) \zeta(s)}{\Gamma\left(\frac{1-s}{2}\right)}. \qquad (9.3)$$

We know that $\Gamma(s)$ has a simple pole at $s = -m$ for $m \in \mathbb{N}$. Thus, for all $m \in \mathbb{N}$,

$$\zeta(-2m) = 0$$

since $1 - s = -2m$ if and only if $s = 2m + 1$, $\zeta(s) \neq 0$ for $\Re(s) > 1$ by the Euler product expansion, and $\Gamma \neq 0$ everywhere. The case $s = 1$ is different: Here the right-hand side has a simple pole in the numerator, too (in ζ), cancelling the one in Γ. Thus $\zeta(s)$ is analytic and nonzero at $s = 0$. By the functional equation, the values of $F(s)$ for $\Re(s) \geqslant 1/2$ determine all of F. We found $\zeta(-2m) = 0$ for all $m \in \mathbb{N}$, and there are no more zeros of ζ with $\Re(s) < 0$ because Γ has no other poles by Equation (9.3). Also, $\zeta(s) \neq 0$ for $\Re(s) > 1$ because of the Euler product expansion (see Remark 1.6).   □

In the course of the proof, we found a set of special values of the zeta function at negative even integers. Later we will see that negative odd integers yield rational values of the zeta function (see Exercise 9.10 on p. 204).

9.2.1 The Riemann Hypothesis

Corollary 9.6 gives another proof that the Riemann zeta function can be continued to the whole plane, where it is analytic apart from a simple pole at $s = 1$. Moreover, any nontrivial zero of ζ must lie in the critical strip defined by $0 \leqslant \Re(s) \leqslant 1$. Riemann stated without proof the following conjecture.

Conjecture 9.7. [THE RIEMANN HYPOTHESIS] All zeros of ζ in the critical strip $0 \leqslant \Re(s) \leqslant 1$ have $\Re(s) = \frac{1}{2}$.

This is still an open problem, and its resolution is viewed as one of the outstanding open problems in mathematics. All the zeros found thus far (the first ten billion are known) lie on the line $\Re(s) = \frac{1}{2}$, and they are all simple. Figure 9.1 shows $\Re(\zeta(\frac{1}{2} + it))$ for $0 \leqslant t \leqslant 60$, which already shows the extraordinary subtlety and complexity of the zeta function along the critical line.

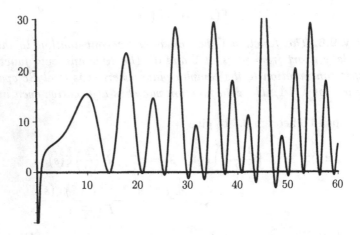

Figure 9.1. The graph of $\Re(\zeta(\frac{1}{2} + it))$ for $0 < t \leqslant 60$.

Just as the Prime Number Theorem is equivalent to a statement about the partial sums of the Möbius function, the Riemann Hypothesis is equivalent to the statement that for every $\varepsilon > 0$

$$\sum_{n \leqslant x} \mu(n) = \mathrm{O}(x^{\varepsilon + 1/2}).$$

Growth properties of the Möbius function are very delicate, and the numerical evidence can be deceptive. A long-standing conjecture of Mertens, supported by a great deal of numerical evidence, was that

$$\left| \sum_{n \leqslant x} \mu(n) \right| < \sqrt{x};$$

this was eventually disproved by Odlyzko and te Riele in 1985.

It is reasonable to ask why certain problems, such as the Riemann Hypothesis, obtain legendary status. Certainly this one has attracted considerable folklore. David Hilbert is reputed to have said that if he were to be awoken in a thousand years, the first question he would ask would be about the status of the Riemann Hypothesis. Many mathematicians believe it must be true, although some great figures have been sceptical. The explanation for its importance is multi-faceted. On the one hand, its statement has great beauty and simplicity, while the many unsuccessful attempts to resolve it have driven forward sophisticated methods in number theory. On the other hand – perhaps more germane to our study – the Riemann Hypothesis is intimately connected to the distribution of the primes. Many results in analytic number theory can be proved in stronger forms if the Riemann Hypothesis is assumed. Less obviously, but perhaps most importantly, the Riemann Hypothesis seems to lie at the heart of future developments in the area of overlap between number theory, geometry and analysis. Workers in this area sometimes need an almost prophetic insight that can lead to layers of conjectures about how hard unsolved problems will eventually be cracked. Much of this has to do with functions that generalize the Riemann zeta function, called L-functions (we will encounter an L-function in the next chapter.) The Riemann Hypothesis seems to be a basic example of a whole series of results that will be needed to make progress in this area. Finally, in addition to its central role in number theory, the Riemann Hypothesis is conjectured to relate to problems in physics – the zeros of the zeta function corresponding to the eigenvalues of an appropriate Hermitian operator.

The Clay Mathematics Institute[1] has offered a million dollars for a proof of the Riemann Hypothesis. The prize is not on offer for a disproof, say by giving a counterexample.

9.3 Fourier Analysis on Schwartz Spaces

For the proof of the functional equation in Theorem 9.5, we will need some Fourier analysis.

Definition 9.8. *The* Schwartz space $\mathcal{S}$ *is the set of functions* $f : \mathbb{R} \to \mathbb{C}$ *that are infinitely differentiable and whose derivatives* $f^{(n)}$ *(including the function itself* $f^{(0)} = f$*) all satisfy*

$$(1 + |x|)^m f^{(n)} = O(1) \tag{9.4}$$

for all $m \in \mathbb{N}$. *The bound in* $O(1)$ *may depend upon* m *and* n.

Example 9.9. The Gaussian function $f(x) = e^{-x^2}$ is in $\mathcal{S}$.

---

[1] On May 24th 2000, the Clay Mathematics Institute established seven Millennium Prize Problems, each worth one million dollars, including the Riemann Hypothesis because "they are important classic questions that have resisted solution over the years."

Notice that S is a complex vector space and that any function $f \in S$ is integrable,

$$\left| \int_{-\infty}^{\infty} f(x)\, dx \right| \leqslant \int_{-\infty}^{\infty} |f(x)|\, dx \leqslant C \int_{-\infty}^{\infty} \frac{1}{(1+|x|)^2}\, dx < \infty,$$

just by taking $n = 0$ and $m = 2$ in Equation (9.4).

Definition 9.10. *For any function* $f \in S$, *the* Fourier transform *of* f *is the function*

$$\widehat{f}(y) = \int_{-\infty}^{\infty} f(x) e^{-2\pi i x y}\, dx.$$

The integral exists for the same reason as before,

$$|\widehat{f}(y)| \leqslant \int_{-\infty}^{\infty} |f(x)|\, dx < \infty,$$

and in fact $\widehat{f} \in S$ again since we may apply Equation (9.4) with $m = n$ to get the bound for $\widehat{f}^{(n)}$.

Thus $f \to \widehat{f}$ is a linear map from S to S. It turns out that this map has a fixed point – a function equal to its Fourier transform. Recall that $\int_{-\infty}^{\infty} e^{-x^2}\, dx = \sqrt{\pi}$.

Lemma 9.11. *If* $f(y) = e^{-\pi y^2}$, *then* $\widehat{f}(y) = f(y)$.

PROOF.

$$\widehat{f}(y) = \int_{-\infty}^{\infty} e^{-\pi x^2} e^{-2\pi i x y}\, dx.$$

The idea is to complete the square,

$$-\pi(x^2 + 2ixy) = -\pi[(x + iy)^2 + y^2],$$

so the Fourier transform of f is

$$\widehat{f}(y) = e^{-\pi y^2} \int_{-\infty}^{\infty} e^{-\pi(x+iy)^2}\, dx.$$

Let

$$I(y) = \int_{-\infty}^{\infty} e^{-\pi(x+iy)^2}\, dx.$$

We know that $I(0) = 1$. What happens if $y \neq 0$? Fix some large N and consider the following paths:

$$\gamma_1 = [-N, N], \qquad\qquad \gamma_2 = [N, N + yi],$$
$$\gamma_3 = [N + yi, -N + yi], \; \gamma_4 = [-N + yi, -N].$$

Put $\gamma = \gamma_1 + \gamma_2 + \gamma_3 + \gamma_4$ (a rectangle). Since $e^{-\pi z^2}$ is an analytic function on the whole of the complex plane, we have, for any $N \geqslant 0$,

$$\int_\gamma e^{-\pi z^2}\, dz = 0.$$

Now, as $N \to \infty$, the integral of $e^{-\pi z^2}$ over γ_1 tends to $I(0) = 1$, the integral over γ_3 tends to $-I(y)$, and the integrals over γ_2 and γ_4 both tend to 0, as $N \to \infty$. This completes the proof of Lemma 9.11. $\qquad\square$

Exercise 9.3. Prove that $\int_N^{N+yi} e^{-z^2}\, dz \to 0$ as $N \to \infty$ for any $y \in \mathbb{R}$.

9.4 Fourier Analysis of Periodic Functions

Fourier analysis is more familiar in the setting of periodic functions.

Definition 9.12. *A function* $g : \mathbb{R} \to \mathbb{C}$ *is periodic with period 1 if*

$$g(x) = g(x+1) \text{ for all } x \in \mathbb{R}.$$

If g *is periodic and piecewise continuous, then its kth* Fourier coefficient *is defined for* $k \in \mathbb{Z}$ *by*

$$c_k = \int_0^1 g(x)e^{-2\pi ikx}\, dx,$$

and its Fourier series *is the function*

$$G(x) = \sum_{k \in \mathbb{Z}} c_k e^{2\pi ikx}.$$

Lemma 9.13. *If* g *is periodic and twice differentiable with continuous second derivative, then there exists a constant* $C > 0$*, depending only upon* g*, such that*

$$|c_k| \leqslant \frac{C}{k^2}$$

for all $k \neq 0$.

PROOF. Integrate by parts:

$$c_k = \left[\frac{-e^{-2\pi ikx}g(x)}{2\pi ik} \right]_0^1 + \int_0^1 \frac{e^{-2\pi ikx}g'(x)}{2\pi ik}\, dx.$$

Now the bracketed term vanishes because g is periodic. Integrate by parts again, so that k^2 appears in the denominator, and then bound the exponential by 1. Finally, put $C = \int_0^1 |g''|\, dx/(4\pi^2)$. $\qquad\square$

Theorem 9.14. *Any function* g *that is periodic and differentiable infinitely often has a Fourier series expansion*

$$g(x) = \sum_{k \in \mathbb{Z}} c_k e^{2\pi i k x}$$

that is uniformly convergent on $\mathbb{R}$.

PROOF. Let G be the Fourier series of g, and apply Lemma 9.13:

$$\left| G(x) - \sum_{k=-n}^{n} c_k e^{2\pi i k x} \right| \leqslant C \sum_{|k|>n} \frac{1}{k^2},$$

where the last sum tends to zero independent of x since the constant C depends only on g. This proves the convergence is uniform.

The equality $g(x) = G(x)$ is not so easy to prove. We first record a few lemmas that are of interest in their own right.

Lemma 9.15. *Consider the sequence of functions* (D_K) *defined by*

$$D_K(x) = \sum_{k=-K}^{K} e^{2\pi i k x}, \qquad \text{for } K \in \mathbb{N},$$

called the Dirichlet kernel. *Then*

$$\int_0^1 D_K(x)\, dx = 1, \tag{9.5}$$

$$D_K(x) = \frac{\sin((2K+1)\pi x)}{\sin(\pi x)}, \tag{9.6}$$

and

$$\int_0^1 g(y+x) D_K(x)\, dx = \sum_{k=-K}^{K} c_k e^{2\pi i k y}, \tag{9.7}$$

where c_k *are the Fourier coefficients of* g *as in Theorem 9.14.*

The functions D_K are useful because they concentrate at the origin and pick out the Fourier coefficients conveniently. The shape of D_K is illustrated in Figure 9.2, that shows the graph of D_{11}.

PROOF OF LEMMA 9.15. Equation (9.5) follows from the fact that

$$\int_0^1 e^{2\pi i k x}\, dx = 0 \text{ for all } k \neq 0.$$

Equation (9.6) is proved by induction on k or directly by summation of a geometric progression. Equation (9.7) follows since

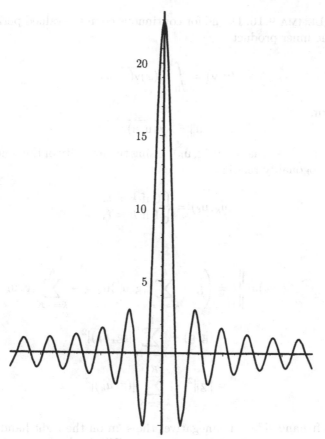

Figure 9.2. The Dirichlet kernel $D_{11}(x)$ for $-\frac{1}{2} \leqslant x \leqslant \frac{1}{2}$.

$$\int_0^1 \mathbf{g}(y+x)D_K(x)\,dx = \int_{y-1}^y \mathbf{g}(z)D_K(z-y)\,dz$$

$$= \int_{-1/2}^{1/2} \mathbf{g}(z)D_K(y-z)\,dz.$$

In the last step, we have used the fact that $\mathbf{g}$ and D_K are periodic functions and that D_K is an even function. At this point, we put in the definition of the D_K, interchange the integral and the sum, and extract a factor $e^{2\pi iky}$ from each summand, which gives the right-hand side of Equation (9.7). $\qquad\square$

Lemma 9.16. [RIEMANN–LEBESGUE LEMMA] *Let* $\mathbf{g}$ *be a continuous periodic function, and let* c_k *be the kth Fourier coefficient of* $\mathbf{g}$. *Then*

$$\lim_{|k|\to\infty} c_k = 0.$$

PROOF OF LEMMA 9.16. Define for continuous complex-valued periodic functions u, v the inner product

$$(u, v) = \int_0^1 u(x)\overline{v(x)}\, dx$$

and the norm

$$\|u\| = \sqrt{(u, u)}.$$

Let $u_k(x) = e^{2\pi ikx}$ so that $c_k = (g, u_k)$. Using the linearity of the inner product and the orthogonality relations

$$(u_k, u_\ell) = \begin{cases} 0 \text{ if } k \neq \ell, \\ 1 \text{ if } k = \ell, \end{cases}$$

we get

$$\left\| g - \sum_{k=-K}^{K} (g, u_k)u_k \right\|^2 = \left(g - \sum_{k=-K}^{K} (g, u_k)u_k, g - \sum_{k=-K}^{K} (g, u_k)u_k \right)$$

$$= (g, g) - \sum_{k=-K}^{K} |(g, u_k)|^2$$

$$= \|g\|^2 - \sum_{k=-K}^{K} |(g, u_k)|^2.$$

Since the left-hand side is nonnegative, the sum on the right-hand side must be bounded independently of K, so the series $\sum_{k=-\infty}^{\infty} |c_k|^2$ converges. In particular, $c_k \to 0$ as $|k| \to \infty$. □

An immediate consequence of the Riemann–Lebesgue Lemma is

$$c_k + c_{-k} = 2i \int_0^1 g(x)\sin(kx)\, dx \longrightarrow 0 \text{ as } k \to \infty. \tag{9.8}$$

Now we are ready to complete the proof of Theorem 9.14. By Lemma 9.15, the partial sums of the Fourier series are given by the left-hand side of Equation (9.7). We manipulate this integral a little, using Equation (9.6):

$$\int_0^1 g(y + x)D_K(x)\, dx = \int_{-1/2}^{1/2} \frac{g(y + x) - g(y)}{\sin(\pi x)} \sin((2K + 1)x)\, dx$$

$$+ \int_{-1/2}^{1/2} g(y)D_K(x)\, dx. \tag{9.9}$$

The last integral in Equation (9.9) simply equals $g(y)$ for all K by the property in Equation (9.5) of the Dirichlet kernel. For the first summand, observe that

$$\frac{g(y + x) - g(y)}{\sin(\pi x)}$$

is a periodic continuous function for $x \in [-1/2, 1/2]$ (the limit in $x = 0$ exists by l'Hôpital's rule). By Equation (9.8), this implies that the first summand tends to zero as K tends to infinity. $\square$

Theorem 9.17. [POISSON SUMMATION FORMULA] *Suppose that* f *belongs to the Schwartz space* S. *Then*

$$\sum_{m \in \mathbb{Z}} f(m) = \sum_{m \in \mathbb{Z}} \widehat{f}(m).$$

PROOF. Let

$$g(x) = \sum_{m \in \mathbb{Z}} f(x + m),$$

which is certainly convergent since $f \in S$. Clearly, g is periodic. Moreover, g is differentiable infinitely often since

$$\left| \sum_{|m| > N} f^{(n)}(x + m) \right| \leqslant \sum_{|m| > N} \left| f^{(n)}(x + m) \right| \leqslant C \sum_{|m| > N} \frac{1}{(1 + |x + m|)^2},$$

where the last series tends to zero for $|x|$ bounded. Therefore the nth derivatives of the partial sums converge uniformly by periodicity for all $n \geqslant 0$. We cannot apply Theorem 8.23 since the functions f_n are not necessarily analytic. However, we can use Lemma 8.25 as follows. Let γ be the real interval $[1, x]$, let G_N be the Nth partial sum of the derivatives f'_n, and use the fundamental theorem of calculus to see that

$$\frac{d}{dx} \int_0^x G_N(t) \, dt = G_N(x) - G_N(0).$$

The integral converges to $g(x) - g(0)$ as $N \to \infty$, and similarly for higher derivatives, using induction. It follows that g is n times differentiable and its nth derivative is the limit of that of the partial sums, so we may do Fourier analysis on g.

Let c_k be the kth Fourier coefficient of g. Then, by Theorem 9.14,

$$g(x) = \sum_{k=-\infty}^{\infty} c_k e^{2\pi i k x}, \qquad g(0) = \sum_{k=-\infty}^{\infty} c_k. \qquad (9.10)$$

On the other hand,

$$c_k = \int_0^1 g(x) e^{-2\pi i k x} \, dx = \int_0^1 \sum_{m=-\infty}^{\infty} f(x + m) e^{-2\pi i k x} \, dx$$

$$= \sum_{m=-\infty}^{\infty} \int_0^1 f(x + m) e^{-2\pi i k x} \, dx.$$

This interchange of sum and integral is justified because the series for g converges uniformly by Lemma 8.25 again. Multiply each summand by a factor of $e^{-2\pi i k m} = 1$ and substitute $x + m$ for x to find

$$c_k = \sum_{m=-\infty}^{\infty} \int_0^1 f(x+m)e^{-2\pi i k(x+m)} \, dx = \int_{-\infty}^{\infty} f(x)e^{-2\pi i k x} \, dx$$

$$= \widehat{f}(k).$$

Now

$$\sum_{m=-\infty}^{\infty} f(m) = g(0) = \sum_{k=-\infty}^{\infty} c_k = \sum_{k=-\infty}^{\infty} \widehat{f}(k)$$

by Equation (9.10). This completes the proof of the Poisson Summation Formula. □

9.5 The Theta Function

Another classical special function we need is the theta function. This satisfies a surprising functional equation, which plays a key role in the proof of the functional equation for the zeta function.

Theorem 9.18. *For real $y > 0$, define the* theta function *by*

$$\theta(y) = \sum_{n=-\infty}^{\infty} e^{-n^2 \pi y}.$$

Then

$$\theta\left(\frac{1}{y}\right) = \sqrt{y}\,\theta(y).$$

PROOF. This relation is far from obvious and looks barely possible. The series defining θ converges uniformly in the range $y > \delta$ for any fixed $\delta > 0$. Fix some real $b > 0$ and define, with $f(y) = e^{-\pi y^2}$ as in Lemma 9.11,

$$f_b(y) = f(by) = e^{-\pi b^2 y^2}.$$

Of course, f_b is in the Schwartz space $\mathcal{S}$, so we may apply the Poisson Summation Formula (Theorem 9.17) to obtain

$$\sum_{n=-\infty}^{\infty} f_b(n) = \sum_{n=-\infty}^{\infty} \widehat{f_b}(n). \tag{9.11}$$

Next, we need to compute $\widehat{f_b}(y)$:

$$\widehat{f_b}(y) = \int_{-\infty}^{\infty} f_b(x)e^{-2\pi i x y} \, dx = \int_{-\infty}^{\infty} f(bx)e^{-2\pi i x y} \, dx. \tag{9.12}$$

Now put $u = bx$, so $dx = \frac{1}{b}du$. Thus, Equation (9.12) becomes

$$\widehat{f_b}(y) = \frac{1}{b} \int_{-\infty}^{\infty} f(u)e^{-2\pi i u \frac{y}{b}} \, du = \frac{1}{b}\widehat{f}\left(\frac{y}{b}\right).$$

Apply Lemma 9.11 to this equation to see that

$$\widehat{f_b}(y) = \frac{1}{b}f\left(\frac{y}{b}\right).$$

Put this result into Equation (9.11), and use the definition of f again to obtain

$$\sum_{n=-\infty}^{\infty} e^{-\pi b^2 n^2} = \frac{1}{b}\sum_{n=-\infty}^{\infty} e^{-\pi n^2/b^2}.$$

Finally, put $b = \sqrt{y}$ and the functional equation for θ emerges. □

We are now ready for the proof of the functional equation of the zeta function.

PROOF OF THEOREM 9.5. We begin with

$$\Gamma\left(\frac{s}{2}\right) = \int_0^{\infty} e^{-x}x^{s/2-1} \, dx = \int_0^{\infty} e^{-x}x^{s/2} \frac{dx}{x},$$

so, in the domain $\Re(s) > 1 + \delta$,

$$F(s) = \pi^{-s/2} \sum_{n=1}^{\infty} \int_0^{\infty} n^{-s}e^{-x}x^{s/2} \frac{dx}{x}.$$

Next, replace x by $\pi n^2 y$ in the integral. This means $\frac{dx}{dy} = \frac{x}{y}$, and after some cancellation we get

$$F(s) = \int_0^{\infty} \sum_{n=1}^{\infty} e^{-\pi n^2 y}y^{s/2} \frac{dy}{y}. \tag{9.13}$$

The interchange of the integral and sum is permitted because the series for the zeta function converges uniformly on $\Re(s) > 1 + \delta$ for any fixed $\delta > 0$. Define

$$g(y) = \sum_{n=1}^{\infty} e^{-\pi n^2 y} = \frac{\theta(y) - 1}{2}.$$

Split the integral in Equation (9.13) into $0 \leqslant y \leqslant 1$ and $1 \leqslant y < \infty$,

$$F(s) = \int_1^{\infty} y^{s/2}g(y) \frac{dy}{y} + \int_0^1 y^{s/2}g(y) \frac{dy}{y}.$$

In the second integral, change y to $z = y^{-1}$, so that it becomes an integral over the region $\infty > z \geqslant 1$, and $\frac{dz}{dy} = -\frac{z}{y}$. Thus

$$F(s) = \int_1^\infty y^{s/2} g(y) \, \frac{dy}{y} + \int_1^\infty z^{-s/2} g(z^{-1}) \, \frac{dz}{z}. \qquad (9.14)$$

The Poisson Summation Formula (Theorem 9.17) gave us Theorem 9.18, which may be applied to give

$$g(y^{-1}) = \frac{\theta(y^{-1}) - 1}{2} = \frac{\sqrt{y}\,\theta(y) - 1}{2}$$

$$= \frac{\sqrt{y}\,(\theta(y) - 1) + \sqrt{y} - 1}{2} = \sqrt{y}\,g(y) + \frac{\sqrt{y} - 1}{2}.$$

Substituting this into Equation (9.14),

$$F(s) = \int_1^\infty y^{s/2} g(y) \, \frac{dy}{y} + \int_1^\infty y^{(1-s)/2} g(y) \, \frac{dy}{y}$$

$$+ \frac{1}{2} \int_1^\infty y^{-s/2} \left(y^{1/2} - 1\right) \frac{dy}{y}. \qquad (9.15)$$

Let J denote the third integral in Equation (9.15). Then

$$2J = \int_1^\infty y^{-(1+s)/2} - y^{-(2+s)/2} \, dy = \left[\frac{y^{(1-s)/2}}{(1-s)/2} - \frac{y^{-s/2}}{-s/2} \right]_1^\infty$$

$$= 2\left(\frac{-1}{1-s} - \frac{1}{s} \right) = 2\left(\frac{1}{s-1} - \frac{1}{s} \right). \qquad (9.16)$$

Lemma 9.19. *For all $z \in \mathbb{C}$, the function*

$$G(z) = \int_1^\infty y^z g(y) \, \frac{dy}{y}$$

is analytic.

Assuming this for the moment, we have, by Equations (9.15) and (9.16),

$$F(s) = G\left(\frac{s}{2}\right) + G\left(\frac{1-s}{2}\right) + \frac{1}{s-1} - \frac{1}{s},$$

so $F(1-s) = F(s)$, and F is analytic for all $s \in \mathbb{C}$ apart from simple poles at $s = 1$ and $s = 0$, completing the proof of Theorem 9.5.   □

All that remains is to prove the lemma.

PROOF OF LEMMA 9.19. Write $G(z) = \sum_{n=1}^\infty G_n(z)$, where

$$G_n(z) = \int_n^{n+1} y^z g(y) \, dy.$$

We will prove that the G_n are analytic functions on all of $\mathbb{C}$ and then use a uniform convergence argument. Consider the difference quotient for $G_n(z)$ (exactly as in the standard proof of Theorem 8.29 for the analytic continuation of the zeta function on p. 176),

$$\frac{1}{h} \int_n^{n+1} y^z \left(y^h - 1 \right) g(y) \, dy = \frac{1}{h} \int_n^{n+1} y^z g(y)(1 + h \log y + \rho(h,y) - 1) \, dy,$$

where $\rho(h,y) = O(h^2)$ for bounded values of y. One may therefore divide by h and take the limit $h \to 0$.

Next, we prove that the partial sums of the G_n converge uniformly on a suitable domain. Consider z in the half-plane $\Re(z) < K$ for some fixed K. There

$$\left| \int_N^\infty y^z g(y) \, dy \right| \leqslant \int_N^\infty y^K |g(y)| \, dy. \tag{9.17}$$

Now we estimate $|g(y)|$:

$$|g(y)| = \sum_{n=1}^\infty e^{-\pi n^2 y} \leqslant \sum_{n=1}^\infty e^{-\pi n y} = \frac{e^{-\pi y}}{1 - e^{-\pi y}}.$$

The denominator is clearly bounded below for $1 < y$, so the right-hand side of the inequality (9.17) is finite for $N = 1$, say. As an immediate consequence, the integrals from N to infinity must tend to zero, and all this was independent of z. Now we may apply Theorem 8.23 and deduce that G is analytic on the half-plane $\Re(z) < K$. Since K was arbitrary, G is analytic on the whole complex plane. $\qquad\square$

9.6 The Gamma Function Revisited

We have seen that the zeta function and the Gamma function go together like Hardy and Wright. We need to know some additional properties of Γ (in particular that $\Gamma(s) \neq 0$ for all $s \in \mathbb{C}$) in order to understand the zeta function better.

Theorem 9.20. [WEIERSTRASS] *Define a function* f *by*

$$f(s) = s e^{\gamma s} \prod_{n=1}^\infty \left[\left(1 + \frac{s}{n} \right) e^{-s/n} \right],$$

where γ is the Euler–Mascheroni constant. Then f *is an analytic function on the whole of the complex plane, and it is zero at $0, -1, -2, -3, \ldots$ only.*

This rather mysterious function turns out to satisfy $f(s) = 1/\Gamma(s)$, giving another formula for the Gamma function and incidentally proving that $\Gamma(s)$

is nonzero for all $s \in \mathbb{C}$ (Theorem 9.4). The argument may appear at first sight an infuriating piece of magic, but it appears more reasonable when thought of as a (functional) factorization. We know that Γ has simple poles at $0, -1, -2, \ldots$, so $1/\Gamma$ must have zeros there. The most naïve approach is to look for a factorization of $1/\Gamma(s)$ in the form

$$Cs(s+1)(s+2)\cdots,$$

but this expression clearly does not converge. Trying to correct the most obvious defect (that the terms do not converge to 1) would lead one to look for expressions such as

$$Cs(1+s)(1+s/2)(1+s/3)\cdots,$$

but this is still not convergent because $\sum_{n=1}^{\infty} \frac{s}{n}$ is not convergent. What is needed is a factorization in which the terms converge to 1 fast enough to guarantee convergence of the infinite product. The argument below gives a quadratic rate of convergence. This kind of adjustment became a standard tool in nineteenth-century analytic number theory, and we will encounter it several times.

PROOF OF THEOREM 9.20. Consider the function

$$g(s) = \sum_{n=1}^{\infty} g_n(s) = \sum_{n=1}^{\infty} \left[\log\left(1 + \frac{s}{n}\right) - \frac{s}{n} \right]. \qquad (9.18)$$

Each g_n is analytic except at $-1, -2, -3, \ldots$. We want to prove that the series Equation (9.18) converges uniformly on $\{s \in \mathbb{C} : |s| < K\}$ for every fixed $K > 0$. Choose $N > 2K$, so for all $n \geqslant N$, $|s/n| \leqslant 1/2$, and therefore

$$\log\left(1 + \frac{s}{n}\right) = \frac{s}{n} - \frac{1}{2}\left(\frac{s}{n}\right)^2 + \frac{1}{3}\left(\frac{s}{n}\right)^3 - \cdots.$$

Thus we can estimate $g_n(s)$ for all these s and n by

$$|g_n(s)| \leqslant \frac{1}{2}\left|\frac{s}{n}\right|^2 + \frac{1}{3}\left|\frac{s}{n}\right|^3 + \cdots$$
$$\leqslant \frac{|s|^2}{n^2} \frac{1}{1 - |s|/n} \leqslant 2\frac{|s|^2}{n^2} < \frac{2K^2}{n^2}.$$

This can be summed from $n = N$ to give

$$\left| \sum_{n=N}^{\infty} g_n(s) \right| \leqslant \sum_{n=N}^{\infty} \frac{2K^2}{n^2},$$

and the latter is arbitrarily small if N is large, as it is the tail end of a convergent series. Thus the series Equation (9.18) is a uniformly convergent sum of analytic functions $g_n(s)$ on $|s| < K$ for any K. By Theorem 8.23, we deduce that the limit $\mathbf{g}(s)$ is analytic for all s not equal to $-1, -2, -3, \ldots$. The same holds for

$$e^{\mathbf{g}(s)} = \prod_{n=1}^{\infty} \left[\left(1 + \frac{s}{n}\right) e^{-s/n} \right].$$

After multiplying this by $se^{\gamma s}$, we see that $\mathbf{f}$ is an analytic function away from $-1, -2, -3, \ldots$. It is clear that $\mathbf{f}$ has zeros at each of these points, so $\log \mathbf{f}$ has a singularity there. Conversely, away from these obvious zeros, we have shown that $\log \mathbf{f}$ is analytic, so $\mathbf{f}$ cannot be zero elsewhere. Finally, for some fixed $m \in \mathbb{N}$, consider the infinite product defining $\mathbf{f}$ without the factor corresponding to $n = m$. The same estimates as above show that the logarithm of this is analytic at $s = -m$, so $\mathbf{f}$ is analytic at $s = -m$ as well. □

Corollary 9.21. *The zeros of $\mathbf{f}$ in Theorem 9.20 are all simple, and the function*

$$\frac{1}{\mathbf{f}(s)} = \frac{1}{s} e^{-\gamma s} \prod_{n=1}^{\infty} \left(1 + \frac{s}{n}\right)^{-1} e^{s/n}$$

is analytic on $\mathbb{C}$ apart from simple poles at $0, -1, -2, \ldots$. The function $1/\mathbf{f}$ has no zeros at all (because $\mathbf{f}$ has no poles).

Theorem 9.22. [EULER] *For all $s \neq 0, -1, -2, \ldots$,*

$$\frac{1}{\mathbf{f}(s)} = \frac{1}{s} \prod_{n=1}^{\infty} \left[\left(1 + \frac{1}{n}\right)^{s} \left(1 + \frac{s}{n}\right)^{-1} \right].$$

PROOF. We use the definition of the Euler–Mascheroni constant γ (see Exercise 1.2 on p. 10 or Theorem 8.3).

$$\mathbf{f}(s) = s \lim_{m \to \infty} e^{s(1 + 1/2 + \cdots + 1/m - \log m)} \lim_{N \to \infty} \prod_{n=1}^{N} \left(1 + \frac{s}{n}\right) e^{-s/n}$$

$$= s \lim_{m \to \infty} e^{s(1 + 1/2 + \cdots + 1/m - \log m)} \prod_{n=1}^{m} \left(1 + \frac{s}{n}\right) e^{-s/n}$$

$$= s \lim_{m \to \infty} m^{-s} \prod_{n=1}^{m} \left(1 + \frac{s}{n}\right). \tag{9.19}$$

Now we pull a rabbit out of the hat: Write m as

$$m = \frac{2}{1} \cdot \frac{3}{2} \cdots \frac{m-1}{m-2} \cdot \frac{m}{m-1}$$

$$= \left(1 + \frac{1}{1}\right) \left(1 + \frac{1}{2}\right) \cdots \left(1 + \frac{1}{m-1}\right), \tag{9.20}$$

where as usual an empty product (the case $m = 1$) is defined to be 1. Substitute this into Equation (9.19) and use the fact that for all s

$$\lim_{m\to\infty} \left(1 + \tfrac{1}{m}\right)^s = 1$$

to see that Equation (9.19) becomes

$$f(s) = s \lim_{m\to\infty} \prod_{n=1}^{m-1} \left(1 + \frac{1}{n}\right)^{-s} \prod_{n=1}^{m} \left(1 + \frac{s}{n}\right)$$

$$= s \lim_{m\to\infty} \left(1 + \frac{1}{m}\right)^s \prod_{n=1}^{m} \left(1 + \frac{1}{n}\right)^{-s} \left(1 + \frac{s}{n}\right)$$

$$= s \lim_{m\to\infty} \prod_{n=1}^{m} \left(1 + \frac{1}{n}\right)^{-s} \left(1 + \frac{s}{n}\right).$$

Now invert both sides, and the proof of Theorem 9.22 is complete.   □

Corollary 9.23. *For all $s \in \mathbb{C}$,*

$$\frac{1}{f(s)} = \lim_{m\to\infty} \frac{1 \cdot 2 \cdots (m-1)m^s}{s(s+1)\cdots(s+m-1)}.$$

PROOF. By Theorem 9.22, for $s \neq 0, -1, -2, \ldots$

$$\frac{1}{f(s)} = \lim_{m\to\infty} \frac{1}{s} \prod_{n=1}^{m-1} \left(1 + \frac{1}{n}\right)^s \left(1 + \frac{s}{n}\right)^{-1}$$

$$= \lim_{m\to\infty} \frac{1}{s} \cdot \frac{(1+1)^s \left(1 + \tfrac{1}{2}\right)^s \cdots \left(1 + \frac{1}{m-1}\right)^s}{\left(1 + \frac{s}{1}\right) \cdots \left(1 + \frac{s}{m-1}\right)}$$

$$= \lim_{m\to\infty} \frac{1}{s} \cdot \frac{(1+1)^s 2 \left(1 + \tfrac{1}{2}\right)^s \cdots (m-1) \left(1 + \frac{1}{m-1}\right)^s}{(1+s)(2+s)\cdots(m-1+s)},$$

where we have just multiplied the numerator and denominator by

$$2 \cdot 3 \cdots (m-1).$$

Now collect the integers in the numerator into one product and the other factors into a product

$$\prod_{n=1}^{m-1} \left(1 + \frac{1}{n}\right)^s = m^s$$

by the identity (9.20). This completes the proof of the corollary.   □

Theorem 9.24. *For all s such that $\Re(s) > 0$,*

$$\frac{1}{f(s)} = \Gamma(s) = \int_0^\infty e^{-t} t^{s-1}\, dt.$$

Thus we have three representations of the Gamma function – Definition 9.1 and the ones given in Theorems 9.20 and 9.22. The ability to move between these different formulations will be very useful.

PROOF OF THEOREM 9.24. For $n \in \mathbb{N}$, define

$$\Gamma_n(s) = \int_0^n \left(1 - \frac{t}{n}\right)^n t^{s-1} dt.$$

Evaluate $\Gamma_n(s)$ using integration by parts. Substitute $t = n\tau$ to give

$$\Gamma_n(s) = n^s \int_0^1 (1-\tau)^n \tau^{s-1} d\tau$$

$$= n^s \left[(1-\tau)^n \frac{\tau^s}{s}\right]_0^1 + \frac{n^s n}{s} \int_0^1 (1-\tau)^{n-1} \tau^s d\tau$$

$$= \frac{n^s n(n-1)}{s(s+1)} \int_0^1 (1-\tau)^{n-2} \tau^{s+1} d\tau = \cdots$$

$$= \frac{n^s n \cdot (n-1) \cdot (n-2) \cdots 2 \cdot 1}{s(s+1) \cdots (s+n)}.$$

Now let n tend to infinity, and use Corollary 9.23, which shows that

$$\lim_{n \to \infty} \Gamma_n(s) = \frac{1}{f(s)}.$$

To complete the proof of Theorem 9.24, we need to prove that

$$\lim_{n \to \infty} \Gamma_n(s) = \Gamma(s).$$

This is plausible because

$$\lim_{n \to \infty} \left(1 - \frac{t}{n}\right)^n = e^{-t} \tag{9.21}$$

for all t. (To prove this, just take logarithms, replace $1/n$ by h, and apply l'Hôpital's rule.) However, to apply Equation (9.21) to our problem, an exchange of limit and integral is required. We must therefore prove that

$$\lim_{n \to \infty} \int_0^n \left[e^{-t} - \left(1 - \frac{t}{n}\right)^n\right] t^{s-1} dt = 0. \tag{9.22}$$

Estimate the integrand in Equation (9.22) by

$$\left| t^{s-1} \left(e^{-t} - \left(1 - \frac{t}{n}\right)^n\right) \right| = t^{\sigma-1} \left| e^{-t} - \left(1 - \frac{t}{n}\right)^n \right|.$$

We need the following estimate:

$$\left| e^{-t} - \left(1 - \frac{t}{n}\right)^n \right| \le \frac{t^2 e^{-t}}{n} \tag{9.23}$$

for all $t \in [0, n]$. Assuming this,

$$\int_0^n t^{\sigma - 1} \left| e^{-t} - \left(1 - \frac{t}{n}\right)^n \right| dt \le \frac{1}{n} \int_0^\infty e^{-t} t^{\sigma + 1} \, dt = \frac{\Gamma(\sigma + 2)}{n},$$

which obviously tends to zero. (Note that the convergence is even uniform for bounded s, although we do not need this here.)   □

Exercise 9.4. Prove the inequality (9.23).

Exercise 9.5. Using logarithmic differentiation on the representation of Γ in Theorem 9.20, prove that

$$\Gamma'(1) = -\gamma. \tag{9.24}$$

Corollary 9.25. *For all $s \in \mathbb{C}$, $s \notin \mathbb{N}$,*

$$\Gamma(s)\Gamma(1 - s) = \frac{\pi}{\sin(\pi s)}.$$

PROOF. By Theorem 9.20,

$$\Gamma(s)\Gamma(-s) = -\frac{1}{s^2} \prod_{n=1}^\infty \left(1 + \frac{s}{n}\right)^{-1} e^{s/n} \prod_{n=1}^\infty \left(1 - \frac{s}{n}\right)^{-1} e^{-s/n}$$

$$= -\frac{1}{s^2} \prod_{n=1}^\infty \left(1 - \frac{s^2}{n^2}\right)^{-1} = -\frac{\pi}{s \sin(\pi s)}$$

using the classical formula

$$\sin(\pi s) = \pi s \prod_{n=1}^\infty \left(1 - \frac{s^2}{n^2}\right). \tag{9.25}$$

The corollary follows because $-s\Gamma(-s) = \Gamma(1 - s)$.   □

Equation (9.25) is another example of an analog of the Fundamental Theorem of Arithmetic in a function-theory context. We know that $\sin(\pi s)$ vanishes at each integer, so we might hope to factorize it in the form

$$cs \prod_{n=1}^\infty (n^2 - s^2).$$

Of course, this does not converge, and attempting to get the terms to converge to 1 fast enough to guarantee convergence of the infinite product plausibly leads one to conjecture Equation (9.25).

Exercise 9.6. Prove the identity (9.25).

Exercise 9.7. Justify the steps in the following argument. The Taylor expansion of the sine function gives

$$\sin(\pi s) = \pi s - \frac{(\pi s)^3}{6} + \cdots . \tag{9.26}$$

By Equation (9.25), this is equal to

$$\pi s \left(1 - s^2 \left(\frac{1}{1} + \frac{1}{4} + \frac{1}{9} + \cdots \right) + \cdots \right)$$

$$= \pi s - \pi s^3 \sum_{n=1}^{\infty} \frac{1}{n^2} + \cdots .$$

Comparing the coefficient of s^3 with that of Equation (9.26) gives

$$\sum_{n=1}^{\infty} \frac{1}{n^2} = \frac{\pi^2}{6} .$$

Exercise 9.8. Prove that $\zeta(2k)$ is a rational multiple of π^{2k} for any $k \geqslant 1$.

Much less is known about the values $\zeta(3), \zeta(5), \ldots$. Apéry proved in 1978 that $\zeta(3) \notin \mathbb{Q}$, and there are some very deep results on the algebraic independence of various values of ζ at odd integers.

Exercise 9.9. This exercise is a more explicit version of the previous one.
(a) Replace s by iz in Equation (9.25) to deduce that

$$\sinh(\pi z) = \pi z \prod_{n=1}^{\infty} \left(1 + \frac{z^2}{n^2} \right). \tag{9.27}$$

(b) Use logarithmic differentiation to prove

$$\frac{\pi z}{e^{\pi z} - 1} + \frac{\pi z}{2} = 1 + \sum_{k=1}^{\infty} \frac{(-1)^{k+1}}{2^{2k-1}} \zeta(2k) z^{2k}. \tag{9.28}$$

(c) Deduce that

$$\zeta(2k) = (-1)^k \pi^{2k} \frac{2^{2k-1}}{(2k-1)!} \left(-\frac{B_{2k}}{2k} \right), \tag{9.29}$$

where B_n denotes the nth Bernoulli number defined by

$$\frac{z}{e^z - 1} = \sum_{n=1}^{\infty} \frac{B_n z^n}{n!}. \tag{9.30}$$

Exercise 9.10. (a) Use Theorem 9.5 and Equation (9.29) to prove that ζ takes rational values at negative odd integers.
(b) Use Equation (9.30) to show that $B_n = 0$ for odd integers $n > 1$.
(c) Deduce that

$$\zeta(-n) = -\frac{B_{n+1}}{n+1} \tag{9.31}$$

for all $n > 0$.

The neatness of Equation (9.31) suggests there might be a more elegant way to prove it. Hurwitz found a beautiful proof using complex analysis.

Exercise 9.11. Use the functional equation together with Equations (9.24) and (8.24) to prove that

$$\frac{\zeta'(0)}{\zeta(0)} = \log(2\pi). \tag{9.32}$$

Prove that $\zeta(0) = -\frac{1}{2}$ and deduce the value of $\zeta'(0)$.

Exercise 9.12. *Prove that $\displaystyle\sum_{n=-\infty}^{\infty} \frac{1}{(4n+1)^k}$ is a rational multiple of π^k for any $k \geqslant 2$.

There are many deep results on the location and distribution of the zeros of the Riemann zeta function, all far beyond our scope.

Theorem 9.26. *Define $N(T)$ to be the number of zeros of the Riemann zeta function in the critical strip up to height T,*

$$N(T) = |\{s \in \mathbb{C} : 0 \leqslant \Re(s) \leqslant 1, \ \zeta(s) = 0, \ 0 < \Im(s) < T\}|.$$

Then there is an asymptotic formula,

$$N(T) = \frac{T}{2\pi} \log\left(\frac{T}{2\pi}\right) - \frac{T}{2\pi} + O(\log T).$$

The proof makes use of Stirling's Formula extended to the complex plane,

$$\log \Gamma(s) = -s + \left(s - \frac{1}{2}\right) \log s + O(1),$$

provided $|\mathrm{Arg}(s)| < \pi - \delta$.

Exercise 9.13. Define a function ν by $\nu(1) = 0$, and $\nu(n)$ is the number of distinct prime divisors of n for $n > 1$.
(a) Prove that $\displaystyle\sum_{n=1}^{\infty} \frac{\nu(n)}{n^s} = \zeta(s) \sum_{p \in \mathbb{P}} \frac{1}{p^s}$.
(b) Prove that $\displaystyle\sum_{n=1}^{\infty} \frac{2^{\nu(n)}}{n^s} = \frac{\zeta^2(s)}{\zeta(2s)}$.

At the start of this chapter, the idea of "factorizing" functions in the way that polynomials are factorized was discussed. Quite apart from the convergence issues that pervade this topic, infinite products may behave in quite surprising ways, as shown by the next exercise.

Exercise 9.14. Using Exercise 8.11, show that, for any x with $|x| < 1$,

$$e^x = \prod_{n=1}^{\infty} (1 - x^n)^{-\mu(n)/n}.$$

The functional equations we have considered in this chapter are analytic properties of known classical functions. The next exercise is (relatively) light relief and is a functional equation in another sense: The unknown solution sought is a function.

Exercise 9.15. *Find the solutions to the functional equation

$$f(xz - y)f(x)f(y) + 3f(0) = 1 + 2f(0)f(0) + f(x)f(y) \text{ for all } x, y, z \in \mathbb{R}.$$

Does the solution change if the identity is only required to hold for all x, y, z in $\mathbb{Z}$?

9.6.1 Factorizing the Riemann Zeta Function

Several times in this chapter, we have seen a function factorize in a meaningful way into an infinite product of "irreducible" terms corresponding to zeros, corresponding to a function-theoretic version of the Fundamental Theorem of Arithmetic. The Riemann Hypothesis itself can be understood in these terms – except that the location of the zeros is not known.

Theorem 9.27. [HADAMARD] *Let Ξ denote the set of zeros of the Riemann zeta function in the critical strip $\{z \mid 0 < \Re(z) < 1\}$. Then*

$$\zeta(s) = \frac{e^{bs}}{2(s-1)\Gamma(\frac{s}{2}+1)} \prod_{\xi \in \Xi} \left(1 - \frac{s}{\xi}\right) e^{s/\xi},$$

where $b = \log(2\pi) - 1 + \frac{\gamma}{2}$.

In this theorem, the zeros of the zeta function outside the critical strip are accounted for by the poles of $\Gamma(\frac{s}{2} + 1)$.

Exercise 9.16. Assuming the statement of Theorem 9.27 for some constant b, show that it must have the stated value by using Exercise 9.11.

NOTES TO CHAPTER 9: For a very interesting discussion of both the mathematics and the history of the type of analysis used in this chapter, and in particular to gain some insight into how Euler came close to the functional equation, see Hardy's monograph [74]. An elegant guide to classical Fourier analysis may be found in Katznelson's book [87]. Apéry's proof that $\zeta(3)$ is irrational appeared in his paper [3]; an accessible account is provided by van der Poorten [118]. More recent results on values of the zeta function at odd integers appear in works by Ball and Rivoal [9] or Rivoal [130] and references therein. The disproof of Merten's conjecture mentioned on p. 186 appears in the paper of Odlyzko and te Riele [114]. A comprehensive guide to many of the analytic arguments here, including Exercises 9.4 and 9.6 is the classic text of Whittaker and Watson [160]. Artin's book [6] is an exceptionally clear account of the main properties of the Gamma function. Deeper properties of the zeta function, emphasizing the role of Poisson summation, may be found in Patterson's book [115]. Several different approaches to the functional equation for the Riemann zeta function appear in the book of Titchmarsh [153]. For a recent overview of the Riemann Hypothesis written by a worker in the field, consult the survey of Conrey [33]. Exercise 9.12 is classical; a proof requiring little background appears in a paper of Beukers, Kolk and Calabi [13] and is discussed in a paper of Elkies [50]. Exercise 9.14 is taken from a paper of Brent [19]. Exercise 9.15 is taken from a paper of Šunik [148].

10

Primes in an Arithmetic Progression

We begin with two elementary results and then give more sophisticated proofs of them, suggesting a general method. The algebraic part of this method concerns characters of Abelian groups, the analytic part is a nonvanishing statement about L-functions. The culmination is Dirichlet's general result, Theorem 10.5 in Section 10.1.

Consider all the primes congruent to 1 modulo 4 (the first row of Figure 10.1) and all the primes congruent to 3 modulo 4 (the second row). We might guess that there are infinitely many primes of each type.

$p \equiv 1 \pmod 4$	5		13 17		29	37 41		53	61		73
$p \equiv 3 \pmod 4$	3	7 11		19 23	31		43 47	59		67 71	

Figure 10.1. The primes modulo 4.

Proposition 10.1. *There are infinitely many primes congruent to 3 modulo 4.*

PROOF. This proceeds like Euclid's proof that there are infinitely many prime numbers. Suppose that the proposition is false, and there are only r such primes $p_1, \ldots, p_r$. Let
$$N = (p_1 \cdots p_r)^2 + 2.$$
Since
$$p_1^2 \equiv \cdots \equiv p_r^2 \equiv 1 \pmod 4,$$
we have $N \equiv 3$ modulo 4. Now N decomposes into prime factors,
$$N = q_1 \cdots q_k,$$
which must all be odd, so they are all congruent to 1 or 3 modulo 4. At least one of the primes q_i must be congruent to 3 since otherwise N would be

congruent to 1. Thus q_i is one of $p_1, \ldots, p_r$ and divides N and $(N-2)$, and hence divides 2, a contradiction.   □

Proposition 10.2. *There are infinitely many primes congruent to 1 modulo 4.*

FIRST PROOF OF PROPOSITION 10.2. This proof is slightly different. Rather than deriving a contradiction, we will show that, for any given $N > 1$, there exists a prime congruent to 1 modulo 4 and greater than N. Given $N > 1$, define

$$M = (N!)^2 + 1. \tag{10.1}$$

Clearly, M is odd. Let p be the smallest prime factor of M. We must have $p > N$ since $N \equiv 1$ modulo q for any prime $q \leqslant N$. We claim that $p \equiv 1$ modulo 4 (which completes the proof since N was arbitrary). To prove the claim, transform Equation (10.1) into

$$(N!)^2 = M - 1 \equiv -1 \quad (\mathrm{mod}\ p). \tag{10.2}$$

Since p divides M, p is odd, and we may raise the congruence (10.2) to the $(\frac{p-1}{2})$th power:

$$(N!)^{p-1} \equiv (-1)^{(p-1)/2}.$$

By Fermat's Little Theorem, $a^{p-1} \equiv 1$ modulo p for all $a \not\equiv 0$ modulo p, so $(p-1)/2$ must be even, proving the claim.   □

Exercise 10.1. Prove that there are infinitely many primes congruent to 1 or to 5 modulo 6.

These results are all very well, but the proofs are awkward and ad hoc. We would like to have a general principle for proving such results.

10.1 A New Method of Proof

Just as the analytic proofs of Theorem 1.2 in the end gave us more information than Euclid's original proof by contradiction, it turns out that the most powerful approach to primes in congruence classes comes from analysis.

SECOND PROOF OF PROPOSITION 10.2. This proof works along the lines of the second proof of Theorem 1.3 on p. 12. Consider for odd $n \in \mathbb{N}$ the function

$$c_1(n) = \frac{1 + (-1)^{(n-1)/2}}{2} = \begin{cases} 1 & \text{for } n \equiv 1 \quad (\mathrm{mod}\ 4), \\ 0 & \text{for } n \equiv 3 \quad (\mathrm{mod}\ 4). \end{cases} \tag{10.3}$$

The function c_1 is a gadget for picking out a particular congruence class. Later, we will generalize this fact using orthogonality relations for characters of Abelian groups. Using the gadget, for real $\sigma > 1$,

$$\sum_{p \equiv 1 \bmod 4} \frac{1}{p^\sigma} = \sum_{p \text{ odd}} \frac{c_1(p)}{p^\sigma}$$

$$= \frac{1}{2} \sum_{p \text{ odd}} \frac{1}{p^\sigma} + \frac{1}{2} \sum_{p \text{ odd}} \frac{(-1)^{(p-1)/2}}{p^\sigma}. \tag{10.4}$$

The rearrangement here is permitted because the series involved converge absolutely. The first summand on the right-hand side of Equation (10.4) tends to infinity as $\sigma \to 1$ (by Theorem 1.3). We claim that the second summand converges for $\sigma \to 1$ and in particular is bounded. (This will be proved below.) This implies that the left-hand side of Equation (10.4) tends to infinity, and we conclude that there must be infinitely many primes over which the summation runs.

For the moment, let us pursue the aim of another proof of Proposition 10.2 since all the essentials of Dirichlet's proof become apparent there already. We still have to prove the convergence claim for the last sum in Equation (10.4). To do this, define two functions $\chi, \chi_0 : \mathbb{N} \to \{-1, 0, 1\}$ by

$$\chi(n) = \begin{cases} 0 & \text{if } n \text{ is even,} \\ (-1)^{(n-1)/2} & \text{if } n \text{ is odd,} \end{cases}$$

$$\chi_0(n) = \begin{cases} 0 & \text{if } n \text{ is even,} \\ 1 & \text{if } n \text{ is odd.} \end{cases}$$

Define a complex function by

$$L(s, \chi) = \sum_{n=1}^{\infty} \frac{\chi(n)}{n^s},$$

and define $L(s, \chi_0)$ similarly. Such functions are called L-functions and are a special kind of Dirichlet series. Clearly, the series defining $L(s, \chi)$ and $L(s, \chi_0)$ converge absolutely for all s with $\Re(s) > 1$.

Lemma 10.3. *The series $L(s, \chi)$ converges for $s = 1$, and*

$$L(1, \chi) = 1 - \frac{1}{3} + \frac{1}{5} - \frac{1}{7} + \cdots = \frac{\pi}{4}.$$

PROOF. Consider the integral

$$\int_0^1 \frac{dt}{1+t^2} = [\tan^{-1}(t)]_0^1 = \frac{\pi}{4}. \tag{10.5}$$

Substitute into this integral the expansion

$$\frac{1}{1+t^2} = \sum_{n=0}^{\infty} \frac{1}{(-t^2)^n}, \tag{10.6}$$

which converges for all $0 \leqslant t < 1$. Fix any $0 < x < 1$, and then the series in Equation (10.6) converges uniformly for $0 \leqslant t \leqslant x$. We therefore have for all x with $0 < x < 1$

$$f(x) = \int_0^x \frac{dt}{1+t^2} = \sum_{n=1}^{\infty} \int_0^x (-t^2)^n \, dt$$

$$= \sum_{n=0}^{\infty} \frac{(-1)^n}{2n+1} x^{2n+1}$$

because there we may interchange integration and summation thanks to the uniform convergence. Now, we may take the limit $x \to 1$ thanks to Abel's Limit Theorem. (This is a useful special feature of power series – see Section 10.6.) For $x \to 1$, we get $L(1, \chi)$ on the right-hand side, and the integral in Equation (10.5), $f(1) = \pi/4$, on the left-hand side. $\square$

Lemma 10.4. *The functions χ and χ_0 are completely multiplicative (see Definition 3.4).*

PROOF. Check all the possible values of m and n modulo 4. $\square$

Now recall the Euler expansion (Theorem 1.5) for the zeta function. Since χ and χ_0 are completely multiplicative, we get in exactly the same way (see Theorem 8.17) an Euler expansion of $L(\sigma, \chi)$ and $L(\sigma, \chi_0)$,

$$L(\sigma, \chi) = \prod_{p \text{ odd}} \left(1 - \frac{\chi(p)}{p^\sigma}\right)^{-1}, \tag{10.7}$$

$$L(\sigma, \chi_0) = \prod_{p \text{ odd}} \left(1 - \frac{1}{p^\sigma}\right)^{-1}. \tag{10.8}$$

Take logarithms of Equation (10.7) and Equation (10.8) to get

$$\log L(\sigma, \chi) = -\sum_{p \text{ odd}} \log\left(1 - \frac{\chi(p)}{p^\sigma}\right) = \sum_{p \text{ odd}} \frac{\chi(p)}{p^\sigma} + O(1), \tag{10.9}$$

$$\log L(\sigma, \chi_0) = -\sum_{p \text{ odd}} \log\left(1 - \frac{1}{p^\sigma}\right) = \sum_{p \text{ odd}} \frac{1}{p^\sigma} + O(1). \tag{10.10}$$

Adding (see Equation (10.4)) gives

$$\log L(\sigma, \chi_0) \cdot L(\sigma, \chi) = \sum_{p \text{ odd}} \frac{1 + \chi(p)}{p^\sigma} + O(1)$$

$$= 2 \sum_{p \equiv 1 \bmod 4} \frac{1}{p^\sigma} + O(1).$$

What is the behavior of the left-hand side as σ tends to 1 from above?

$$L(\sigma, \chi) \to \frac{\pi}{4} \neq 0,$$

$$L(\sigma, \chi_0) = \left(1 - \frac{1}{2^\sigma}\right) \sum_n \frac{1}{n^\sigma} \longrightarrow \infty.$$

The terms $O(1)$ in Equations (10.9) and (10.10) are still $O(1)$ for $\sigma \to 1$, so we conclude that

$$\sum_{p \equiv 1 \bmod 4} \frac{1}{p^\sigma} \longrightarrow \infty$$

as $\sigma \to 1$ from above. This completes the second proof of Proposition 10.2. $\square$

Had we subtracted Equations (10.9) and (10.10) instead of adding them, we would have proved that

$$\sum_{p \equiv 3 \bmod 4} \frac{1}{p^\sigma}$$

diverges as $\sigma \to 1$ from above and hence would have found another proof of Proposition 10.1.

Exercise 10.2. Use the gadget

$$c_3(n) = \frac{1 - (-1)^{(n-1)/2}}{2} = \begin{cases} 0 \text{ for } n \equiv 1 \pmod 4 \\ 1 \text{ for } n \equiv 3 \pmod 4 \end{cases} \tag{10.11}$$

to prove that there are infinitely many primes $p \equiv 3$ modulo 4.

The biggest payoff of this more sophisticated approach is that the argument can be made to work in complete generality. At the end of this chapter, we will have proved the following theorem.

Theorem 10.5. [DIRICHLET] *If $a \in \mathbb{N}$ and $q \in \mathbb{N}$ are coprime, then there are infinitely many primes p such that*

$$p \equiv a \pmod q.$$

Note that this is the most general result we could hope for: If a and q are not coprime, then every number $n \equiv a$ modulo q will be divisible by $\gcd(a, q) > 1$, so there can only be finitely many such primes.

10.2 Congruences Modulo 3

To understand the ingredients necessary to prove Dirichlet's Theorem, we repeat the argument above for primes congruent to 1 or 2 modulo 3.

Consider the functions

$$\chi_0(n) = \begin{cases} 1 \text{ if } 3 \nmid n \\ 0 \text{ if } 3 \mid n, \end{cases}$$

$$\chi(n) = \begin{cases} 1 \text{ if } n \equiv 1 \pmod 3 \\ -1 \text{ if } n \equiv 2 \pmod 3 \\ 0 \text{ if } n \equiv 0 \pmod 3. \end{cases}$$

As in the previous example, the functions c_1 and c_2 picking out a particular congruence class can be rewritten using χ and χ_0 as

$$c_1(n) = \frac{1}{2}(\chi_0(n) + \chi(n)) = \begin{cases} 1 \text{ if } n \equiv 1 \pmod 3 \\ 0 \text{ otherwise}, \end{cases}$$

and

$$c_2(n) = \frac{1}{2}(\chi_0(n) - \chi(n)) = \begin{cases} 1 \text{ if } n \equiv 2 \pmod 3 \\ 0 \text{ otherwise}. \end{cases}$$

Define the associated L-functions

$$L(\sigma, \chi) = \sum_{n=1}^{\infty} \frac{\chi(n)}{n^\sigma}$$

and similarly $L(\sigma, \chi_0)$. As in the previous example, χ and χ_0 are completely multiplicative and hence $L(\sigma, \chi)$ and $L(\sigma, \chi_0)$ have Euler product expansions. Moreover,

$$L(\sigma, \chi_0) = \sum_{3 \nmid n} \frac{1}{n^\sigma} = \left(1 - \frac{1}{3^\sigma}\right) \zeta(\sigma)$$

tends to infinity as $\sigma \to 1$. We have all the ingredients to repeat the analog of the second proof of Proposition 10.2 in this case except one: We do not yet know whether

$$L(1, \chi) \neq 0. \tag{10.12}$$

If we knew this, we could proceed exactly as before; the key step is to notice that

$$\log L(\sigma, \chi_0) \cdot L(\sigma, \chi) = 2 \sum_{p \equiv 1 \bmod 3} \frac{1}{p^\sigma} + O(1).$$

As long as we do not know the inequality (10.12), the left-hand side might have a limit as σ tends to 1. We will prove the inequality (10.12) in Section 10.5 as part of a general result.

Thus if we want to prove results such as Proposition 10.2 along the lines of the second proof, then there are two things we need to get to grips with:

1. A mechanism for pulling out a particular congruence class via multiplicative functions (see Sections 10.3 and 10.4).
2. A nonvanishing statement about L-functions at $\sigma = 1$ (see Section 10.5).

10.3 Characters of Finite Abelian Groups

In this section, we want to deal with the first problem from the preceding section. Consider the example $n = 5$. Define the functions

$$\chi_0(n) = \begin{cases} 1 \text{ if } n \not\equiv 0 \pmod{5} \\ 0 \text{ if } 5 | n, \end{cases}$$

$$\chi(n) = \begin{cases} i \text{ if } n \equiv 2 \pmod{5} \\ -1 \text{ if } n \equiv 4 \pmod{5} \\ -i \text{ if } n \equiv 3 \pmod{5} \\ 1 \text{ if } n \equiv 1 \pmod{5} \\ 0 \text{ if } n \equiv 0 \pmod{5}. \end{cases}$$

Now check that

$$\frac{1}{4}(\chi_0(n) + \chi(n) + \chi^2(n) + \chi^3(n)) = c_1(n) = \begin{cases} 1 \text{ if } n \equiv 1 \pmod{5} \\ 0 \text{ otherwise}. \end{cases}$$

What if you want to pull out the congruence class $n \equiv 2$ modulo 5? Is an ingenious ad hoc. argument needed each time? We clearly need a general setup.

Recall that, in the ring $\mathbb{Z}/n\mathbb{Z}$, the units are

$$U(\mathbb{Z}/n\mathbb{Z}) = \{k \pmod{n} \mid \gcd(k, n) = 1\}.$$

The units form a group under multiplication.

Example 10.6. Let $n = 5$, so $U(\mathbb{Z}/5\mathbb{Z}) = \{1, 2, 3, 4\}$ is a cyclic group, and 2 is a generator. The multiplication table of $U(\mathbb{Z}/5\mathbb{Z})$ is

	1	2	3	4
1	1	2	3	4
2	2	4	1	3
3	3	1	4	2
4	4	3	2	1

so $U(\mathbb{Z}/5\mathbb{Z}) \cong \{1, i, -1, -i\}$.

Definition 10.7. *Let G be a finite Abelian group. A character of G is a homomorphism*

$$\chi : G \to (\mathbb{C}^*, \cdot).$$

The multiplicative group $\mathbb{C}^*$ is $\mathbb{C} \backslash \{0\}$ equipped with the usual multiplication. By convention, we will write all finite groups multiplicatively in this section – hence the identity will be written as 1_G or 1. For any group, the map

$$\chi_0 : G \to \mathbb{C}^*, \ \chi_0(g) = 1,$$

is a character called the trivial character.

Lemma 10.8. *Let G be a finite Abelian group, and let χ be a character of G. Then $\chi(1_G) = 1$ and $\chi(g)$ is a root of unity for any $g \in G$. In particular, $|\chi(g)| = 1$. Thus $\chi(g)$ lies on the unit circle in $\mathbb{C}$.*

PROOF. Clearly

$$\chi(1_G) = \chi(1_G \cdot 1_G) = \chi(1_G)\chi(1_G),$$

so $\chi(1_G) = 1$ since $\chi(1_G) \neq 0$. As to the second statement, we use the fact that for every $g \in G$ there exists $n \in \mathbb{N}$ such that $g^n = 1_G$. This implies that

$$\chi(g)^n = \chi(g^n) = \chi(1_G) = 1.$$

$\square$

Example 10.9. Let $G = C_k = \langle g \rangle$, a cyclic group of order k. Now

$$g^k = 1,$$

so

$$\chi(g)^k = 1$$

and therefore $\chi(g)$ must be a kth root of unity. Any of the k different kth roots of unity can occur as $\chi(g)$, and of course $\chi(g)$ determines all the values of χ on G since G is generated by g, so there are k distinct characters of G. We can label the characters of G with labels $0, 1, \ldots, n-1$ as follows: χ_j is determined by $\chi_j(g) = e^{2\pi i j / k}$, so $\chi_j(g^m) = e^{2\pi i j m / k}$.

Theorem 10.10. *Let G be a finite Abelian group. Then the characters of G form a group with respect to the multiplication*

$$(\chi \cdot \psi)(g) = \chi(g)\psi(g),$$

denoted $\widehat{G}$. The identity in $\widehat{G}$ is the trivial character. The group $\widehat{G}$ is isomorphic to G. In particular, any finite Abelian group G of order n has exactly n distinct characters.

This theorem is the first intimation of an entire dual world, a mirror image to the familiar world of finite Abelian groups. This duality extends to a larger class of Abelian groups and in that wider class takes subgroups to quotient groups, quotient groups to subgroups, and products to sums.

Exercise 10.3. What happens if the same construction is made for other groups?

(a) Describe the group

$$\widehat{\mathbb{Z}} = \{\text{homomorphisms } \mathbb{Z} \to \mathbb{S}^1\}.$$

(b) For nondiscrete groups G, we need to restrict to continuous characters. Find

$$\widehat{\mathbb{S}^1} = \{\text{continuous homomorphisms } \mathbb{S}^1 \to \mathbb{S}^1\}.$$

(c)*A more challenging problem is to describe the group $\widehat{\mathbb{Q}}$.

PROOF OF THEOREM 10.10. Use the structure theorem for finite Abelian groups, which says that G is isomorphic to a product of cyclic groups,

$$G \cong \prod_{j=1}^{k} C_{n_j}.$$

Choose a generator g_j for each of the factors C_{n_j} and define characters on G by

$$\chi^{(j)}(*,\ldots,*,g_j,*,\ldots,*) = e^{2\pi i/n_j},$$

that is, ignore all entries except the jth, and there use the same definition as in Example 10.9. Then the characters $\chi^{(1)},\ldots,\chi^{(k)}$ generate a subgroup of $\widehat{G}$ that is isomorphic to G: Each $\chi^{(j)}$ generates a cyclic group of order n_j, and this group has a trivial intersection with the span of all the other $\chi^{(i)}$s since all characters in the latter have value 1 at g_j. Likewise, for any given character of G, it is easy to write down a product of powers of the $\chi^{(j)}$ that coincides with χ on the generators g_j and hence on all of G.   □

Corollary 10.11. *Let G be a finite Abelian group. For any $1 \neq g \in G$, there exists $\chi \in \widehat{G}$ such that $\chi(g) \neq 1$.*

PROOF. Looking again at the proof of Theorem 10.10, we may write

$$g = (*,\ldots,*,g_j^r,*,\ldots,*)$$

with some entry $g_j^r \neq 1, 0 < r < n_j$. Then $\chi^{(j)}(g) = e^{2\pi i r/n_j} \neq 1$.   □

Theorem 10.12. *Let G be a finite Abelian group. Then, for any element $h \in G$ and any character $\psi \in \widehat{G}$,*

$$\sum_{g \in G} \psi(g) = \begin{cases} |G| & \text{if } \psi = \chi_0 \\ 0 & \text{if } \psi \neq \chi_0, \end{cases} \tag{10.13}$$

$$\sum_{\chi \in \widehat{G}} \chi(h) = \begin{cases} |G| & \text{if } h = 1 \\ 0 & \text{if } h \neq 1. \end{cases} \tag{10.14}$$

These identities are known as the *orthogonality relations for finite Abelian group characters.*

PROOF. Consider Equation (10.13) first. The case $\psi = \chi_0$ is trivial, so assume $\psi \neq \chi_0$. There is an element $h \in G$ such that $\psi(h) \neq 1$. Then

$$\psi(h) \sum_{g \in G} \psi(g) = \sum_{g \in G} \psi(gh) = \sum_{g \in G} \psi(g)$$

because multiplication by h only permutes the summands. This equation can only be true if $\sum_{g \in G} \psi(g) = 0$.

For Equation (10.14), assume $h \neq 1$. By Corollary 10.11, there exists some character $\psi \in \widehat{G}$ such that $\psi(h) \neq 1$. We now use the dual of the argument above,

$$\psi(h) \sum_{\chi \in \widehat{G}} \chi(h) = \sum_{\chi \in \widehat{G}} (\psi \cdot \chi)(h) = \sum_{\chi \in \widehat{G}} \chi(h)$$

since multiplication by ψ only permutes the elements of $\widehat{G}$, and again this can only be true if $\sum_{\chi \in \widehat{G}} \chi(h) = 0$. $\qquad\square$

Corollary 10.13. *For all $g, h \in G$, we have*

$$\sum_{\chi \in \widehat{G}} \chi(g)\overline{\chi(h)} = \begin{cases} |G| & \text{if } g = h \\ 0 & \text{if } g \neq h. \end{cases}$$

PROOF. Note that
$$\chi(h^{-1}) = \chi(h)^{-1} = \overline{\chi(h)}$$
since $\chi(h)$ is on the unit circle in $\mathbb{C}$. Then use Theorem 10.12 with gh^{-1} in place of h. $\qquad\square$

This is the gadget in its ultimate form. Character theory allows us to construct functions that will extract any desired residue class. As an example, take $G = U(\mathbb{Z}/5\mathbb{Z}) \cong C_4$. Table 10.1 shows all the characters on G.

Table 10.1. Characters on $U(\mathbb{Z}/5\mathbb{Z})$.

	χ_0	χ_1	χ_2	χ_3
1	1	1	1	1
2	1	i	-1	$-i$
4	1	-1	1	-1
3	1	$-i$	-1	i

Note that we have written the elements of $U(\mathbb{Z}/5\mathbb{Z})$ in Table 10.1 in an unusual ordering $2^0, 2^1, 2^2, 2^3$, adapted to the generator 2. The character values behave likewise. Note also $\chi_1^2 = \chi_2$ and $\chi_1^3 = \chi_3 = \chi_1^{-1}$. We used earlier

$$\chi_0(n) + \chi_1(n) + \chi_2(n) + \chi_3(n) = 4c_1(n),$$

which is just the case $h = 1$ of Corollary 10.13. We asked then, "What about $c_2(n)$, which is 1 if n is congruent to 2 and 0 otherwise?" The corollary suggests that we take $h = 2$, and we get

$$\chi_0(n) - i\chi_1(n) - \chi_2(n) + i\chi_3(n) = 4c_2(n).$$

This can be checked simply by going through the possible cases.

If you compare the ideas used here with Fourier analysis, much is familiar. The expression

$$\frac{1}{|G|} \sum_{h \in G} f(h)\overline{g(h)}$$

is an inner product on the vector space of all functions on G, and the characters form a complete orthonormal set. There are no difficulties about convergence because the group is finite. In particular, *any* complex function on G can be written as a linear combination of the characters.

10.4 Dirichlet Characters and L-Functions

Definition 10.14. *Given* $1 < q \in \mathbb{N}$, *let* $G = U(\mathbb{Z}/q\mathbb{Z})$ *and fix a character* χ *in* $\widehat{G}$. *Extend* χ *to a function* X *on* $\mathbb{N}$ *by setting*

$$X(n) = \begin{cases} \chi(n \bmod q) & \text{if } n \text{ is coprime to } q, \\ 0 & \text{otherwise.} \end{cases}$$

The function X *is called a* Dirichlet character modulo q.

This is a slight abuse of language – characters are functions on groups, and $\mathbb{N}$ certainly is not a group. We will even write χ instead of X for the Dirichlet character associated with χ. In the same way, for any $a \in G$, we can extend the function

$$c_a(b) = \begin{cases} 1 & \text{if } b = a, \\ 0 & \text{otherwise,} \end{cases} \tag{10.15}$$

to a periodic function on $\mathbb{N}$, which will also be written as c_a. Finally, associate to each Dirichlet character χ the L-function

$$L(s, \chi) = \sum_{n=1}^{\infty} \frac{\chi(n)}{n^s},$$

which is called the L-function of χ.

Example 10.15. Take the trivial character χ_0 of $U(\mathbb{Z}/4\mathbb{Z})$. The associated Dirichlet character is just the function χ_0 that we used in the second proof of Proposition 10.2. The functions c_1 and c_3 that we used there are extensions of functions on $U(\mathbb{Z}/4\mathbb{Z})$ as in Equation (10.15). The same holds for the corresponding L-functions – the notation has been carefully chosen to be consistent.

Theorem 10.16. *A Dirichlet character is completely multiplicative, and the associated L-function therefore has an Euler product expansion.*

PROOF. Let χ be a Dirichlet character modulo q. If two integers m, n are given, and at least one of them is not coprime to q, then neither is the product mn. Thus, $\chi(mn) = 0 = \chi(m)\chi(n)$. If on the other hand, both m and n are coprime to q, then $(m \bmod q)\cdot(n \bmod q) = (mn \bmod q)$ by definition, and because χ in the original sense is a group character, we have $\chi(mn) = \chi(m)\chi(n)$. The existence of an Euler product expansion then follows directly from Theorem 8.17. Since $\chi(p) = 0$ for all p dividing q, we get

$$L(s,\chi) = \prod_p \left(1 - \frac{\chi(p)}{p^s}\right)^{-1} = \prod_{p \nmid q} \left(1 - \frac{\chi(p)}{p^s}\right)^{-1}. \qquad (10.16)$$

$\square$

Clearly, the L-functions converge for $\Re(s) > 1$ by comparison with the Riemann zeta function. Now let us see how these L-functions can be marshaled to prove Dirichlet's Theorem about primes in an arithmetic progression.

By Theorem 8.17, which gives the Euler product expansion, $L(s,\chi) \neq 0$ for $\Re(s) > 1$, so we may take logarithms in Equation (10.16) and expand the logarithm in a Taylor expansion

$$\log L(s,\chi) = -\sum_{p \nmid q} \log\left(1 - \frac{\chi(p)}{p^s}\right) = \sum_{p \nmid q} \sum_{m=1}^{\infty} \frac{1}{m} \frac{\chi(p^m)}{p^{sm}}$$

$$= \sum_{p \nmid q} \frac{\chi(p)}{p^s} + O(1),$$

as before. For a given congruence class $a \bmod q$, with a coprime to q, multiply both sides by $\overline{\chi(a)}$ and sum over all $\chi \in \widehat{U(\mathbb{Z}/q\mathbb{Z})}$ (the associated Dirichlet characters). We get

$$\sum_\chi \overline{\chi(a)} \log L(s,\chi) = \sum_\chi \overline{\chi(a)} \sum_{p \nmid q} \frac{\chi(p)}{p^s} + O(1).$$

Since the series on the right converges absolutely, we may interchange summations. By Corollary 10.13,

$$\sum_\chi \overline{\chi(a)}\chi(p) = \begin{cases} \phi(q) & \text{if } p = a \pmod q \\ 0 & \text{otherwise}, \end{cases}$$

where $\phi(q) = |U(\mathbb{Z}/q\mathbb{Z})|$ by definition of the Euler function. We have proved

$$\sum_\chi \overline{\chi(a)} \log L(s,\chi) = \phi(q) \sum_{p \equiv a \bmod q} \frac{1}{q^s} + O(1). \qquad (10.17)$$

Now let $s \to 1$ from above. We claim the following.

1. The L-function $L(s, \chi_0)$ has a simple pole at $s = 1$.
2. For all $\chi \neq \chi_0$, the L-function $L(s, \chi)$ has a nonzero limit as $s \to 1$.

Once these claims have been proved, we know that the left-hand side in Equation (10.17) tends to infinity as $s \to 1$. For the right-hand side, this means that there must be infinitely many summands, which will complete the proof of Dirichlet's Theorem.

The first claim is quite easy to prove,

$$L(s, \chi_0) = \prod_{p \nmid q} \left(1 - \frac{1}{p^s}\right)^{-1} = \prod_{p \mid q} \left(1 - \frac{1}{p^s}\right) \zeta(s),$$

and we know that ζ has a simple pole at $s = 1$. The second claim – the nonvanishing of the L-function – is very difficult to prove. Thus far, nothing that we have done has required the L-function to be defined for complex values of s. It is the proof of nonvanishing that requires the complex variable methods.

A last remark before we embark on this. Looking back at Figure 10.1, one might have guessed that both congruence classes of primes modulo 4 contain about the same number of primes up to a given bound. This is true in complete generality: For a coprime to q,

$$\frac{|\{p \in \mathbb{P} \mid p \equiv a \pmod{q}, \ p \leqslant T\}|}{|\{p \in \mathbb{P} \mid p \leqslant T\}|} \longrightarrow \frac{1}{\phi(q)} \text{ as } T \to \infty.$$

In particular, the limit is independent of a. This can be proved by a slight refinement of the methods given in this chapter.

10.5 Analytic Continuation and Abel's Summation Formula

In this section, we will not only complete the proof of Dirichlet's Theorem 10.5. On the way, we will see Abel's Summation Formula and the analytic continuation of L-functions to the half-plane $\Re(s) > 0$.

Theorem 10.17. [ABEL] *Let* a *be an arithmetic function, and define*

$$\mathsf{A}(x) = \sum_{n \leqslant x} \mathsf{a}(n).$$

Let $\mathsf{f} : [x, y] \to \mathbb{C}$ *be differentiable with a continuous derivative. Then*

$$\sum_{x < n \leqslant y} \mathsf{a}(n)\mathsf{f}(n) = \mathsf{A}(y)\mathsf{f}(y) - \mathsf{A}(x)\mathsf{f}(x) - \int_x^y \mathsf{A}(t)\mathsf{f}'(t) \, dt. \tag{10.18}$$

PROOF. Assume $x, y \in \mathbb{N}$ are integral for simplicity, $m = y$ and $k = x$, so the left-hand side of Equation (10.18) is

$$\sum_{n=k+1}^{m} \mathsf{a}(n)\mathsf{f}(n) = \sum_{n=k+1}^{m} (\mathsf{A}(n) - \mathsf{A}(n-1))\mathsf{f}(n)$$

$$= -\sum_{n=k+1}^{m} \mathsf{A}(n-1)(\mathsf{f}(n) - \mathsf{f}(n-1)) + \mathsf{A}(m)\mathsf{f}(m) - \mathsf{A}(k)\mathsf{f}(k).$$

This is rather like integration by parts, using sums instead of integrals and taking differences instead of differentiating. There are some terms coming from the boundary of the summation interval, too, which complicates things a little. Notice that we have not used the hypothesis that f be differentiable up to now. Using $\mathsf{A}(t) = \mathsf{A}(n)$ for all $t \in [n, n+1)$, we get

$$\mathsf{A}(n-1)(\mathsf{f}(n) - \mathsf{f}(n-1)) = \mathsf{A}(n-1)\int_{n-1}^{n} \mathsf{f}'(t)\, dt = \int_{n-1}^{n} \mathsf{A}(t)\mathsf{f}'(t)\, dt.$$

Sum this from $n = k+1$ to $n = m$ to get the statement of the theorem. □

We apply this formula to $\mathsf{a}(n) = \chi(n)$ (for χ a Dirichlet character modulo q, but not the trivial character) and $\mathsf{f}(t) = t^{-s}$. Notice that $\sum_{n=k}^{k+q-1} \chi(n) = 0$ for all $k \in \mathbb{N}$, so $\mathsf{A}(x) = O(1)$; in fact $|\mathsf{A}(x)| \leqslant \phi(q)$ for all x. Abel's Summation Formula gives

$$\sum_{1 < n \leqslant y} \frac{\chi(n)}{n^s} = \frac{\mathsf{A}(y)}{y^s} - 1 + s\int_{1}^{y} \frac{\mathsf{A}(t)}{t^{s+1}}\, dt. \tag{10.19}$$

The integral on the right-hand side of Equation (10.19) can be split into integrals from 1 to 2, from 2 to 3, and so on. The series of these integrals converges uniformly for $\Re(s) > \delta$, with any fixed $\delta > 0$, so we may let y go to ∞. Each of these integrals is an analytic function of s for $\Re(s) > 0$. Hence the function $L(s, \chi)$ is analytic in this domain by Theorem 8.23. This argument is in fact the same as the one we used in one of the proofs of the analytic continuation of the zeta function.

PROOF OF NONVANISHING OF L-FUNCTIONS. We will now finally prove the last step in Dirichlet's Theorem by showing that $L(s, \psi) \neq 0$ for $s = 1$. Consider first the case that ψ is a non-real Dirichlet character (non-real meaning that not all values of $\psi(n)$ are ± 1). Consider, for $\sigma > 1$,

$$\log \prod_{\chi} L(\sigma, \chi) = \sum_{\chi} \log L(\sigma, \chi) = -\sum_{\chi} \sum_{p \nmid q} \log \left(1 - \frac{\chi(p)}{p^s}\right)$$

$$= \sum_{\chi} \sum_{p \nmid q} \sum_{m=1}^{\infty} \frac{\chi(p)^m}{m p^{\sigma m}}. \tag{10.20}$$

Suppose $L(1, \psi) = 0$. Then we must also have $L(1, \bar{\psi}) = 0$. By hypothesis, $\psi \neq \bar{\psi}$, and both ψ and $\bar{\psi}$ appear in the product over all characters in Equation (10.20). As σ tends to 0, the simple pole of $L(s, \chi_0)$ is doubly cancelled by the zeros in $L(s, \psi)$ and $L(s, \bar{\psi})$, and hence the product must tend to 0 and the logarithm in Equation (10.20) must go to $-\infty$. But the right-hand side of Equation (10.20) is always nonnegative. This follows from the fact that $\sum_\chi \chi(p^m) = 0$ or $\phi(q)$ by Theorem 10.12. This is a contradiction, and we have proved $L(1, \psi) \neq 0$ in the case that ψ is not real.

The case that ψ is real, so $\psi(n) = \pm 1$ for all $n \in \mathbb{N}$, is rather more complicated. Suppose again that $L(1, \psi) = 0$. Then $\zeta(s)L(s, \psi)$ must be analytic on the half-plane $\Re(s) > 0$. Write $F(s) = \zeta(s)L(s, \psi)$ as a Dirichlet series (see Exercise 8.15 on p. 166),

$$F(s) = \sum_{n=1}^{\infty} \frac{f(n)}{n^s},$$

where the function $f = \psi * u$ is defined by

$$f(n) = (\psi * u)(n) = \sum_{d|n} \psi(d).$$

Lemma 10.18. *Define another arithmetic function g by*

$$g(n) = \begin{cases} 1 & \text{if } n \text{ is a square;} \\ 0 & \text{otherwise.} \end{cases}$$

Then $f(n) \geqslant g(n)$ for all $n \in \mathbb{N}$.

PROOF. Note that both f and g are multiplicative arithmetic functions, so it is enough to consider the case $n = p^k$, a prime power. We have

$$f(p^k) = 1 + \psi(p) + \cdots + \psi(p)^k = \begin{cases} 1 & \text{if } \psi(p) = 0, \\ k+1 & \text{if } \psi(p) = 1, \\ 0 & \text{if } \psi(p) = -1 \text{ and } k \text{ is odd}, \\ 1 & \text{if } \psi(p) = -1 \text{ and } k \text{ is even.} \end{cases}$$

Clearly $f(n) \geqslant 0$ for all n. This settles already the claim of the lemma in the case that n is not a square. If n is a square, the exponent of each prime in n is even, and we get $f(n) \geqslant 1$ by looking at the preceding equation. This completes the proof of Lemma 10.18. □

Returning to the main proof, fix a number r with $0 < r < 3/2$. Now F is analytic on the half-plane $\sigma > 0$, so we may consider the Taylor expansion of F about $s = 2$,

$$F(2 - r) = \sum_{\nu=1}^{\infty} \frac{F^{(\nu)}(2)}{\nu!}(-r)^\nu,$$

where the νth derivative $F^{(\nu)}(2)$ is given by

$$F^{(\nu)}(2) = \sum_{n=1}^{\infty} f(n)\frac{(-\log n)^\nu}{n^2}. \tag{10.21}$$

We will prove that the Dirichlet series for F converges uniformly for $\sigma > 0$ so that we may indeed differentiate term by term. Now consider a general summand of the Taylor expansion,

$$\frac{F^{(\nu)}(2)}{\nu!}(-r)^{\nu} = \frac{r^{\nu}}{\nu!}\sum_{n=1}^{\infty}\frac{f(n)(\log n)^{\nu}}{n^2} \geqslant \frac{r^{\nu}}{\nu!}\sum_{n=1}^{\infty}\frac{g(n)(\log n)^{\nu}}{n^2}$$

$$= \frac{r^{\nu}}{\nu!}\sum_{n=1}^{\infty}\frac{1\cdot(\log n^2)^{\nu}}{n^4} = \frac{(-2r)^{\nu}}{\nu!}\sum_{n=1}^{\infty}\frac{(-\log n)^{\nu}}{n^4}$$

$$= \frac{(-2r)^{\nu}}{\nu!}\zeta^{(\nu)}(4).$$

Use this inequality for all terms of the Taylor expansion of F to deduce that

$$F(2-r) \geqslant \sum_{\nu=0}\frac{(-2r)^{\nu}}{\nu!}\zeta^{(\nu)}(4) = \zeta(4-2r). \tag{10.22}$$

Now let r converge to $3/2$ from below. The right-hand side of Equation (10.22) tends to infinity since ζ has a pole at $s = 1$. The left-hand side is bounded because $F(s)$ is analytic for $s > 0$, a contradiction.

We still have to prove our claim that the Dirichlet series for F converges uniformly for all $s > 0$. Look again at Equation (10.21) and substitute it into the Taylor series for F about $s = 2$ to get

$$F(2-r) = \sum_{\nu=0}^{\infty}\frac{1}{\nu!}r^{\nu}\sum_{n=1}^{\infty}f(n)\frac{(\log n)^{\nu}}{n^2}.$$

Note that the minus sign of $-r$ cancels with that in the derivative so that all terms are positive. Hence we may interchange the summations,

$$F(2-r) = \sum_{n=1}^{\infty}\frac{f(n)}{n^2}\sum_{\nu=0}^{\infty}\frac{1}{\nu!}(r\log n)^{\nu}, \tag{10.23}$$

and this sum converges for all r with $|r| < 2$ because by our assumption $F(s)$ is analytic on the whole half-plane $\Re(s) > 0$. The inner sum in Equation (10.23) is just $e^{r\log n} = n^r$, so Equation (10.23) becomes

$$F(2-r) = \sum_{n=1}^{\infty}\frac{f(n)}{n^{2-r}}.$$

The right-hand side converges for all $r < 2$, and if we substitute $s = 2-r$, we get just the Dirichlet series for F back again, which converges for all $s > 0$. Since any Dirichlet series convergent for all $s > s_0$ converges *uniformly* in that domain, the proof is complete. □

Exercise 10.4. Locate the steps in the preceding proof that required $L(\chi, s)$ to be a function of a complex variable s rather than a real one.

10.6 Abel's Limit Theorem

We used the following result on p. 210 in the proof of Lemma 10.3.

Theorem 10.19. [ABEL] *Given a real power series*

$$f(x) = \sum_{n=0}^{\infty} a_n x^n$$

that converges for all x with $0 < x < x_0$, suppose that the limit

$$L = \sum_{n=0}^{\infty} a_n x_0^n = \lim_{N \to \infty} \sum_{n=0}^{N} a_n x_0^n$$

exists. Then the limit of $f(x)$ exists as x tends to x_0 from below, and

$$\lim_{x \to x_0^-} f(x) = L.$$

PROOF. For any $\epsilon > 0$, we will show that, for all x sufficiently close to x_0 and for all N sufficiently large,

$$\left| \sum_{n=0}^{N} a_n (x_0^n - x^n) \right| \leqslant \epsilon. \tag{10.24}$$

To do this, rewrite the sum in the inequality (10.24) as

$$\sum_{n=0}^{N} a_n (x_0^n - x^n) = \sum_{n=0}^{N} a_n \left(1 - \left(\frac{x}{x_0} \right)^n \right) x_0^n.$$

Let $y = x/x_0$ and use the geometric series expansion

$$1 - y^n = (1 - y)(1 + y + y^2 + \cdots + y^{n-1})$$

to obtain

$$\sum_{n=0}^{N} a_n (x_0^n - x^n) = (1 - y) \sum_{n=1}^{N} a_n \sum_{k=0}^{n-1} y^k x_0^n. \tag{10.25}$$

We may interchange the summations since these sums are both finite, giving

$$\sum_{n=0}^{N} a_n (x_0^n - x^n) = (1 - y) \sum_{k=0}^{N} y^k \sum_{n=k+1}^{N} a_n x_0^n. \tag{10.26}$$

The coefficients of y^k in Equation (10.26) form a Cauchy sequence since the corresponding series converges. Hence they are bounded in absolute value by some b_k,

$$\left| \sum_{n=k+1}^{N} a_n x_0^n \right| \leqslant b_k,$$

which depends only on k, not on N. Moreover, we know that

$$b_k \longrightarrow 0 \text{ as } k \longrightarrow \infty$$

so we may choose K such that $b_k < \epsilon$ for all $k \geqslant K$. Then we can estimate the sum in Equation (10.25) by splitting it,

$$\left| \sum_{n=0}^{N} a_n(x_0^n - x^n) \right| \leqslant (1-y) \sum_{k=0}^{K-1} y^k b_k + (1-y) \sum_{k=K}^{\infty} y^k \epsilon. \qquad (10.27)$$

As x tends to x_0, y tends to 1. This means that the first summand on the right-hand side of the inequality (10.27) becomes arbitrarily small. The second summand is equal to $y^K \epsilon$, and hence tends to ϵ, and our estimates no longer depend on N. $\square$

NOTES TO CHAPTER 10: There are accounts of this material in many of the references. Monsky's paper [109] discusses one of many possible simplifications of the proof of Dirichlet's Theorem. Theorem 10.19 is a result of *Abelian* type; typically the converse of such a result is false but becomes true under an additional assumption. The converse theorems obtained in this way are called *Tauberian* and are generally deeper; see Hardy [74, Chapter VII]. The survey paper [68] of Gelbart and Miller discusses the historical development of L-functions, starting with Riemann's original paper and explaining how L-functions may be associated with groups. The solution to Exercise 10.3(c) is most naturally given in terms of *adeles* – see Ramakrishnan and Valenza [121], Tate's thesis [150], or Weil [158].

11

Converging Streams

In Chapters 2 and 4, and again in Chapters 8–10, we developed two different approaches to number theory. The first viewed the study of numbers in more algebraic terms, the second in more analytic terms. One of the great early achievements of algebraic number theory was a reconciliation of these two approaches in the class number formula, discussed in Section 11.1. The name algebraic number theory is a little unfortunate – it is the study of algebraic numbers (solutions of polynomial equations) and the fields and rings they generate, but it uses an enormous range of techniques, including analysis. We will need to discuss the evaluation of Gauss sums as part of this study. This too shows that an idea which appears to belong within elementary number theory can only be understood, apparently, by deep methods.

Similarly, in Chapters 5–7 we developed the arithmetic of elliptic curves and this again appeared to be somewhat distinct in flavour from the other topics. A second unifying theme presented informally in this chapter is the conjecture of Birch and Swinnerton-Dyer, which is a profound connection between the arithmetic and analytic properties of elliptic curves. The bridge connecting this to the other chapters is the theory of L-functions. It is no surprise that some of the deepest and most important unsolved problems in number theory for the new millennium are concerned with these mysterious objects.

11.1 The Class Number Formula

We will state the class number formula for quadratic fields and then look at some aspects of the proof. There is a more general formulation of the class number formula for algebraic number fields. Dirichlet originally proved the quadratic case in 1837 using quadratic forms rather than quadratic fields, as part of his proof concerning primes in arithmetic progressions. He showed that the associated L-function, for real characters, does not vanish at 1 because it is equal to a recognizably nonzero expression. Thus the results in this chapter could be taken as another approach to the nonvanishing of the L-function for a real character.

In essence, the class number formula gives the class number as a finite sum of easily computable quantities. In special cases, it gives an unexpected relation between units of the ring of algebraic integers that depends upon the class number. To this day, the only convincing proofs of this result use nontrivial analysis.

Jacobi made many fundamental discoveries about the class number, and Dirichlet was able to build upon these in his formulation and eventual proof of the class number formula. Jacobi's name is attached to a generalization of the Legendre symbol (defined on p. 65) which will be needed later. This symbol is sometimes called the quadratic symbol.

Definition 11.1. *Let $n \geqslant 1$ be an odd integer with prime factorization*

$$n = p_1^{\alpha_1} \cdots p_r^{\alpha_r}.$$

Then the Jacobi symbol $\left(\dfrac{\cdot}{n}\right) : \mathbb{Z} \to \{0, \pm 1\}$ is defined to be

$$\left(\frac{a}{n}\right) = \left(\frac{a}{p_1}\right)^{\alpha_1} \cdots \left(\frac{a}{p_r}\right)^{\alpha_r},$$

where $\left(\dfrac{a}{p_i}\right)$ denotes the Legendre symbol.

Clearly, $\left(\dfrac{a}{n}\right) = 0$ unless $\gcd(n, a) = 1$. Also, if a prime p does not divide n, then

$$\left(\frac{p^2 a}{n}\right) = \left(\frac{a}{n}\right)$$

and

$$\left(\frac{a}{p^2 n}\right) = \left(\frac{a}{n}\right).$$

Thus, in order to evaluate Jacobi symbols, we need only consider a and n square-free and coprime. Note that the symbol only depends upon a modulo n.

The Quadratic Reciprocity Law extends to the Jacobi symbol in the following form.

Theorem 11.2. *Suppose a and n are positive, nonzero, coprime integers. Then*

$$\left(\frac{a}{n}\right) = (-1)^{(a-1)/2 \cdot (n-1)/2} \left(\frac{n}{a}\right) \quad \text{if a and n are odd;}$$

$$\left(\frac{2}{n}\right) = (-1)^{(n^2-1)/8} \quad \text{if n is odd.}$$

Exercise 11.1. Prove Theorem 11.2. (Hint: Consider the primes $p \equiv 3$ modulo 4 dividing a and n.)

Notice that the Jacobi symbol does not characterize the property of being a quadratic residue in the way the Legendre symbol does. Certainly if a is a quadratic residue modulo n with $\gcd(a, n) = 1$, then $\left(\frac{a}{n}\right) = 1$. The converse does not hold, as the next example shows.

Example 11.3. Since 2 is not a quadratic residue modulo 3, it is not a quadratic residue modulo 15. On the other hand,

$$\left(\frac{2}{15}\right) = \left(\frac{2}{3}\right)\left(\frac{2}{5}\right) \text{ by definition}$$
$$= (-1)(-1) \text{ by Theorem 11.2}$$
$$= 1.$$

Definition 11.4. *The* Kronecker symbol, *also written* $\left(\frac{a}{n}\right)$, *is an extension of the Jacobi symbol to all $n \neq 0$. It is defined by the properties:*

- $\left(\frac{a}{n}\right) = 0$ *if* $\gcd(a, n) > 1$;

- $\left(\frac{a}{-1}\right) = \begin{cases} 1 \text{ if } a > 0, \\ -1 \text{ if } a < 0; \end{cases}$

- $\left(\frac{a}{2}\right) = \begin{cases} 1 \text{ if } a \equiv \pm 1 \pmod{8}, \\ -1 \text{ if } a \equiv \pm 3 \pmod{8}; \end{cases}$

- $\left(\frac{ab}{cd}\right) = \left(\frac{a}{c}\right)\left(\frac{b}{c}\right)\left(\frac{a}{d}\right)\left(\frac{b}{d}\right).$

Exercise 11.2. Let D be the discriminant of the quadratic field $\mathbb{Q}(\sqrt{d})$ for a square-free integer d. Show that $\chi : \mathbb{Z} \to \{0, \pm 1\}$ defined by $\chi(n) = \left(\frac{D}{n}\right)$ is a Dirichlet character modulo $|D|$.

Theorem 11.5. [CLASS NUMBER FORMULA] *Let D denote the discriminant of a quadratic field $\mathbb{K} = \mathbb{Q}(\sqrt{d})$. Then the class number h of $O_{\mathbb{K}}$ is given by*

$$h = \begin{cases} -\dfrac{1}{2 \log u} \displaystyle\sum_{a=1}^{D-1} \chi(a) \log \sin(a\pi/D) & \text{for } D > 0, \\[4mm] -\dfrac{w}{2|D|} \displaystyle\sum_{a=1}^{|D|-1} \chi(a) a & \text{for } D < 0, \end{cases}$$

where w denotes the number of roots of unity in $\mathbb{K}$, u is the fundamental unit of $\mathbb{K}$, and χ denotes the associated character (see Exercise 11.2).

The fundamental unit u is introduced in Exercise 4.4 on p. 85.

The following theorem is a special case of the class number formula, yet it already makes an amazing claim, giving a totally unexpected relation between units in a real quadratic number field.

Theorem 11.6. *Let $q > 0$ denote a prime congruent to 1 modulo 4. Then the quantity*

$$v = \prod_{a=1}^{(q-1)/2} \sin(a\pi/q)^{-\left(\frac{a}{q}\right)},$$

is a unit in $\mathbb{Z}[(1 + \sqrt{q})/2]$, the ring of algebraic integers in $\mathbb{Q}(\sqrt{q})$. This unit is related to the fundamental unit u by the equation

$$v = u^h,$$

where h denotes the class number of the field $\mathbb{Q}(\sqrt{q})$.

Exercise 11.3. Prove that

$$\sin(\pi/5) = \frac{\sqrt{10 - 2\sqrt{5}}}{4},$$

and deduce that

$$\sin(2\pi/5) = \frac{\sqrt{10 + 2\sqrt{5}}}{4}.$$

Use this to show that Theorem 11.6 applied to the prime $q = 5$ constructs the unit $v = \frac{1+\sqrt{5}}{2}$.

Exercise 11.4. Show that $h = 1$ when $-D$ is 3, 4, 7, 8, 11, 19, 43, 67, or 163.

Exercise 11.5. Use Exercise 4.5 on p. 85 to find the class number of

$$\mathbb{Q}(\sqrt{2}), \quad \mathbb{Q}(\sqrt{3}), \quad \mathbb{Q}(\sqrt{5}) \quad \text{and} \quad \mathbb{Q}(\sqrt{7}).$$

Gauss, using the equivalent notion for quadratic forms, conjectured that the only values of $-D$ with $D > 0$ for which $h = 1$ are those given in Exercise 11.4. This was eventually proved[1] in the second half of the twentieth-century by Heegner, and then by Baker and Stark. Baker and Stark verified the result independently using different methods. They also solved the class number two problem.

---

[1] Kurt Heegner was a school teacher in Berlin. In 1952 he published a proof of the class number one problem, an old and famous problem. For several reasons – minor errors in the work, Heegner's refusal to give seminar presentations of his work, and perhaps some reluctance by professional mathematicians to accept that an "amateur" had solved such an important problem – his proof was not generally accepted. Heegner's proof was finally accepted after Alan Baker and Harold Stark independently proved the result in 1967.

Exercise 11.6. Show that $h = 2$ when $-D$ is 15, 20, 24, 35, 40, 51, 52, 88, 91, 115, 123, 148, 187, 232, 235, 267, 403, or 427.

Exercise 11.7. Using a computer algebra package, find the class number of the imaginary quadratic field $\mathbb{Q}(\sqrt{-d})$ for $1 \leqslant d \leqslant 100$.

We are going to prove the class number formula up to sign. Then we will consider the sign of the Gauss sum in a separate section. Firstly we develop some machinery: These ideas are so important, they are worth considering just for their own sake.

11.2 The Dedekind Zeta Function

A quadratic field has a complex function associated with it that generalizes the Riemann zeta function associated with $\mathbb{Q}$. The theory of ideals proved to be ideal as a way of defining such a function. Euler's momentous observation about the product formula for the Riemann zeta function requires the Fundamental Theorem of Arithmetic in the integers, which does not hold at the level of elements in the ring of algebraic integers in a quadratic field. Instead, one defines the *Dedekind zeta function* as follows. If I is an ideal in $O_{\mathbb{K}}$, then write $N(I)$ for the norm of I defined on p. 87. Now define

$$\zeta_{\mathbb{K}}(s) = \sum_I N(I)^{-s}$$

for $s \in \mathbb{C}$, where the sum runs over all the nonzero ideals in $O_{\mathbb{K}}$. If $\mathbb{K} = \mathbb{Q}$, then the Dedekind zeta function $\zeta_{\mathbb{K}}$ coincides with the Riemann zeta function ζ.

Notice that our results about the recovery of the Fundamental Theorem of Arithmetic at the level of ideals (see Section 4.3) means the Dedekind zeta function admits an Euler product expansion

$$\zeta_{\mathbb{K}}(s) = \prod_P \left(1 - \frac{1}{N(P)^s}\right)^{-1},$$

where the product is taken over all prime ideals of $O_{\mathbb{K}}$.

Exercise 11.8. Let $\mathbb{K} = \mathbb{Q}(\sqrt{d})$ for a square-free integer d. Show that

$$\zeta_{\mathbb{K}}(s) = \zeta(s)L(s, \chi) \text{ for } \Re(s) > 1,$$

where ζ is the Riemann zeta function and $L(\cdot, \chi)$ is the L-function for the character χ from Exercise 11.2. (Hint: Use Exercise 4.18 on p. 91.) Using the results from earlier chapters, deduce the analytic continuation of $\zeta_{\mathbb{K}}$ to a larger half-plane.

Exercise 11.8, when combined with the results in Chapter 10, gives information about the nature of $\zeta_K(s)$ near $s = 1$. The L-function is analytic at $s = 1$ so $\zeta_K(s)$ inherits only a simple pole from the Riemann zeta function. The class number formula eventually follows because it is possible to directly compute the nature of the singularity in two different ways and then equate them. The residue of ζ_K at the pole $s = 1$ is given by the following theorem.

Theorem 11.7. *Let ζ_K denote the Dedekind zeta function of a quadratic number field with discriminant D. Let h denote the class number of O_K, let w denote the number of units in O_K if $D < 0$ and let u denote a fundamental unit of O_K^* if $D > 0$. Then*

$$\lim_{s \to 1}(s-1)\zeta_K(s) = \rho_K = \begin{cases} \dfrac{2h \log u}{\sqrt{D}} & \text{if } D > 0, \\[2ex] \dfrac{2\pi h}{w\sqrt{|D|}} & \text{if } D < 0. \end{cases}$$

Dirichlet proved this using the language of quadratic forms. Nowadays, Hecke's proof using the language of fields and ideals tends to be preferred. The proof sketched below is a variation of Hecke's and uses some complex analysis.

OUTLINE PROOF OF THEOREM 11.7. The following proof varies from that in most of the textbooks; references are provided in the notes at the end of the chapter.

The two cases $D > 0$ and $D < 0$ vary ultimately but begin in the same way. The idea is to sum over each ideal class, so fix an ideal class $\mathfrak{C}$. There is a *fixed ideal* J belonging to the inverse class with the property that $IJ = (b_I)$ is a principal ideal generated by the quadratic integer b_I which depends on I. Now $b_I \in J$ so sum over the elements of J, up to multiplication by units, and consider

$$\sum_{0 \neq b \in J} \left(\frac{N(J)}{N(b)} \right)^s = N(J)^s \sum_{0 \neq b \in J} \frac{1}{N(b)^s}. \tag{11.1}$$

In this sum it is understood that no pair of distinct elements b and b' have b/b' equal to a unit. The technical details thus become much easier in the imaginary quadratic case because there are only finitely many units, so we assume for now that $D < 0$. After choosing a basis $\{b_1, b_2\}$ for the ideal J, $N(b)$ becomes a positive-definite quadratic form $Q(x, y)$ in the variables x, y, where $b = b_1 x + b_2 y$.

The sum in Equation (11.1) can be compared with the corresponding integral

$$\int \int_{\mathfrak{A}} Q(x, y)^{-s} dx dy,$$

where x and y now become continuous variables. The range of integration $\mathfrak{A}$ is an infinite cone, with a small region around $(0, 0)$ removed to take account of the fact that $b = 0$ is excluded from the summation. The shape of the cone depends on the number of units in O_K – if there are w units in O_K then the angle of the cone is $2\pi/w$. The difference between the sum and the integral is analytic on the region $\Re(s) > 0$. The method of proof for this is an extension of the methods used in the second proof of Theorem 8.29 on p. 176. Around each integral point (x, y) there is a square of area 1. The integral of Q^{-s} over that square differs from $Q(x, y)^{-s}$ by an amount that may be computed using the Taylor expansion. The sum over the integral points near the boundary of the cone contributes a function which is analytic on $\Re(s) > \frac{1}{2}$ so we can ignore it.

Exercise 11.9. Prove that this sum is analytic, by comparing the sum with an integral along an infinite strip with parallel sides.

It follows that the singularity of the sum can be calculated from that of the integral.

The integral is easily integrated because Q is a positive-definite quadratic form. A linear substitution with Jacobian $N(J)\sqrt{|D|}$ reduces the integral to

$$\frac{1}{N(J)\sqrt{|D|}} \int \int_{\mathfrak{B}} (X^2 + Y^2)^{-s}\, dX dY,$$

where the region $\mathfrak{B}$ is a cone of angle $2\pi/w$ with a small region around the origin removed. Using polar coordinates (r, θ), the region around the origin can be taken to be $r < 1$. The resulting integral has a simple pole at $s = 1$ with residue

$$\frac{2\pi}{wN(J)\sqrt{|D|}}.$$

When evaluating the residue in the sum in Equation (11.1) there are two things to be borne in mind. The first is that the factor $N(J)$ cancels because of the factor $N(J)^s$ in Equation (11.1), which is $N(J)$ when $s = 1$. The second is that we sum over all the ideal classes. This explains the appearance of the term h and the shape of the residue in Theorem 11.7.

When $D > 0$ a complication is added to the proof because there are infinitely many units in the ring of algebraic integers O_K. Thus greater care is needed in counting the elements b of J. Roughly the same technique is used as in the $D < 0$ case – the sum can be written as in Equation (11.1) with the same proviso about elements differing by unit multiplication. An additional assumption will be inserted however; assume that $b > 0$ and work with the subgroup of positive units. Write b and b^* for the conjugates of $b \in J$ and fix a basis for J as above. Writing $b = b_1 x + b_2 y$ and extending the coordinates x and y to continuous real variables, we now switch to the continuous, positive real variables $X = b_1 x + b_2 y$ and $Y = b_1^* x + b_2^* y$. The transformation has Jacobian $N(J)\sqrt{D}$ exactly as before. A factor of 2 will be inserted below because of the assumption about the positive unit group. The sum equals (up to an analytic function on $\Re(s) > \frac{1}{2}$) the following integral in a half-plane containing $s = 1$:

$$\frac{2}{N(J)\sqrt{D}} \int\int_{\mathfrak{C}} (XY)^{-s} \, dX dY,$$

where $\mathfrak{C}$ is a region we will now describe. Essentially, the region is a fundamental domain under the action of the positive part of the unit group. This can be expressed neatly in terms of X and Y as follows: The map $b \mapsto bu$ sends $c = b^*/b$ to c/u^2. Hence each element b in the sum can be represented by one with

$$1 \leqslant \frac{b^*}{b} < u^2.$$

This imposes the following constraints upon X and Y:

$$1 \leqslant \frac{Y}{X} < u^2. \tag{11.2}$$

One final piece of bookkeeping requires a small region around $(0,0)$ to be removed, to take account of the fact that $b = 0$ is not included in the sum: Perform the double integral, first over Y satisfying the inequality (11.2) and then over the interval $1 < X < \infty$ – the lower bound ensuring that a suitable region has been removed around the origin. Integrating over Y gives

$$\frac{2}{N(J)\sqrt{D}} \int_1^\infty X^{-s} \left[\int_{1 \leqslant \frac{Y}{X} < u^2} Y^{-s} \, dY \right] dX = \frac{2(u^{2(1-s)} - 1)}{(1-s)N(J)\sqrt{D}} \int_1^\infty X^{1-2s} \, dX.$$

Integrating over X yields the following closed formula for the value of the integral:

$$\mathfrak{J} = \frac{1 - u^{2-2s}}{N(J)\sqrt{D}(1-s)^2}.$$

Although this might appear to yield a double pole at $s = 1$, notice that $s = 1$ is a zero of the numerator. The Taylor series of the numerator about $s = 1$ begins

$$1 - 1 + 2(s-1) \log u + \cdots.$$

The term $N(J)$ disappears just as it did before. Now the claims made about the nature of the singularity and the residue in Theorem 11.7 follow after summing over the classes. $\square$

A version of Theorem 11.7 exists which is expressed in terms of the counting of norms of ideals.

Exercise 11.10. Let $\mathbb{K}$ denote a quadratic field with discriminant D. For positive real T, let $R(T)$ denote the ideal counting function defined as follows:

$$R(T) = |\{I \mid N(I) < T\}|,$$

where I denotes an ideal of the ring of algebraic integers of $\mathbb{K}$. Prove that

$$\frac{R(T)}{T} \to \rho_{\mathbb{K}} \quad \text{as} \quad T \to \infty,$$

where $\rho_{\mathbb{K}}$ is defined by the formulas in Theorem 11.7, depending upon the sign of D.

11.3 Proof of the Class Number Formula

Before we can prove Theorem 11.5, we record some basic lemmas to allow the proof to proceed unhindered.

Lemma 11.8. *Let $N > 1$ denote a positive integer and suppose $1 \leqslant a < N$. Then*

$$-\log\left(1 - e^{2\pi ai/N}\right) = -\log\left(2\sin\frac{\pi a}{N}\right) + \frac{\pi i}{2}\left(1 - \frac{2a}{N}\right).$$

PROOF. This is elementary, relying only upon the definition of the principal branch of the complex logarithm and some manipulation with the half-angle formulas. □

Let χ denote the character from Exercise 11.2. We will use a form of Gauss sum (compare this definition with the one in Equation (3.11)).

Lemma 11.9. *Let $\zeta = e^{2\pi i/|D|}$ and define*

$$G = \sum_{a=1}^{|D|-1} \chi(a)\zeta^a. \tag{11.3}$$

Then

$$G^2 = D. \tag{11.4}$$

Moreover, for every n,

$$\chi(n) = \frac{1}{G}\sum_{a=1}^{|D|-1} \chi(a)\zeta^{an}. \tag{11.5}$$

PROOF. The claim about G^2 is proved in exactly the same way as Equation (3.12) on p. 70 was proved. The product G^2 may be written

$$\sum_{\substack{a=1;\\(a,|D|)=1}}^{|D|-1} \chi(a)\zeta^a \sum_{\substack{r=1;\\(r,|D|)=1}}^{|D|-1} \chi(ar)\zeta^{ar} = \sum_{\substack{a=1;\\(a,|D|)=1}}^{|D|-1} \sum_{\substack{r=1;\\(r,|D|)=1}}^{|D|-1} \chi(r)\zeta^{(1+r)a}.$$

Instead of adding zero in the form of Equation (3.14), here we add zero in the two forms

$$\sum_{\substack{a=1; \\ (a,|D|)=1}}^{|D|-1} \sum_{\substack{r=1; \\ (r,|D|)>1}}^{|D|-1} \chi(r)\zeta^{(1+r)a}$$

and

$$\sum_{\substack{a=1; \\ (a,|D|)>1}}^{|D|-1} \sum_{r=1}^{|D|-1} \chi(r)\zeta^{(1+r)a}.$$

The rest of the proof proceeds as before. Equation (11.5) is proved as follows. For $(n, D) = 1$,

$$\sum_{a=1}^{|D|-1} \chi(a)\zeta^{an} = \sum_{a=1}^{|D|-1} \chi(an^2)\zeta^{an} = \chi(n) \sum_{a=1}^{|D|-1} \chi(an)\zeta^{an} = \chi(n)G,$$

while for $(n, D) > 1$ it is clear. $\qquad\qquad\qquad\qquad\qquad\qquad\qquad\qquad$ $\square$

Notice that G lies in a quadratic field. Which quadratic field it lies in depends upon the sign of D. If $D > 0$ then it follows from Equation (11.5) that G is real; hence $G = \pm\sqrt{D}$ by Equation (11.4).

When $D < 0$, it follows that

$$\overline{G} = -G,$$

so $G = \pm i\sqrt{|D|}$ using Equation (11.4). Understanding which sign occurs is not trivial – during the proof below we will fudge this issue, essentially giving a proof of the class number formula up to sign. In Section 11.4 we will show how a simple application of Fourier Analysis can be used to determine the sign of the simplest Gauss sum.

PROOF OF THEOREM 11.5. We begin with the formal evaluation of $L(\chi, 1)$, worrying about convergence later. By definition,

$$L(\chi, 1) = \sum_{n=1}^{\infty} \frac{\chi(n)}{n}.$$

Apply Equation (11.5) and rearrange to give

$$L(\chi, 1) = \frac{1}{G} \sum_{a=1}^{|D|-1} \chi(a) \sum_{n=1}^{\infty} \frac{1}{n} e^{2\pi a i n/|D|}.$$

The inner sum is $-\log\left(1 - e^{2\pi a i/|D|}\right)$ so invoke Lemma 11.8 with $N = |D|$ to obtain

$$L(\chi, 1) = -\frac{1}{G} \sum_{a=1}^{|D|-1} \chi(a) \left[\log\left(2\sin\frac{\pi a}{|D|}\right) + \frac{\pi i}{2}\left(1 - \frac{2a}{|D|}\right)\right]. \qquad (11.6)$$

At this point, the sign of D brings about a dichotomy. If $D > 0$ then only the logarithm terms in Equation (11.6) survive. We know that $L(\chi, 1)$ must be real and since the Gauss sum G is real, the imaginary part of Equation (11.6) must cancel. We obtain

$$L(\chi, 1) = -\frac{1}{G} \sum_{a=1}^{|D|-1} \chi(a) \log \sin \frac{\pi a}{D}.$$

By Theorem 11.7 this is equal to $2h \log u / \sqrt{|D|}$. Since $G = \pm\sqrt{|D|}$, cancellation occurs and the theorem is proved up to sign.

When $D < 0$ it is the logarithm terms in Equation (11.6) which cancel. The Gauss sum G is purely imaginary so no real part of Equation (11.6) can survive. The cancelling leaves

$$L(\chi, 1) = -\frac{\pi i}{G|D|} \sum_{a=1}^{|D|-1} \chi(a)a.$$

By Theorem 11.7 this is equal to $2\pi h / w \sqrt{|D|}$. In this case, cancellation occurs because $G = \pm i\sqrt{|D|}$ and once again the formula is proved up to sign.

Finally, the rearrangement can be justified using Abel's Summation Formula, Theorem 10.17. All that is needed to use this is the uniform boundedness of $\sum_{a \leqslant x} \chi(a)$.

Alternatively, one could work backwards from the Taylor Series for the sine function, whose convergence properties are known to be adequate.   □

PROOF OF THEOREM 11.6. In this special case the class number formula comes out as

$$\sum_{a=1}^{(q-1)/2} \left(\frac{a}{q}\right) \log \sin \left(\frac{a\pi}{q}\right) = h \log u,$$

which proves the claim about v as well as giving the proof of its relation with u.   □

Of particular note is the way that so much of our earlier material goes into the class number formula. It suggests that the formula lies very deep in the fabric of number theory, and it has long been recognized as a profound relationship. We hope this material might persuade a reader to look into more advanced topics in the area of overlap between algebra and analysis.

11.4 The Sign of the Gauss Sum

We are only going to consider a simple example, which should be compared with the Gauss sum defined by Equation (3.11) on p. 70. Let q denote an odd prime number and $\left(\frac{\cdot}{q}\right)$ the Legendre symbol.

Write

$$G = \sum_{a=1}^{q-1} \left(\frac{a}{q}\right) \zeta^a$$

where $\zeta = e^{2\pi i/q}$. It is important to recognize, as Gauss himself did, that G is sensitive to the choice of qth root of unity. In particular, replacing ζ by some other primitive qth root of unity could change the sign of G. In the following, we will use Dirichlet's method[2] to evaluate G, based on Fourier analysis.

Exercise 11.11. Generalize Exercise 5.11 to show that

$$\sum_{a=1}^{q-1} e^{2\pi i a/q} = -1.$$

By Exercise 11.11 we may write

$$G = \sum_{a=1}^{q-1} \left(\frac{a}{q}\right) e^{2\pi i a/q} + \sum_{a=1}^{q-1} e^{2\pi i a/q} + 1$$

$$= 2 \sum_{\substack{a=1; \\ \left(\frac{a}{q}\right)=1}}^{q-1} e^{2\pi i a/q} + 1$$

$$= \sum_{b=0}^{q-1} e^{2\pi i b^2/q}.$$

Dirichlet's method works for a more general class of sums.

Theorem 11.10. [DIRICHLET] *Let N denote a positive integer and define*

$$H = \sum_{k=0}^{N-1} e^{2\pi i k^2/N}. \tag{11.7}$$

Then

$$H = \begin{cases} (1+i)\sqrt{N} & \text{if } N \equiv 0 \pmod 4, \\ \sqrt{N} & \text{if } N \equiv 1 \pmod 4, \\ 0 & \text{if } N \equiv 2 \pmod 4, \\ i\sqrt{N} & \text{if } N \equiv 3 \pmod 4, \end{cases}$$

where $\sqrt{N}$ denotes the positive square root.

---

[2] Dirichlet contributed significantly to the theory of Fourier analysis. For example, in 1829, he became the first person to give a rigorous proof of the Poisson Summation Formula. He obtained Theorem 11.10 in 1835 as an application.

PROOF. The functions $\{x \mapsto e^{2\pi i n x/N}\}_{n\in\mathbb{Z}}$ form an orthonormal family with respect to the inner product

$$< f, g >= \frac{1}{N} \int_0^N f(t)\overline{g(t)}\, dt.$$

The Fourier expansion[3] of the map $x \mapsto e^{2\pi i x^2/N}$ with respect to the orthonormal family shows that

$$e^{2\pi i k^2/N} = \sum_{n=-\infty}^{\infty} \left(\frac{1}{N} \int_0^N e^{2\pi i x^2/N} e^{-2\pi i n x/N}\, dx \right) e^{2\pi i n k/N},$$

so

$$H = \sum_{k=0}^{N-1} \sum_{n=-\infty}^{\infty} \left(\frac{1}{N} \int_0^N e^{2\pi i x^2/N} e^{-2\pi i n x/N}\, dx \right) e^{2\pi i n k/N}$$

$$= \sum_{n=-\infty}^{\infty} \left(\frac{1}{N} \int_0^N e^{2\pi i x^2/N} e^{-2\pi i n x/N}\, dx \right) \sum_{k=0}^{N-1} e^{2\pi i n k/N}. \qquad (11.8)$$

The orthogonality relations Theorem 10.12 applied to the group $\mathbb{Z}/N\mathbb{Z}$ show that

$$\sum_{k=0}^{N-1} e^{2\pi i n k/N} = \begin{cases} N & \text{if } N \mid n, \\ 0 & \text{if } N \nmid n. \end{cases}$$

Thus Equation (11.8) simplifies to give

$$H = \sum_{n=-\infty}^{\infty} \int_0^N e^{2\pi i x^2/N} e^{-2\pi i n x}\, dx$$

$$= \sum_{n=-\infty}^{\infty} \int_0^N e^{2\pi i (x^2 - N n x)/N}\, dx$$

$$= N \sum_{n=-\infty}^{\infty} e^{-\pi i n^2 N/2} \int_{-n/2}^{1-n/2} e^{2\pi i N v^2}\, dv \qquad (11.9)$$

where $v = \frac{x}{N} - \frac{n}{2}$. Now

---

[3] The function which is being expanded here is defined as follows. Let

$$f(x) = e^{2\pi i x^2/N} \text{ for } 0 \leqslant x < N,$$

and then extend f by requiring that $f(x + N) = f(x)$ for all x. It is clear that $f(0) = \lim_{x \to N} f(x)$, so the resulting function is continuous and piecewise twice continuously differentiable. It follows that the Fourier series for f converges pointwise to f everywhere.

$$e^{-\pi \mathrm{i} n^2 N/2} = \begin{cases} 1 & \text{if } n \text{ is even,} \\ \mathrm{i}^{-N} & \text{if } n \text{ is odd} \end{cases}$$

because odd squares are congruent to 1 modulo 4 so the sum in Equation (11.9) may be split into sums over $n = 2m + 1$ and $n = 2m$, giving

$$H = N \sum_{m=-\infty}^{\infty} \int_{-m}^{1-m} e^{2\pi \mathrm{i} N v^2}\, dv + N\mathrm{i}^{-N} \sum_{m=-\infty}^{\infty} \int_{-m-1/2}^{-m+1/2} e^{2\pi \mathrm{i} N v^2}\, dv.$$

Recombining the integrals shows that

$$H = N(1 + \mathrm{i}^{-N}) \int_{-\infty}^{\infty} e^{2\pi \mathrm{i} N v^2}\, dv$$

$$= \sqrt{N}(1 + \mathrm{i}^{-N}) \int_{-\infty}^{\infty} e^{2\pi \mathrm{i} w^2}\, dw \qquad (11.10)$$

where $w = v\sqrt{N}$. To compute the integral, notice that Equation (11.10) holds for all N, in particular for $N = 1$. When $N = 1$, $H = 1$ by Equation (11.7) and it follows that

$$\int_{-\infty}^{\infty} e^{2\pi \mathrm{i} w^2}\, dw = \frac{1}{1 + \mathrm{i}^{-1}}.$$

Substituting this value gives

$$H = \frac{1 + \mathrm{i}^{-N}}{1 + \mathrm{i}^{-1}} \sqrt{N},$$

and checking the possible congruence classes modulo 4 completes the proof of Theorem 11.10. $\square$

Exercise 11.12. Evaluate the sum G in Equation (11.3).

11.5 The Conjectures of Birch and Swinnerton-Dyer

The group law on an elliptic curve introduced in Chapter 5 involves rational functions only so, as pointed out in Section 5.3, the theory of elliptic curves makes sense over a finite field $\mathbb{F}_q$ of characteristic p as long as we avoid division by p. Exercises 5.20 and 5.21 on p. 109 began our study of elliptic curves over finite fields.

11.5.1 The Hasse Theorem

Consider the elliptic curve E defined by the affine equation

$$E : y^2 = x^3 + ax + b, \qquad (11.11)$$

with a and b integral. Recall that the non-degeneracy condition on the curve E is defined in terms of

$$\Delta = 4a^3 + 27b^2.$$

Fix a prime p. If $p \nmid \Delta$, then the curve defined over $\mathbb{F}_p$ obtained by reducing Equation (11.11) modulo p is an elliptic curve; let $N_p = N_p(E)$ denote the number of points on this curve. How large is N_p?

There is a trivial bound: The projective plane $\mathbb{P}^2(\mathbb{F}_p)$ is defined by

$$\mathbb{P}^2(\mathbb{F}_p) = \{(x, y, z) \in \mathbb{F}_p^3 \mid (x, y, z) \neq (0, 0, 0)\}/ \sim,$$

where $(x, y, z) \sim (x', y', z')$ if and only if there is a $\lambda \in \mathbb{F}_p^*$ with

$$(x, y, z) = \lambda(x', y', z').$$

There are $(p^3 - 1)$ choices for the triple (x, y, z), and each equivalence class under $\sim$ has $|\mathbb{F}_p^*| = (p-1)$ elements. It follows that there are $\frac{p^3-1}{p-1} = p^2+p+1$ points in $\mathbb{P}^2(\mathbb{F}_p)$, so certainly

$$|N_p| \leqslant p^2 + p + 1. \tag{11.12}$$

This estimate ignores the fact that the points we are counting lie on the curve defined by Equation (11.11), so it is hardly surprising that much more can be done. The curve in projective coordinates is defined by

$$\{[x, y, z] \in \mathbb{P}^2(\mathbb{F}_p) \mid y^2 z = x^3 + axz^2 + bz^3\}.$$

For each of the $(p^2 - p)$ possible values of (x, z) with $z \neq 0$, we are trying to solve an equation of the form $y^2 = f(x, z)$ for y. This has at most two possible solutions. If $z = 0$, then $x = 0$ also, so $y \neq 0$, and there is (projectively) one choice for y. Thus

$$|N_p| \leqslant 2(p^2 - p)/(p - 1) + (p - 1)/(p - 1) = 2p + 1, \tag{11.13}$$

a dramatic improvement over the inequality (11.12). However, we know that not all numbers are quadratic residues modulo p. Indeed, for $p \neq 2$, exactly half of the elements of $\mathbb{F}_p^*$ are quadratic residues.

Exercise 11.13. Let p denote an odd prime, and assume that $\gcd(a, p) = 1$. Find the exact number of solutions to $y^2 = ax + b$ over $\mathbb{F}_p$.

Exercise 11.14. Let p denote a prime congruent to 2 modulo 3. Show that for an elliptic curve of the form $E : y^2 = x^3 + b$, $N_p = p + 1$.

Exercise 11.15. Let p denote a prime congruent to 3 modulo 4. Show that for an elliptic curve of the form $E : y^2 = x^3 - x$, $N_p = p + 1$.

Our problem is a little more subtle. Working in projective coordinates as before, when $z \neq 0$ there are two choices (or one choice if $p = 2$) for y if

$$(x^3 + axz^2 + bz^2)/z$$

is a quadratic residue modulo p and no possible choices for y if

$$(x^3 + axz^2 + bz^2)/z$$

is a quadratic nonresidue modulo p. If $z = 0$, then $x = 0$, so there is one choice for y. Thus for an odd prime p, we expect

$$|N_p| = 2|\{x \in \mathbb{F}_p \mid x^3 + ax + b \text{ is a quadratic residue} \pmod{p}\}|$$
$$+ |\{x \in \mathbb{F}_p \mid x^3 + ax + b \equiv 0 \pmod{p}\}|$$
$$+ 1.$$

It is not clear if this is an improvement over the inequality (11.13), but it is nonetheless suggestive. For now, we ignore the second term since no more than three values of x in $\mathbb{F}_p$ will have $x^3 + ax + b \equiv 0$ modulo p. If $x^3 + ax + b$ is no more or less likely than x to be a quadratic residue, then we would expect the first term to contribute p to the total. This suggests that we write

$$N_p = (p+1) - a_p, \tag{11.14}$$

where a_p encodes the information about the extent to which the polynomial $x^3 + ax + b$ fails to distribute its values fairly between quadratic residues and nonresidues. If the polynomial behaves reasonably well, then we expect the "error" a_p defined by Equation (11.14) to be small relative to the prime p.

This turns out to be the case. The next theorem was conjectured by Artin in his thesis and proved by Hasse. We will not prove it here; proofs may be found in several of the references at the end of the chapter.

Theorem 11.11. [HASSE'S THEOREM] *Let N_p denote the number of points in $\mathbb{F}_p$ on an elliptic curve defined over $\mathbb{F}_p$. Then*

$$|N_p - (p+1)| \leqslant 2\sqrt{p}. \tag{11.15}$$

Notice that the hypothesis (of being an elliptic curve over $\mathbb{F}_p$) requires that $p \nmid \Delta$. This theorem gives a precise bound for the size of the error term a_p in Equation (11.14).

Birch and Swinnerton-Dyer carried out extensive calculations on the numbers N_p and combined this with deep theoretical insights into the arithmetic of elliptic curves. One of the resulting conjectures is that one of the "global" measures of the complexity of $E(\mathbb{Q})$ – the rank of the curve – should be reflected in the extent to which the "local" quantities N_p exceed $(p+1)$. Numerical experiments led to the following conjecture.

Conjecture 11.12. There is a constant C, depending only on the curve E, with the property that

$$\prod_{p \in \mathbb{P}, p < X} \frac{N_p}{p} \sim C(\log X)^r$$

where r is the rank of the group $E(\mathbb{Q})$.

11.5.2 The L-function Attached to an Elliptic Curve

The error term a_p defined in Equation (11.14) as p varies carries subtle information about the elliptic curve. For $p \mid \Delta$, the reduction of the curve modulo p is not an elliptic curve. This phenomena is called *bad reduction*.

In order to understand some of the possibilities in the more familiar setting of curves over the reals, we recall two examples. The point $(0,0)$ on $y^2 = x^3$ (mentioned on p. 54) is a *cusp* – a point where two tangents coincide. The point $(1,0)$ on $y^2 = x^2(x+1)$ is also singular but for a different reason: there is a pair of distinct tangents (see Exercise 5.2).

Returning to the primes of bad reduction, the associated number a_p is defined according to the type of reduction as follows, using formal derivatives and tangents over finite fields.

- If the curve reduced modulo p has a cusp (a point where two tangents coincide), then set a_p to be 0. This is called *additive reduction*.
- If the curve has a double point as its only singularity, with tangents having rational slopes over $\mathbb{F}_p$, then set a_p to be 1. This is called *split multiplicative reduction*.
- The remaining possibility is that the curve has a double point as its only singularity over $\mathbb{F}_p$, with tangents having slopes defined over a quadratic extension of $\mathbb{F}_p$ but not over $\mathbb{F}_p$. In that case set a_p to be -1. This is called *non-split multiplicative reduction*.

With the notation above, the L-function attached to the curve E is defined as an Euler product to be

$$L_E(s) = \prod_{p \mid \Delta} \left(1 - a_p p^{-s}\right)^{-1} \prod_{p \nmid \Delta} \left(1 - a_p p^{-s} + p^{1-2s}\right)^{-1} \qquad (11.16)$$

Exercise 11.16. Use the inequality (11.15) to show that the Euler product (11.16) converges when $\Re(s) > \frac{3}{2}$.

Exercise 11.17. If L_E is expanded as a Dirichlet series

$$L_E(s) = \sum_{n=1}^{\infty} \frac{c_n}{n^s},$$

prove that for each prime p, $c_p = a_p$.

The Euler product defining L_E does not converge at $s = 1$, but for any X notice that Equation (11.14) shows that

$$\prod_{p < X} \frac{p}{N_p} = \prod_{p < X} \left(1 - a_p/p + 1/p\right)^{-1}.$$

Thus

$$\prod_{\substack{p \mid \Delta, \\ p < X}} \left(1 - a_p p^{-1}\right)^{-1} \prod_{\substack{p \nmid \Delta, \\ p < X}} \left(1 - a_p p^{-1} + p^{1-2}\right)^{-1} = A \times \prod_{p < X} \frac{p}{N_p},$$

where A is a term depending only on the finitely many primes dividing Δ. What this means is that if the L-function can be extended to include the point $s = 1$, we expect the analytic properties at $s = 1$ of the extension to carry information about the numbers N_p. The conjectures of Birch and Swinnerton-Dyer are a very precise formulation of this idea. In order to state some of them, we need several preliminary results and some new definitions. Some details, even of the definitions, are omitted.

The first of the results needed is a conjecture usually attributed to Hasse, Weil, and Deuring, which was proved by work of Wiles and of Taylor and Wiles in a special case. The full result was obtained by a strengthening of Wiles' method due to Breuil, Conrad, Diamond, and Taylor. The *arithmetic conductor* N_E of the elliptic curve E, referred to in Theorem 11.13, is a refined version of Δ. It has the same prime divisors as Δ but reflects more precisely the behavior of the curve under reduction modulo each prime.

Theorem 11.13. *The L-function L_E extends to an entire function on the whole complex plane, and the extended function satisfies a functional equation of the form*

$$\Lambda_E(s) = \pm \Lambda_E(2 - s),$$

where

$$\Lambda_E(s) = \frac{1}{(2\pi)^s} \Gamma(s) N_E^{s/2} L_E(s).$$

There are several aspects to the conjecture of Birch and Swinnerton-Dyer, two of which we present below in ascending order of strength.

Conjecture 11.14. [BIRCH AND SWINNERTON-DYER] The extended L-function has the following properties.

1. $L_E(1) \neq 0$ if and only if the rank of E is zero.
2. If the expansion of L_E at $s = 1$ has the form

$$L_E(s) = b_g(s - 1)^g + \cdots$$

with $g \geqslant 0$ and $b_g \neq 0$, then the rank of E is g.

The third part of the conjecture is the analog of the formula in Equation (11.7); it gives an exact formula for b_g in terms of data associated with the elliptic curve. It is too complicated to state here but may be found in the references. Remarkably, each of the ingredients in Theorem 11.7 has an elliptic analog.

The conjecture of Birch and Swinnerton-Dyer is one of the seven Millennium Prize Problems announced by the Clay Mathematics Institute.

11.5.3 Tunnell's Theorem

Theorem 5.8 hints at a connection between elliptic curves and congruent numbers. It is very far from a characterization of congruent numbers. A deep result due to Tunnell exploits the connection between elliptic curves, their L-functions and congruent numbers at a more profound level, and gives a form of characterization of congruent numbers. Part of his result may be stated as follows.

Theorem 11.15. [TUNNELL'S THEOREM] *Let n be an odd natural number with no square factor. Say that n has property T if the number of integer triples (x, y, z) satisfying the equation $2x^2 + y^2 + 8z^2 = n$ is twice the number of integer triples (x, y, z) satisfying $2x^2 + y^2 + 32z^2 = n$. Then*

$$n \text{ is congruent} \implies n \text{ has property } T.$$

If (part of) the Birch and Swinnerton-Dyer conjecture holds, then

$$n \text{ has property } T \implies n \text{ is congruent}.$$

NOTES TO CHAPTER 11: The material in Sections 11.1 to 11.4 is classical, and complete treatments may be found in many of the references, including Davenport [40], Hasse [76], Lang [96], and Weil [158]. Davenport's book [40] approaches the class number formula via quadratic forms, which is closer to Dirichlet's original formulation. The results on the class number one problem mentioned appear in papers of Baker [7], Heegner [78], and Stark [146]. We followed Davenport [40, Chapter 2] closely in the proof of Theorem 11.10. The equivalence between ideal counting (Exercise 11.10) and Theorem 11.7 is explained, in the general case, in Lang [96, Chapters VI and XV]. Hasse's Theorem is proved in many places including Chahal [28] and Silverman [139]; Silverman and Tate [143] give a particularly accessible treatment. For more on Conjecture 11.12 and its connection to the conjecture of Birch and Swinnerton-Dyer, see the paper of Goldfeld [69]. The work of Wiles and others on Theorem 11.13 is discussed in many places, including the notes of Darmon [39], which contains references to the original papers; the analytic continuation as we have stated it is proved in the paper of Breuil, Conrad, Diamond and Taylor [23]. The conjectures of Birch and Swinnerton-Dyer, and the proofs of special cases, are discussed in (relatively) accessible form in a paper by Wiles [161] on the Clay Institute Web site and in a paper of Zagier [166]. They were originally formulated in the papers [16] of Birch and Swinnerton-Dyer. Tunnell's Theorem (Theorem 11.15) is shown in his paper [155]; an accessible treatment is in Koblitz [89].

Computational Number Theory

Expressions such as "too slow" or "computationally infeasible" have been used to describe methods for finding large primes and verifying their primality. For example, we dismissed the sieve of Eratosthenes on these grounds. In this chapter, we are going to discuss objective tests on algorithms to try to quantify statements about how fast or slowly they may run. This comprises a brief introduction to a field of growing importance. Three important themes taken up in the books cited at the end of the chapter are the following. How can advances in the speed of electronic computing devices be exploited by number theorists? How can number theory be applied to improve the speed of calculations? How can number theory contribute to the search for secure methods of communication?

12.1 Complexity of Arithmetic Computations

To begin, let us analyze the basic arithmetic operations. A computer may seem to take no time at all to carry out a simple calculation, but in fact computer calculations run in a time that is approximately proportional to the number of bit operations required. Computers work in binary so instead of expressing integers in decimal notation, we use their binary equivalents: Just as the string of decimal digits $a_k a_{k-1} \cdots a_0$ is used to denote the integer

$$a_0 + 10a_1 + \cdots + 10^k a_k,$$

a string of binary (zero–one) digits $b_k b_{k-1} \cdots b_0$ denotes the integer

$$b_0 + 2b_1 + \cdots + 2^k b_k.$$

The first few binary numbers are shown in Table 12.1.

Arithmetic in binary is carried out as usual. For example, the long multiplication $120 \cdot 30 = 3600$ becomes

Table 12.1. Binary numbers.

Decimal number	Binary equivalent
1	1
2	10
3	11
4	100
5	101
6	110
7	111

$$1111000 \times 11110 = 111000010000$$

by the calculation shown in Figure 12.1.

Notice that this only involves the repetition of a simple basic operation: Add a binary digit 0 or 1 to a binary digit 0 or 1 and output a 0 or a 1 with a carry of a 0 or a 1 to the next column. This bit operation (bit is an abbreviation of binary digit),

$$0 + 0 \to 0, \quad 1 + 0 \to 1, \quad 0 + 1 \to 1, \quad 1 + 1 \to 0 \text{ carry } 1,$$

is the building block from which many arithmetic operations can be built. It will be useful to allow a carried digit to be included; for example, 1 plus 1 with a carried digit 1 outputs 1 and a carried 1 in one operation.

```
        1 1 1 1 0 0 0
          1 1 1 1 0 ×
      _____

      1 1 1 1 0 0 0
        1 1 1 1 0 0 0      +
          1 1 1 1 0 0 0    +
            1 1 1 1 0 0 0  +
      _____

  1 1 1 0 0 0 0 1 0 0 0 0
```

Figure 12.1. Binary multiplication.

Definition 12.1. *A bit operation is the process of adding or subtracting two binary digits, taking account of any carried or borrowed digits, and outputting an answer and a carry or a borrow.*

Now consider the problem of adding two numbers m and n presented in decimal form, outputting the sum in binary. There are two steps to carry out.

1. Convert m and n to binary. Without loss of generality, assume that m has k bits and n has ℓ bits with $k \geqslant \ell$.

2. Add the two numbers. Notice that no more than $(k+1)$ bit operations will be required.

Definition 12.2. *The complexity* C *of an arithmetic calculation is the number of bit operations required to carry it out.*

The complexity of a calculation is an upper bound on the time it will take to run on a computer.

Thus

$$\text{C(add a } k\text{-bit number to an } \ell\text{-bit number)} \leqslant k+1 \text{ if } k \geqslant \ell.$$

Example 12.3. The calculation $23 + 7 = 30$ in binary takes five bit operations:

$$
\begin{array}{r}
1\,0\,1\,1\,1 \\
1\,1\,1\,+ \\
\hline
1\,1\,1\,1\,0
\end{array}
$$

Recall the following notation for functions $f, g : \mathbb{N} \to \mathbb{R}$. We say that

$$f = O(g) \text{ or } f(x) = O(g(x))$$

if there is a constant $C \geqslant 0$ with

$$f(x) \leqslant Cg(x) \text{ for all } x \in \mathbb{N}.$$

This is particularly convenient for complexity calculations because it is only an upper bound: $C = O(g)$ is a true statement as soon as we find an algorithm that runs in g bit operations, even if we have missed an ingenious algorithm that runs much faster. Finding *lower* bounds for the complexity of a calculation is a more subtle problem that we do not touch on here.

Lemma 12.4. *For $k \geqslant \ell$,*

$$\text{C(add a } k\text{-bit number to an } \ell\text{-bit number)} = O(k).$$

Subtracting one number from another has the same complexity.

Lemma 12.5. *For $k \geqslant \ell$,*

$$\text{C(subtract an } \ell\text{-bit number from a } k\text{-bit number)} = O(k).$$

Lemma 12.6. *For $k \geqslant \ell$,*

$$\text{C(multiply a } k\text{-bit number by an } \ell\text{-bit number)} = O(k^2).$$

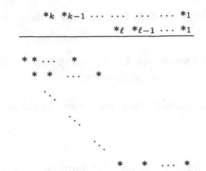

Figure 12.2. Multiplying a k-bit number by an ℓ-bit number.

PROOF. Notice that all we need is an upper bound for the number of bit operations required. The long multiplication can be done in the shape shown in Figure 12.2, with arbitrary bits denoted $*$. We have the following upper bounds. There are no more than $(k+2)$ operations needed to add the lowest two rows, resulting in an integer with no more than $(k+2)$ bits. Adding this to the third lowest row requires no more than $(k+3)$ bit operations, resulting in an integer with no more than $(k+3)$ bits, and so on. The total number of bit operations does not exceed

$$(k+2) + (k+3) + \cdots + (\ell + k + 1) \leqslant k(\ell + k + 1)$$
$$\leqslant k(2k+1) = O(k^2).$$

$\square$

Dividing one integer by another results in a rational number. In arithmetic, we are primarily interested in integer operations, so division means finding the quotient and remainder.

Lemma 12.7. *For $k \geqslant \ell$,*

$$C(\text{divide a } k\text{-bit number by an } \ell\text{-bit number}) = O(k^2).$$

PROOF. This calculation amounts to repeatedly subtracting an ℓ-bit number from a k-bit number and testing to see if the answer is less than the ℓ-bit number. By Lemma 12.5, this takes no more than $k\,O(k) = O(k^2)$ bit operations.

$\square$

12.1.1 Improving Complexity Estimates

We have been considering only the most obvious methods for doing arithmetic. It is possible to speed up these basic algorithms considerably. For example, Lemma 12.6 says that

$$C(\text{multiply a } k\text{-bit number by a } k\text{-bit number}) = O(k^2).$$

Using a different algorithm to do the multiplying allows complexity

$$O(k \cdot \log k \cdot \log \log k \cdot \log \log \log k).$$

We will not prove any sophisticated results of this kind, but some indication of how even simple algorithms can be improved is given by the following result.

Theorem 12.8. $C(\text{multiply a } k\text{-bit number by a } k\text{-bit number}) = O(k^{1.59})$.

PROOF. This relies on chopping up the integers m and n to be multiplied in a clever way (as all these methods do). Assume for simplicity that k is even. If m has k bits, then we can write

$$m = a \cdot 2^{k/2} + b,$$

where a and b have at most $k/2$ bits. Similarly,

$$n = c \cdot 2^{k/2} + d,$$

where c and d have at most $k/2$ bits.
Then

$$mn = ac2^k + (bc + ad)2^{k/2} + bd$$
$$= ac2^k + [ac + bd - (a - b)(c - d)]2^{k/2} + bd.$$

We have computed ac, bd, $(a - b)(c - d)$, all of which are approximately half the bit length of m and n. Notice that multiplying by 2^k really constitutes a shift of binary data and so can be ignored in complexity terms.

Write $C(k)$ for the complexity of multiplying two k-bit numbers this way. Then

$$C(k) \leqslant 3C(k/2) + O(k).$$

This means that there is a constant C_1 with

$$C(k) \leqslant 3C(k/2) + C_1 k.$$

We want to iterate this to find an estimate for $C(k)$; the danger is that we cannot add $O(\cdot)$ terms carelessly. If we abuse notation slightly to also write $C(k)$ for an upper bound of the functions satisfying the relation for real values of k as well as integral values, then

$$C(k) \leqslant 3C(k/2) + C_1 k$$
$$\leqslant 3\left(3C(k/2^2) + C_1(k/2)\right) + C_1 k$$
$$\leqslant 3^2 C(k/2^2) + (1 + 3/2)C_1 k$$
$$\vdots$$
$$\leqslant 3^r C(k/2^r) + (1 + 3/2 + \cdots + (3/2)^{r-1}) \text{ for all } r \geqslant 1,$$

so

$$C(k) \leqslant 3^r C(k/2^r) + C_2 k(3/2)^r \text{ for all } r \geqslant 1.$$

Choose r so large that $k/2^r \leqslant 1$; that is, $r = \log k / \log 2$. It follows that

$$C(k) = O\left(3^{\log k / \log 2}\right)$$
$$= O\left(k^{\log 3 / \log 2}\right)$$
$$= O(k^{1.59}).$$

□

12.1.2 Polynomial Complexity

All the complexity calculations so far have involved the number of bits in the numbers. Sometimes it is useful to relate this to the size of the numbers in the usual sense. If n has bit length k, then

$$2^{k-1} \leqslant n \leqslant 1 + 2 + \cdots + 2^{k-1} = 2^k - 1 \leqslant 2^k,$$

so

$$(k - 1) \log 2 \leqslant \log n \leqslant k \log 2.$$

This means that $\log n$ is interchangeable with k in all O-estimates. For example,

$$C(\text{multiply } m \text{ by } n) = O(\log m \cdot \log n).$$

The complexity estimates above have been logarithmic in the input variables. In general, a computation involving numbers $n_1, \ldots, n_r$ is said to have *polynomial complexity* (or to be a *polynomial time* calculation) if its complexity satisfies

$$C = O\left((\log n_1)^{d_1} \cdots (\log n_r)^{d_r}\right),$$

where the d_i are integers. As a rule of thumb, polynomial complexity estimates are considered computationally feasible for large values of the variables. Of course, in practice, the constant hidden in the O notation will influence the running time. For practical implementation, it is desirable to make this as small as possible. We do not touch on such issues here.

There are several quite basic arithmetic problems that are believed not to be solvable in polynomial time, and this complexity is used in cryptography.

Example 12.9. [THE KNAPSACK PROBLEM] Given natural numbers $n_1, \ldots, n_d$ and N, decide if there is a subset I of the index set $\{1, 2, \ldots, d\}$ with the property that

$$\sum_{i \in I} n_i = N.$$

The most naïve approach is to simply try all the subsets, and there are 2^d of them. No algorithm is known to decide this problem in polynomial time.

12.2 Public-key Cryptography

Part of the great interest in computational number theory comes about because the arithmetic properties of large integers lie at the heart of all modern methods of secure electronic communication. This is a large and active field of research, and we will simply describe one very simple example, the RSA cryptosystem. This is a method for *encrypting* a message, in such a way that only the intended recipient can feasibly *decrypt* it. It also allows for *digital signatures* that authenticate that you are who you claim to be. The scheme is named after Ronald Rivest, Adi Shamir, and Leonard Adleman who developed it in 1977. What we will describe below is a method for parties to send each other large numbers (that may be used to encode any form of message) in a secure fashion. Real implementations of RSA are a great deal more complicated than the description here, but the basic idea of connecting the difficulty of an arithmetic operation to secure communication should become clear.

12.2.1 The RSA algorithm

Fix two large primes p and q, and compute their product $n = pq$. The number n is called the *modulus* of the resulting scheme. Choose a number e, the *public exponent*, with

$$1 \leqslant e < n;$$
$$\gcd(e, (p-1)(q-1)) = 1.$$

Next compute d, the *private exponent*, with the property that

$$ed \equiv 1 \pmod{(p-1)(q-1)}.$$

There is such a d since the requirement that $\gcd(e, (p-1)(q-1)) = 1$ means that $e \in (\mathbb{Z}/pq\mathbb{Z})^*$ (see Corollary 1.25 on p. 37). Now publish openly the *public key*, which is the pair of numbers (n, e). The *private key* (n, d) will be used to decode messages. The original primes p and q are no longer needed – but must be kept secret.

The security of the system is based on the following: It is very difficult to compute d given e and n, unless you know p and q. Finding the inverse of e modulo $(p-1)(q-1)$ is easy if you know the value of $\phi(pq) = (p-1)(q-1)$, but here we know the value of pq and can only work out $(p-1)(q-1)$ by finding p and q. A fast method to factorize large numbers would render the method insecure. We will see later that the calculations involved in encryption and decryption can be done very quickly (see Exercise 12.5).

Encrypting a message

Suppose that Albertina wants to send an integer $m < n$ to Bill. Albertina converts m into an encrypted number c (usually called the *cyphertext*) by computing

$$c = m^e \pmod{n},$$

where the pair (n, e) is Bill's published public key. She then sends the number c to Bill over an open channel.

When Bill receives the encrypted message c, he uses his private key (n, d) to compute

$$m = c^d \pmod{n}. \tag{12.1}$$

Exercise 12.1. Show that Equation (12.1) correctly recovers the original message m.

Exercise 12.2. In order to use the RSA algorithm, a ready supply of primes of specified size is needed. Use the Prime Number Theorem (Theorem 8.1) to estimate the number of primes with N decimal digits for a large N.

Anyone other than Bill will only have access to the public key (n, e). In order to recover the message m, they would have to somehow compute d, which involves factorizing n.

Digital Signatures

Suppose now that Albertina wants to send a message m to Bill in such a way that Bill is assured that the message has not been substituted by another message, and that the message comes from Albertina and not from someone else. Both of these problems are real because the public key allows anyone to generate messages that appear authentic. In order to do this, both Albertina and Bill create public and private keys as above.

Albertina creates a *digital signature* s by computing

$$s_A = m^{d_A} \pmod{n_A},$$

where (n_A, d_A) is her private key. She sends m (using the system above) and the signature s_A to Bill. To verify the signature, Bill checks that the message m is recovered by

$$m = s_A^{e_A} \pmod{n_A},$$

where (n_A, e_A) are Albertina's public key.

Exercise 12.3. Explain what happens if either the encrypted message or the signature is tampered with.

This approach allows both the encryption of the message and the signing to be done without any exchange of private keys. Each party only needs to know the other's public key and their own private key. Only someone with the correct private key can decrypt received messages or sign outgoing messages.

12.3 Primality Testing: Euclidean Algorithm

The two most natural arithmetic problems of computational interest are the following. Firstly, to decide if a given integer is a prime. Secondly, to factorize an integer known to be composite.

Lucas proved that $M_{67} = 147573952589676412927$ is composite without exhibiting any factors. In fact, M_{67} is a difficult number to factorize in the sense that it has no small factors:

$$2^{67} - 1 = 193707721 \cdot 761838257287$$

is the prime factorization of M_{67}. Being able to prove that a number is composite without exhibiting any of its factors seems unlikely, but as we have already seen in Section 1.5, Fermat's Little Theorem sometimes allows this. For example, the fact that $2^{90} \equiv 64$ modulo 91 implies that 91 must be a composite number.

This idea – exploiting known properties of primes as a test for primality – leads to very efficient tests for primality that only rarely give the wrong answer. The first ingredients needed are complexity estimates for modular arithmetic and the Euclidean Algorithm.

Lemma 12.10. $C(\text{multiply } a \text{ modulo } m \text{ by } b \text{ modulo } m) = O\big((\log m)^2\big)$.

PROOF. If m has k bits, then the residues of a and b modulo m are numbers with no more than k bits. Computing $a \cdot b$ takes $O(k^2)$ bit operations by Lemma 12.6, and $a \cdot b$ has no more than $2k$ bits. Finding the residue of $a \cdot b$ modulo m takes $O(k^2)$, giving a total time of $O(k^2) + O(k^2) = O(k^2)$. $\square$

The Euclidean Algorithm finds the greatest common divisor of two integers without factorizing them. Given integers $a > b > 0$, the algorithm proceeds by the following successive divisions:

$$a = bq_1 + r_1, \quad 0 < r_1 < b$$
$$b = r_1 q_2 + r_2, \quad 0 < r_2 < r_1$$
$$\vdots$$
$$r_{n-2} = r_{n-1}q_n + r_n, \quad 0 < r_n < r_{n-1}$$
$$r_{n-1} = r_n q_{n+1}$$

and then

$$\gcd(a, b) = r_n.$$

Theorem 12.11. $C(\text{find } \gcd(a, b)) = O\big((\log a)^3\big)$.

PROOF. Each step involves a simple division of integers that are less than a, so each step has complexity $O\big((\log a)^2\big)$. We need to estimate how many steps are taken before the algorithm terminates. Since the remainders r_i are nonnegative integers, the number of steps can be bounded if we know that the remainders r_i are decreasing rapidly. We claim that

$$r_{i+2} \leqslant \tfrac{1}{2} r_i \text{ for } i = 1, 2, \dots. \tag{12.2}$$

There are two ways to see this. The claim follows by noticing that every two steps, the bit length of the remainders must go down by at least 1. This becomes clear with an explicit approach to long division; if after one step the bit length has not gone down, then the next step involves dividing two binary integers of the same bit length, and hence the quotient is 1 and the remainder must be shortened.

To prove this in a more rigorous manner, consider the following possibilities for a fixed i. If

$$r_{i+1} \leqslant \tfrac{1}{2} r_i, \tag{12.3}$$

then $r_{i+2} < r_{i+1} \leqslant \tfrac{1}{2} r_i$, as required. If the inequality (12.3) does not hold, then

$$r_{i+1} > \tfrac{1}{2} r_i. \tag{12.4}$$

Now

$$r_i = r_{i+1} q_{i+2} + r_{i+2},$$

so if $q_{i+2} \geqslant 2$ then $r_i \geqslant 2r_{i+1} + r_{i+2} \geqslant 2r_{i+1}$, which contradicts the inequality (12.4). It follows that $q_{i+2} = 1$ (as we predicted), so

$$r_{i+2} = r_i - r_{i+1} < r_i - \tfrac{1}{2} r_i = \tfrac{1}{2} r_i,$$

proving the inequality (12.2).

The inequality (12.2) gives the following rapid decay in the size of the remainders:

$$r_3 < \tfrac{1}{2} r_1,$$
$$r_5 < \tfrac{1}{2} r_3 < \tfrac{1}{4} r_1,$$
$$\vdots$$
$$r_{2n+1} < \tfrac{1}{2^n} r_1.$$

It follows that

$$n > \frac{\log a}{\log 2} \implies \frac{1}{2^n} r_1 < \frac{1}{2^n} a < 1 \implies r_{2n+1} = 0,$$

so the number of steps is $n = \mathrm{O}(\log a)$.

Thus, the total time taken to complete the algorithm is $\mathrm{O}\big((\log a)^3\big)$. □

Corollary 12.12. *Let $a > b > 0$ be integers. Then integers m, n can be found with $|m| \leqslant b$, $|n| \leqslant a$, and*

$$am + bn = \gcd(a, b)$$

in complexity $\mathrm{O}\big((\log a)^3\big)$.

Corollary 12.13. *Let $m \geqslant 2$ be an integer and let a, $1 \leqslant a \leqslant m$, be an integer with $\gcd(m, a) = 1$. Then the inverse of a modulo m can be found with complexity $\mathrm{O}\left((\log m)^3\right)$.*

Example 12.14. Let $m = 31$ and $a = 12$. Since $\gcd(31, 12) = 1$, there is an x with $12x \equiv 1$ modulo 31. Apply the Euclidean Algorithm:

$$31 = 12 \cdot 2 + 7,$$
$$12 = 7 \cdot 1 + 5,$$
$$7 = 5 \cdot 1 + 2,$$
$$5 = 2 \cdot 2 + 1.$$

It follows that

$$1 = 5 - 2 \cdot 2$$
$$= 5 - 2(7 - 5) = 3 \cdot 5 - 2 \cdot 7$$
$$= 3(12 - 7) - 2 \cdot 7 = 3 \cdot 12 - 5 \cdot 7$$
$$= 3 \cdot 12 - 5(31 - 2 \cdot 12) = 13 \cdot 12 - 5 \cdot 31,$$

so $12^{-1} \equiv 13$ modulo 31.

We will exploit the speed of the Euclidean Algorithm in what follows. The first goal is to extend our use of Fermat's Little Theorem to construct a primality test. This involves exponentiation modulo m, so the first step is to find a way to compute b^n modulo m for large values of m, n, and b. Certainly, in a calculation such as $2^{90} \equiv 64$ modulo 91 the huge intermediate number 2^{90} should not be computed – all the intermediate steps should be done modulo m. On the face of it, this still leaves n possible multiplications and reductions modulo m of numbers as large as m. It turns out that there is an additional simplification that reduces the order of complexity dramatically: Carry out the multiplications using the *Repeated Squaring algorithm*.

Theorem 12.15. *Given $m \geqslant 2$ and $b, n \in \mathbb{N}$, $0 \leqslant b < m$,*

$$\mathsf{C}(\text{compute } b^n \text{ modulo } m) = \mathrm{O}\left((\log m)^2 \log n\right).$$

PROOF. Assume that n is presented as a binary number

$$n = n_0 + 2n_1 + \cdots + 2^k n_k$$

with bits $n_i \in \{0, 1\}$. Now carry out the following calculations (all modulo m).

Step 1: compute $b_0 = b^{n_0}$;

Step 2: compute b^2;

Step 3: compute $b_1 = b_0 (b^2)^{n_1}$;

Step 4: compute $b^4 = (b^2)^2$;

Step 5: compute $b_2 = b_1 (b^{2^2})^{n_2}$;

Step 6: compute $b^8 = (b^{2^2})^2$;

Step 7: compute $b_3 = b_2 (b^8)^{n_3}$;

$$\vdots$$

Step $2k + 1$: compute $b_n = b^n$.

The total number of steps is $2k + 1 = O(k)$. The complexity of each step is $O\big((\log m)^2\big)$ because it involves multiplying two integers modulo m. It follows that the total complexity is

$$O(k) \cdot O\big((\log m)^2\big) = O\big(k(\log m)^2\big) = O\big((\log m)^2 \log n\big).$$

$\square$

We also need a general result about solutions to simultaneous linear congruences, the Chinese Remainder Theorem (see p. 61). We know how to solve linear congruencies of the form

$$ax \equiv b \pmod{m}$$

if $\gcd(a, m) = 1$ using the Euclidean Algorithm. It is often useful to be able to solve simultaneous congruences.

Theorem 12.16. *Suppose that $m_1, \ldots, m_r$ are integers with $\gcd(m_i, m_j) = 1$ for all $i \neq j$. Then the simultaneous congruences*

$$x \equiv a_1 \pmod{m_1}$$
$$x \equiv a_2 \pmod{m_2}$$
$$\vdots$$
$$x \equiv a_r \pmod{m_r}$$

have a solution, and this solution is unique modulo $M = m_1 \cdots m_r$.

The uniqueness means that if x and y are integers solving all the congruences, then $x \equiv y$ modulo M.

PROOF. If x and y both satisfy the congruences, then

$$x - y \equiv 0 \pmod{m_i} \text{ for } i = 1, \ldots, r.$$

Since the m_i are all coprime, this means $(x - y)$ is divisible by M as required. We show that there is a solution by constructing one. Let

$$M_i = \frac{M}{m_i} = m_1 \cdots m_{i-1} m_{i+1} \cdots m_r \text{ for } i = 1, \ldots, r.$$

Then $\gcd(M_i, m_i) = 1$ for all i, so by the Euclidean Algorithm there are integers N_i, $0 \leqslant N_i \leqslant m_i$, with

$$M_i N_i \equiv 1 \pmod{m_i} \text{ for } i = 1, \ldots, r.$$

Let

$$x = \sum_{i=1}^{r} a_i M_i N_i.$$

Then x satisfies all the congruences. □

Example 12.17. Consider the simultaneous congruences

$$x \equiv 2 \pmod 3,$$
$$x \equiv 3 \pmod 4,$$
$$x \equiv 4 \pmod 5.$$

The Chinese Remainder Theorem predicts a solution that is unique modulo 60. Working through the proof gives

$$M_1 = 20 \text{ so } 20 N_1 \equiv 1 \pmod 3 \text{ and } N_1 = 2,$$
$$M_2 = 15 \text{ so } 15 N_2 \equiv 1 \pmod 4 \text{ and } N_2 = 3,$$
$$M_3 = 12 \text{ so } 12 N_3 \equiv 1 \pmod 5 \text{ and } N_3 = 3.$$

Hence a solution is

$$x = a_1 M_1 N_1 + a_2 M_2 N_2 + a_3 M_3 N_3$$
$$= 2.20.2 + 3.15.3 + 4.12.3$$
$$= 80 + 135 + 144$$
$$= 359 \equiv 59 \pmod{60}$$

Exercise 12.4. Find an estimate for C(compute $(n - 1)!$ modulo n) (see the discussion in Section 1.5 concerning the practicality of using al-Haytham's Theorem for primality testing).

Exercise 12.5. Estimate the complexity of the steps involved in the RSA algorithm from Section 12.2

12.4 Primality Testing: Pseudoprimes

Given an integer n, it is easy enough to construct a flawless test that will certify the primality of n. If n is not prime, then it must have a prime factor no larger than $\sqrt{n}$, so a test for the primality of n is to see if n is divisible by any prime smaller than $\sqrt{n}$. This is completely impractical as soon as n is large. To see why, consider what is involved in applying the method to a relatively small number close to 10^{16}.

1. We need to know or find all the primes up to about 10^8.
2. Even if we have a list of all those primes, there are about $\frac{N}{\log N}$ primes up to N, so we would have to trial divide by about $5 \cdot 10^6$ numbers.

This approach will never give complexity that is polynomial in n.

   In general, any approach that requires knowledge of all the primes up to a bound that is relatively large in relation to n in order to test the primality of n is doomed to be impractical for large values of n. This includes Eratosthenes for example, although this is useful for small values of n.

   Let us go back to Fermat's Little Theorem to construct a primality testing algorithm free of this basic weakness. Suppose n is a large positive integer (with 80 digits, say). Consider the following algorithm.

(1) Choose an integer b, $1 < b < n$.
(2) Compute $\gcd(b, n)$ using the Euclidean Algorithm.
(3) If $\gcd(b, n) > 1$, then we have found a nontrivial factor of n so n is not prime.
(4) If $\gcd(b, n) = 1$, then compute b^{n-1} modulo n using the Repeated Squaring algorithm.
(5) If $b^{n-1} \not\equiv 1$ modulo n, then n is not prime because it does not satisfy Fermat's Little Theorem.

Example 12.18. Let $n = 91$ and choose $b = 6$. We check that $\gcd(6, 91) = 1$, and find that $6^{90} \equiv 63$ modulo 91, so 91 is not prime.

   Suppose this algorithm is run several times, and for each of the chosen values of b with $\gcd(b, n) = 1$ we find that

$$b^{n-1} \equiv 1 \pmod{n}. \tag{12.5}$$

How convinced should we be that n is prime?

Definition 12.19. *A composite integer n with $b^{n-1} \equiv 1$ modulo n for some b with $\gcd(b, n) = 1$ is called a* pseudoprime *to base b.*

Example 12.20. The fact that 91 is not a prime was readily detected by this method. To see how recalcitrant some numbers can be, consider $n = 561$. The first few numbers b with $\gcd(b, 561) = 1$ are $2, 4, 5, 7, 8, 10$, and we can easily find that

$$2^{560} \equiv 1 \pmod{561},$$
$$4^{560} \equiv 1 \pmod{561},$$
$$5^{560} \equiv 1 \pmod{561},$$
$$7^{560} \equiv 1 \pmod{561},$$
$$8^{560} \equiv 1 \pmod{561},$$

and

$$10^{560} \equiv 1 \pmod{561}.$$

Despite this prime-like behavior, 561 is composite: $561 = 3 \cdot 11 \cdot 17$. We have already seen that this number is a stubborn mimic of primality: It satisfies Equation (12.5) for *every* base b with $\gcd(b, 561) = 1$ (see Exercise 1.25 on p. 34).

In order to start to understand how prevalent pseudoprimality is, we need to look at the algebraic properties of pseudoprimes. It will be useful to write

$$\bar{x} \in \{0, 1, \ldots, n-1\}$$

for the residue modulo n of an integer x.

Theorem 12.21. *Suppose that n is odd.*

(1) *n is a pseudoprime to base b if and only if the multiplicative order of $\bar{b}$ in $(\mathbb{Z}/n\mathbb{Z})^*$ divides $(n-1)$.*

(2) *If n is a pseudoprime to bases b_1 and b_2, then n is a pseudoprime to the bases $b_1 b_2$, b_1^{-1}, and b_2^{-1} modulo n.*

(3) *If $b^{n-1} \not\equiv 1$ modulo n for some base b with $\gcd(n, b) = 1$, then $c^{n-1} \not\equiv 1$ modulo n for at least half of all possible bases c.*

PROOF. (1) The congruence $b^{n-1} \equiv 1$ modulo n means that $\bar{b}^{n-1}$ is the identity in the group $(\mathbb{Z}/n\mathbb{Z})^*$, which holds if and only if the order of $\bar{b}$ divides $(n-1)$.

(2) This holds because $(b_1 b_2)^{n-1} \equiv b_1^{n-1} . b_2^{n-1}$ modulo n (and similarly for the inverses).

(3) Let $B = \{b_1, \ldots, b_s\}$ denote the set of bases with respect to which n is a pseudoprime, and let b be a base for which n is not a pseudoprime. By (2) the set $\{bb_1, \ldots, bb_s\}$ consists of bases for which n is not a pseudoprime. Thus B contains no more elements than its complement. Since there are $\phi(n)$ possible bases, we must have

$$s < \frac{1}{2}\phi(n).$$

$\square$

It is tempting to argue as follows: If we find a base b for which n is a pseudoprime (that is, $b^{n-1} \equiv 1$ modulo n), then the probability that n is prime is at least $\frac{1}{2}$.

Unfortunately, this does not make sense unless we know that a composite number n will always have a base b for which $b^{n-1} \not\equiv 1$ modulo n. For numbers n that do have such witnesses to their non-primality, it does make sense to say that if n is a pseudoprime with respect to k different bases chosen randomly,[1] then the probability that n is prime exceeds $1 - 2^{-k}$.

For relatively modest values of k, this probability is so close to 1 that the probability that we have passed a composite number as prime is comparable with the probability of a numerical error in the computer itself.

However, if our candidate number n is composite but has the property that $b^{n-1} \equiv 1$ modulo n for all b with $\gcd(n, b) = 1$, then this test will always fail.

12.5 Carmichael Numbers

Definition 12.22. *A composite integer n with the property that $b^{n-1} \equiv 1$ modulo n for all b with $\gcd(n, b) = 1$ is called a* Carmichael number.

Example 12.23. Let $n = 561$. We saw in Example 12.20 that 561 is a pseudoprime with respect to the bases $2, 4, 5, 7, 8$, and 10. Knowing that $561 = 3 \cdot 11 \cdot 17$, we can use the Chinese Remainder Theorem to argue as follows. Let b be any integer with $\gcd(b, 561) = 1$. Then

$$\gcd(b, 3) = 1 \implies b^2 \equiv 1 \pmod 3 \implies b^{560} = (b^2)^{280} \equiv 1 \pmod 3,$$
$$\gcd(b, 11) = 1 \implies b^{10} \equiv 1 \pmod{11} \implies b^{560} = (b^{10})^{56} \equiv 1 \pmod{11},$$
$$\gcd(b, 17) = 1 \implies b^{16} \equiv 1 \pmod{17} \implies b^{560} = (b^{16})^{35} \equiv 1 \pmod{17},$$

so by the Chinese Remainder Theorem

$$b^{560} \equiv 1 \pmod{561}.$$

Thus 561 is a Carmichael number.

We shall see later that a Carmichael number cannot have a square factor. A great deal is known about the structure of Carmichael numbers: They must have at least three prime factors, for example, and 561 is the smallest Carmichael number. A striking result of Alford, Granville and Pomerance from 1993 is that there are infinitely many Carmichael numbers, indeed

$$|\{n \mid n \leqslant X \text{ and } n \text{ is a Carmichael number}\}| > X^{2/7} \qquad (12.6)$$

---

[1] We are assuming here that the bases can be chosen "randomly" and in particular independently of the property of pseudoprimality.

asymptotically as $X \to \infty$ (see the notes at the end of the chapter for the reference).

For a prime p, $\mathbb{Z}/p\mathbb{Z}$ is a finite field, so its multiplicative group $(\mathbb{Z}/p\mathbb{Z})^*$ is cyclic. Despite the fact that the ring $\mathbb{Z}/p^2\mathbb{Z}$ is not a field (it has zero-divisors and should not be confused with the Galois field $\mathbb{F}_{p^2}$), its multiplicative group is also cyclic.

Lemma 12.24. *If p is a prime, then $G = (\mathbb{Z}/p^2\mathbb{Z})^*$ is a cyclic group.*

PROOF. For $p = 2$, the invertible elements of $\mathbb{Z}/4\mathbb{Z}$ are the residues 1 and 3, so in this case G is a cyclic group with two elements. We may thus assume that p is an odd prime.

We first claim that if $\gcd(a,p) = 1$, then $1 + ap$ has order p in G. To see this, notice that

$$(1+ap)^p = 1 + ap^2 + \binom{p}{2}(ap)^2 + \cdots + \binom{p}{p-1}(ap)^{p-1} + (ap)^p$$

$$\equiv 1 \pmod{p^2}.$$

Since p is prime, this means that the order of $1 + ap$ is either 1 (which is impossible, as $1 + ap \not\equiv 1$ modulo p^2) or p.

We next claim that there is an element in G of order $(p-1)$. Since $(\mathbb{Z}/p\mathbb{Z})^*$ is cyclic, we can find an integer g, $1 < g < p$, with

$$g^n \equiv 1 \pmod{p} \text{ for } 1 < n \leqslant p - 1$$

only if $n = p - 1$. If $g^{p-1} \equiv 1$ modulo p^2, then the order of g modulo p^2 will still be $(p - 1)$. If not, then $g^{p-1} = 1 + bp$ modulo p^2 with $\gcd(b,p) = 1$ and we claim that $g_1 = g(1 + bp)$ has order $(p - 1)$ modulo p^2. The order of g_1 cannot be less than $(p - 1)$ since $g_1 \equiv g$ modulo p, so it is enough to check that

$$g_1^{p-1} = (g(1+bp))^{p-1} = g^{p-1}(1+bp)^{p-1}$$

$$\equiv (1+bp)(1+b(p-1)p) \pmod{p^2}$$

$$\equiv (1+bp)(1-bp) \pmod{p^2}$$

$$\equiv 1 \pmod{p^2}.$$

Thus G has $p(p - 1)$ elements and contains an element of order p and an element of order $(p - 1)$; since $\gcd(p, p - 1) = 1$, the product of these two elements has order $p(p - 1)$, so G is cyclic. $\qquad\square$

In general, $(\mathbb{Z}/m\mathbb{Z})^*$ is cyclic if $m = p^r$ or $2p^r$ for an odd prime p. When $(\mathbb{Z}/m\mathbb{Z})^*$ is cyclic, any generator for it is called a *primitive root* modulo m.

Theorem 12.25. *If n is a Carmichael number, then n is square-free.*

PROOF. Let n be a Carmichael number with $p^2 | n$ for some prime p. Let n' denote the p-primary part of n – that is, n' is np^{-r} if n is divisible by p exactly r times. Let g, $1 < g < p^2$, be a generator of $(\mathbb{Z}/p^2\mathbb{Z})^*$. By the Chinese Remainder Theorem, the congruences

$$b \equiv g \pmod{p^2},$$
$$b \equiv 1 \pmod{n'}$$

have a solution. Since

$$\gcd(b, p) = \gcd(b, n') = 1,$$

we must have $\gcd(b, n) = 1$. Now n is a Carmichael number, so $b^{n-1} \equiv 1$ modulo p^2 which implies that $g^{n-1} \equiv 1$ modulo p^2 since $b \equiv g$ modulo p^2. The order of g modulo p^2 is $p(p-1)$, so $p(p-1) | (n-1)$. This is impossible because $n - 1 \equiv -1$ modulo p, which completes the proof.   □

12.6 Probabilistic Primality Testing

Recall the Jacobi symbol defined in Definition 11.1. If p is an odd prime, then Euler's Criterion says that

$$\left(\frac{a}{p}\right) \equiv a^{(p-1)/2} \pmod{p}. \tag{12.7}$$

Suppose now that n is an odd integer that we wish to test for primality. Choose an integer a at random between 1 and n and compute $\gcd(a, n) = d_a$.

If $1 < d_a < n$, then n has a proper divisor and so is not prime.

If $\gcd(a, n) = 1$, compute $a^{(n-1)/2}$ modulo n using the Repeated Squaring algorithm.

Next compute the Jacobi symbol $\left(\frac{a}{n}\right)$; since $\gcd(a, n) = 1$, this symbol is either $+1$ or -1.

We have computed two numbers, $a^{(n-1)/2}$ modulo n and $\left(\frac{a}{n}\right)$. If they *differ* then n is *not* prime. (If n is prime, they cannot differ by Euler's criterion Equation (12.7).)

As with Fermat's Little Theorem, the important question is what it means if the number n passes this test in the sense that the two numbers agree? It turns out that there is no analog of the problem caused by Carmichael numbers.

Theorem 12.26. *If $n > 1$ is an odd composite number, then there is an integer b, $1 < b < n$, with $\gcd(b, n) = 1$ and*

$$\left(\frac{b}{n}\right) \neq b^{(n-1)/2} \pmod{n}.$$

PROOF. Assume first that there is a prime p with $p^2 \mid n$, and let $b = 1 + \frac{n}{p}$. By the multiplicative property of the Jacobi symbol,

$$\left(\frac{b}{n}\right) = \left(\frac{b}{p}\right)\left(\frac{b}{n/p}\right).$$

Now $\left(\frac{b}{p}\right) = 1$ because $b \equiv 1$ modulo p. On the other hand, $b \equiv 1$ modulo n/p, so $\left(\frac{b}{n/p}\right) = 1$ and therefore $\left(\frac{b}{n}\right) = 1$.

By the Binomial Theorem, for $j \geq 2$,

$$b^j = \left(1 + \frac{n}{p}\right)^j = 1 + \frac{n}{p}j + \left(\frac{n}{p}\right)^2 \binom{j}{2} + \cdots + \left(\frac{n}{p}\right)^j$$

$$\equiv 1 + \frac{n}{p}j \pmod{n}$$

since $p^2 \mid n$ implies that $(\frac{n}{p})^2, (\frac{n}{p})^3, \ldots, (\frac{n}{p})^j$ are all congruent to 0 modulo n. Taking $j = \frac{n-1}{2}$ gives

$$b^{(n-1)/2} \equiv 1 + \frac{n}{p}\left(\frac{n-1}{2}\right) \not\equiv 1 \pmod{n}$$

since $\frac{n}{p}\left(\frac{n-1}{2}\right) \not\equiv 0$ modulo n, so we have found an integer b with $\gcd(b, n) = 1$ and

$$\left(\frac{b}{n}\right) \not\equiv b^{(n-1)/2} \pmod{n}.$$

If there is no prime p with $p^2 \mid n$, then n is square-free. Suppose that p is an odd prime with $p \mid n$. Let a be a quadratic nonresidue modulo p, and consider the congruences

$$x \equiv a \pmod{p},$$
$$x \equiv 1 \pmod{n/p}.$$

Notice that $\gcd(p, n/p) = 1$ because n is square-free, so by the Chinese Remainder Theorem there is a solution b with $1 \leq b \leq n$. Now $\gcd(a, p) = 1$, so $\gcd(b, p) = 1$ and therefore $\gcd(b, n) = 1$. Notice that

$$\left(\frac{b}{n}\right) = \left(\frac{b}{p}\right)\left(\frac{b}{n/p}\right)$$

$$= \left(\frac{a}{p}\right)\left(\frac{1}{n/p}\right)$$

$$= (-1) \cdot 1$$

$$= -1.$$

Since $b \equiv 1$ modulo $\frac{n}{p}$, $b^{(n-1)/2} = 1 + \frac{dn}{p}$ for some $d \in \mathbb{Z}$. If $b^{(n-1)/2} \equiv -1$ modulo n, then we may write $b^{(n-1)/2} = -1 + en$. Now

$$1 + \frac{dn}{p} = -1 + en$$
$$\implies 2p = n(ep - d),$$

which contradicts the fact that n is an odd composite number, so

$$b^{(n-1)/2} \equiv 1 \pmod{n}$$

and we have again found an integer b with $\gcd(b, n) = 1$ and

$$\left(\frac{b}{n} \right) \neq b^{(n-1)/2} \pmod{n}.$$

$\square$

Once we know there is some witness b to non-primality, there must be many such witnesses.

Lemma 12.27. *If $n > 1$ is an odd composite number, then at least half of all the integers b, $1 \leqslant b \leqslant n$, with $\gcd(b, n) = 1$ will satisfy*

$$\left(\frac{b}{n} \right) \neq b^{(n-1)/2} \pmod{n}.$$

PROOF. This is essentially the same as the proof of Theorem 12.21(3). If b_1 and b_2 satisfy $\gcd(b_1, n) = \gcd(b_2, n) = 1$, but

$$\left(\frac{b_1}{n} \right) = b_1^{(n-1)/2} \pmod{n}$$

and

$$\left(\frac{b_2}{n} \right) \neq b_2^{(n-1)/2} \pmod{n}$$

then

$$\left(\frac{b_1 b_2}{n} \right) \neq (b_1 b_2)^{n-1/2} \pmod{n}.$$

The rest of the proof proceeds as before. $\square$

This gives a probabilistic algorithm for primality testing. Make k random choices $b_1, \ldots, b_k$ of distinct integers with $1 \leqslant b_i \leqslant n$ and $\gcd(b_i, n) = 1$. If

$$\left(\frac{b_i}{n} \right) = b_i^{(n-1)/2} \pmod{n},$$

then we accept that n is probably prime in that the probability that a composite number would pass all those tests is $\frac{1}{2^k}$. In order to decide how practical this is, we need to estimate the complexity of the ingredient steps. We know the complexity of exponentiation modulo n using the Repeated Squaring algorithm; the only step we do not know is the complexity of computing the Jacobi symbol.

Theorem 12.28. *Let b and n be integers with $1 \leqslant b \leqslant n$. Then*

$$C\left(\text{calculate the Jacobi symbol } \left(\frac{b}{n}\right)\right) = O\left((\log n)^3\right).$$

PROOF. We may assume that $\gcd(b,n) = 1$. Choose b_1 with $b_1 \equiv b$ modulo n and $-\frac{n}{2} \leqslant b_1 \leqslant \frac{n}{2}$. Then

$$\left(\frac{b}{n}\right) = \left(\pm\frac{b_2}{n}\right) \text{ with } 0 < b_2 \leqslant \frac{n}{2}$$

$$= \left(\pm\frac{1}{n}\right)\left(\frac{b_2}{n}\right).$$

If $b_2 = 2$, then we can use the $\left(\frac{2}{n}\right)$ formula. If b_2 is odd, then apply quadratic reciprocity to see that

$$\left(\frac{b}{n}\right) = \left(\pm\frac{1}{n}\right)\left(\frac{n}{b_2}\right)$$

$$= \left(\pm\frac{1}{n}\right)\left(\frac{n \pmod{b_2}}{b_2}\right)$$

$$= \left(\pm\frac{1}{n}\right)\left(\frac{b_3}{b_2}\right) \text{ with } 1 \leqslant b_3 < b_2 \leqslant \frac{n}{2}.$$

At each repetition of this basic step, the denominator is reduced by a factor of 2 so there are at most $O(\log n)$ steps. Each step involves finding

$$b, b_i, b_j \leqslant n \pmod 4$$

in $O(\log n)$ operations and finding $b_i \leqslant n$ modulo $b_j \leqslant n$ in $O((\log n)^2)$ steps. Thus the total complexity is

$$O(\log n) \cdot O\left((\log n)^2\right) = O\left((\log n)^3\right).$$

□

This test is known as the Solovay–Strassen primality test and it has several refinements; the Miller–Rabin test is even faster and is widely used in computer algebra packages. Typically, when such a package verifies the primality of an integer, it only guarantees that the probability that the integer is composite is very small (usually less than 10^{-6}). Given the random element to choosing bases, repeating the test multiplies the probability – by this stage it is pretty well certain the number is prime.

Exercise 12.6. This exercise introduces the Miller–Rabin test. Given $n \in \mathbb{N}$, write $n - 1 = 2^s t$ with t odd. Given b, $0 < b < n$, $\gcd(b,n) = 1$, n is called a *strong pseudoprime to base b* if $b^t \equiv 1$ modulo n or there is an r, $0 \leqslant r < s$ such that $b^{2^r t} \equiv -1$ modulo n. Prove that if n is an odd composite integer then n is a strong pseudoprime for at most one quarter of the possible bases b with $0 < b < n$ and $\gcd(b,n) = 1$.

We recommend using the primality testing command on a computer with a number theory package. It is almost trivial now to construct an integer with thousands of decimal digits that passes the Miller–Rabin test and is thus almost certainly prime. Numbers which are constructed in this way are often called *industrial primes*. For all practical purposes they behave like primes and indeed are probably so with probability very close to 1. For example, when the RSA cryptosystem is implemented, in order to produce a public key, it relies upon the construction of an integer which is the product of two large primes. The primes used are industrial primes in the above sense.

In 1983, Adleman, Pomerance, and Rumely found a sophisticated deterministic algorithm with complexity

$$(\log n)^{O(\log \log \log n)}.$$

The exponent $\log \log \log n$ grows very slowly in n, and this algorithm is in practice very fast. Other modern algorithms use elliptic curves and Abelian varieties. In 1992 Adleman and Huang gave a probabilistic algorithm with polynomial running time that after k iterations either gives a definitive answer or gives no answer. The probability of no answer is 2^{-k}. This algorithm will never give a wrong answer but may with low probability fail to give an answer.

In Section 12.7.1 we will gave an account of a deterministic version of the Solovay–Strassen test. This test relies upon a hard unproven hypothesis. Until very recently, there was general frustration at the lack of a deterministic, polynomial time algorithm that does not rely upon unproven hypotheses. In 2003 there was great excitement when Agrawal, Kayal, and Saxena announced an ingenious approach to primality testing that gives such an algorithm.

12.7 The Agrawal–Kayal–Saxena Algorithm

There are attractive and readable accounts of this brilliant work and its later refinements in the references at the end of the chapter. All the methods considered so far begin with some theoretical characterization of primality that turns out to be implementable in some practical way. The Agrawal–Kayal–Saxena algorithm is no exception. We have seen that Fermat's Little Theorem gives a property of prime numbers that is shared by some composite numbers. A similar property in polynomials does give a complete characterization of prime numbers.

Lemma 12.29. *Let n denote a positive integer and suppose a is an integer with $\gcd(a, n) = 1$. Then n is prime if and only if the following congruence holds for polynomials*

$$(x - a)^n \equiv x^n - a \pmod{n}.$$

In other words, if and only if the equation

$$(x - a)^n = x^n - a$$

holds in the ring $\mathbb{Z}/n\mathbb{Z}[x]$.

Exercise 12.7. Prove Lemma 12.29. (Hint: It is not much harder than the proof of Fermat's Little Theorem. However, on this occasion, we suggest the use of congruences and the Binomial Theorem rather than Lagrange.)

Although this lemma is simple, the problem as it stands is that it cannot be checked with polynomial complexity. The neat idea that unlocked this beautiful result was to consider the congruence not just modulo n, but also modulo a polynomial of the form $x^r - 1$ for some prime r. However then a also needs to vary in order to keep the integrity of the test. What Agrawal–Kayal–Saxena proved is that r and various a can be chosen in such a way as to yield a primality test whose complexity is polynomial in $\log n$.

Here is a version of their algorithm as refined by Bernstein, taken from an expository article by Bornemann.

Theorem 12.30. *Suppose $n \in \mathbb{N}$ and $s \leqslant n$. Suppose primes q and r are chosen with the properties that $q \mid (r - 1)$, $n^{(r-1)/q} \not\equiv 0, 1$ modulo r, and*

$$\binom{q + s - 1}{s} \geqslant n^{2\lfloor \sqrt{r} \rfloor}.$$

If, for all a with $1 \leqslant a < s$,

(1) $\gcd(a, n) = 1$, and
(2) $(x - a)^n = x^n - a$ modulo $(x^r - 1, n)$ in the ring of polynomials $\mathbb{Z}[x]$,

then n is a prime power.

This gives a version of the Agrawal–Kayal–Saxena algorithm.

(1) Decide if n is a power of a natural number. If it is, go to (5).
(2) Choose integers q, r, and s satisfying the hypotheses of Theorem 12.30.
(3) For $a = 1, \ldots, s - 1$, do two checks. If $a \mid n$, go to (5). If $(x - a)^n = x^n - a$ modulo $(x^r - 1, n)$, go to (5).
(4) If you have reached this step, then n is prime.
(5) If you have reached this step, then n is composite.

The complexity of the original algorithm was a little over $O\big((\log n)^{12}\big)$. In 2003 Lenstra and Pommerance reduced this to a little over $O\big((\log n)^6\big)$ and it can be reduced further for special types of integers. Despite the apparent simplicity, proving that the algorithm does indeed test for primality and that the choices can all be made in polynomial complexity requires considerable ingenuity. However, the mathematics involved is not much beyond the scope of this book. We recommend that an interested reader follow the references in the notes to this chapter.

In one sense this algorithm resolves an outstanding problem. The existence of a polynomial time algorithm for determining primality is certainly a theoretical result of major interest and importance. However, something of a cloud hangs over the implementation of the algorithm as it stands. In practice, on current understanding, it is rather slow, so there is a rather interesting kind of trade-off. The probabilistic algorithms we discussed are very fast and very easy to implement, but give an uncertain answer, whereas the deterministic algorithm is currently too slow to implement. This is an area where much remains to be done and it is likely to see significant development. To further complicate things, the next section discusses some powerful methods that work under the assumption of generalized versions of the Riemann Hypothesis.

12.7.1 Deterministic Primality Testing

We will now show that the Solovay–Strassen primality test does have a deterministic form, but only under the assumption of an unproven hypothesis related to the Riemann Hypothesis. We have already seen a remarkable phenomenon at work in the area of primality testing; new results often draw upon earlier classical theorems from the literature, proven themselves simply for interest's sake. Here comes another example.

If you compose a list of quadratic residues and nonresidues for the first few prime numbers, you are likely to wonder whether the distribution of the quadratic residues obeys any predictable patterns. Since 1 is always a quadratic residue, the most basic question concerns the smallest predictable value of a quadratic nonresidue. This is a difficult problem: In particular, the bounds that are *provable* seem much weaker than the data suggests. On the other hand, a measure of the strength of such bounds is their applicability. The strongest known bound is due to Ankeny, but it relies on the *Extended Riemann Hypothesis*. This is stated in several different ways in the literature; one form is the following.

Conjecture 12.31. [EXTENDED RIEMANN HYPOTHESIS] Let $L(\cdot, \chi)$ denote an L-function associated with a Dirichlet character χ. Then all the solutions of $L(s, \chi) = 0$ in the critical strip $0 \leqslant \Re(s) \leqslant 1$ have $\Re(s) = 1/2$.

The following is a refined form of Ankeny's original theorem.

Theorem 12.32. [ANKENY] *Let p denote a prime and assume the Extended Riemann Hypothesis holds for the L-function associated with the character given by the Legendre symbol for p. Then the smallest quadratic residue modulo p is no larger than $2(\log p)^2$.*

We are not going to prove Ankeny's Theorem. His theorem has an easy generalization for a non-prime modulus which is used in the following result, which is also not going to be proved here.

Theorem 12.33. *Suppose $n > 1$ denotes any odd integer and the Extended Riemann Hypothesis holds for $L(\cdot, \chi)$ where χ denotes the Jacobi symbol $\left(\frac{\cdot}{n}\right)$. Then either an integer a exists with $a < 2(\log n)^2$ such that $\gcd(a, n) \neq 1$, or*

$$\left(\frac{a}{n}\right) \neq a^{(n-1)/2} \pmod{n}.$$

Using this latter result it is easy to extend the Solovay–Strassen test to deterministic polynomial time form; one simply has to run the test for each integer $a > 1$ below $2(\log n)^2$. The deterministic version of Solovay–Strassen is an excellent test that is easy to implement and runs very quickly. The one thing against it is that it relies upon an unproven hypothesis. Whether that hypothesis will be proved or – contrary to expectation – disproved, soon, nobody can tell. Thus, for the moment, primality testing lies in a state of flux and we urge the reader to watch for developments.

12.8 Factorizing

Given an integer n, the Fundamental Theorem of Arithmetic guarantees that n can be factorized. A natural question is to ask if we can find polynomial complexity algorithms to do this, and the answer seems to be no. Much hangs upon this question; the success of the RSA cryptosystem relies upon the apparent intractability of finding a fast factorizing algorithm. There are even large rewards for finding factors of some apparently difficult numbers. RSA Laboratories offers rewards of up to $200000 for factorizing certain *RSA numbers*, the name given to numbers with just two distinct prime factors. There are methods that are substantially faster than trial division for factorizing while not having polynomial complexity. On the other hand, they can be difficult to analyze; we simply describe three approaches whose implementation is easily understood and that will enable the reader to be able to enter into the literature with a good grasp of basic principles.

To give an idea about the way that primality testing and factorizing differ in practice, it might be helpful to consider the relative sizes of the integers to which the known methods can be applied with a hope of success. A well-chosen integer with approximately 200 decimal digits has a good chance of resisting factorization today, even using the best method, the *number field sieve*. (It will become clear later what is meant by the term "well-chosen".) By contrast, modern primality testing techniques installed on a PC can verify (with the earlier probabilistic caveat) the primality of integers with tens of thousands of digits in a matter of seconds. The *Cunningham Project* is an online attempt to tabulate factorizations of integers of the shape $b^n \pm 1$, where b is small (up to 12) and n is very large – the Fermat and Mersenne numbers, for example. The tables give a good indication of the successes, as well as the current limitations of factoring techniques.

In case this sounds pessimistic, it should be acknowledged how rapid developments in the field have been. Carl Pomerance has pointed out that in the 1970s, even 20-digit integers were difficult to factorize. By 1980, the factorization of 50-digit integers was becoming commonplace and by 1990, the record stood at 116-digits (in each case, these are difficult numbers, constructed as products of two large primes). In 1994, a 129-digit integer was factorized, which was remarkable because in an article in 1976, this number was predicted to be safe for 40 quadrillion years.

12.8.1 The Rho Method

This method generates test numbers as candidates to be factors of the given number in a particular way, and the logical structure of the algorithm resembles the Greek letter ρ (rho), hence the name.

Start with a map $f : \mathbb{Z}/n\mathbb{Z} \to \mathbb{Z}/n\mathbb{Z}$, typically a polynomial of degree greater than 1. Pick a starting value $x_0 \in \mathbb{Z}/n\mathbb{Z}$ and compute differences between iterates of f applied to x_0 as follows. Let $x_1 = f(x_0)$, $x_2 = f(x_1)$ and so on, and then compute $\gcd(x_1 - x_0, n)$, $\gcd(x_2 - x_1, n)$, and so on. The hope is that the iterates of f are randomly distributed among the residue classes of proper divisors of n, so among the calculations of $\gcd(x_r - x_{r-1}, n)$ we hope to quickly find a factor of n. The name "rho" is given to this method because the algorithm is said to run in the shape of a letter ρ.

Example 12.34. Let $n = 91$ and $f(x) = x^2 + 1$. Then take $x_0 = 1$ and compute

$$x_1 = 2, \ x_2 = 5, \ x_3 = 26, \ x_4 = 40,$$

and so on. (Remember that these are residues modulo n.) We find

$$\gcd(x_3 - x_2, n) = \gcd(26 - 5, 91) = \gcd(21, 91) = 7.$$

Thus we have found a factor of n after carrying out relatively few calculations.

Example 12.35. Let $n = 323$ and $f(x) = x^2 + 1$. Take $x_0 = 1$, then

$$x_1 = 2, \ x_2 = 5, \ x_3 = 26, \ x_4 = 31.$$

We find that

$$\gcd(x_5 - x_4, n) = \gcd(316 - 31, 323) = \gcd(285, 323) = 19,$$

so 323 is divisible by 19.

Exercise 12.8. Apply this method with $n = 437$, $f(x) = x^2 + 1$, and $x_0 = 1$. Confirm that $\gcd(x_5 - x_4, n)$ finds a nontrivial factor of n.

Several questions present themselves immediately. How do you know which iterates to compare? How do you pick f and x_0? In the literature, the following iterates are often compared:

$$x_1 - x_0,\ x_4 - x_3,\ x_8 - x_7,\ x_{16} - x_{15},\ x_{32} - x_{31},\ \ldots$$
$$x_2 - x_1,\ x_5 - x_3,\ x_9 - x_7,\ x_{17} - x_{15},\ x_{33} - x_{31},\ \ldots$$
$$x_3 - x_1,\ x_6 - x_3,\ x_{10} - x_7,\ x_{18} - x_{15},\ x_{34} - x_{31},\ \ldots$$
$$x_7 - x_3,\ x_{11} - x_7,\ x_{19} - x_{15},\ x_{35} - x_{31},\ \ldots$$
$$x_{12} - x_7,\ x_{20} - x_{15},\ x_{36} - x_{31},\ \ldots$$
$$\vdots \qquad\qquad x_{21} - x_{15},\ x_{37} - x_{31},\ \ldots$$
$$x_{15} - x_7, \qquad \vdots \qquad x_{38} - x_{31},\ \ldots$$
$$x_{31} - x_{15}, \qquad \vdots \quad \ldots$$
$$x_{63} - x_{31},\ \ldots$$

and so on. This appears to leave gaps but, in fact, the gaps can be accounted for.

Example 12.36. Let $n = 4087$ and $f(x) = x^2 + x + 1$. Pick $x_0 = 2$ and compute

$$\gcd(x_1 - x_0, n) = \gcd(7 - 2, 4087) = 1,$$
$$\gcd(x_2 - x_1, n) = \gcd(57 - 7, 4087) = 1,$$
$$\gcd(x_3 - x_1, n) = \gcd(3307 - 7, 4087) = 1,$$
$$\gcd(x_4 - x_3, n) = \gcd(2745 - 3307, 4087) = 1,$$
$$\gcd(x_5 - x_3, n) = \gcd(1343 - 3307, 4087) = 1,$$
$$\gcd(x_6 - x_3, n) = \gcd(2626 - 3307, 4087) = 1,$$
$$\gcd(x_7 - x_3, n) = \gcd(3734 - 3307, 4087) = 61;$$

we have found a factor of 4087.

Frustratingly, this method seems to work when it chooses. Deciding when and how it will work – and how quickly it will work when it does – involves a complicated and unsatisfactory argument. There are several reasons for this difficulty.

1. It is difficult to make the "spreading" property (through the residue classes modulo n) of f precise. Even having done so, it is difficult to check the property for concrete functions.
2. Even if you satisfy yourself about the spreading property, the complexity estimate is well away from being polynomial in $\log n$.
3. It is a probabilistic approach in character (but this is not such a serious difficulty).

12.8.2 The Factor Base Method

This method has been quite successful in the sense that the largest nontrivial factorizations have been performed using it, often using many computers operating in tandem. It is not really viable in general, however, as we will see. On the other hand, the *number field sieve* (often referred to as NFS), extensively refined by many workers, is a successful factoring method based on this method. The complexity of the NFS can be shown to be $O(\exp[(\log n)^{1/3}(\log\log n)^{2/3}])$.

The basic idea goes as follows. Suppose a large integer n can be expressed in the form $n = s^2 - t^2$ with $s, t \in \mathbb{Z}$. Then obviously $(s + t)$ and $(s - t)$ are factors of n. Of course, there is no hope of doing this in general, and even if such a representation exists, it is not clear how to find it. If we suppose further that $(s + t)$ and $(s - t)$ are approximately equal (relative to the size of n), then we could hope to find them. Start with $x = \lfloor \sqrt{n} \rfloor$ and then check

$$(x + 1)^2 - n, (x + 2)^2 - n, (x + 3)^2 - n,$$

and so on to see if the result is a square.

This is still not of much practical use because the hypothesis is much too strong. In the RSA cryptosystem, the security depends on the practical impossibility (at present) of finding the factors of a product $pq = n$, where p and q are very large distinct primes. The user of such a system guarantees that p and q are not close, giving no hope of expressing n in the form

$$n = s^2 - t^2$$

in a reasonable amount of time. However, the basic idea becomes quite workable if we replace equality by congruences. We try to solve

$$s^2 - t^2 \equiv 0 \pmod{n}.$$

This is much easier to solve, and if we have a solution pair (s, t), then we can compute $\gcd(s \pm t, n)$ and hope this will yield a nontrivial factor. The factor base method relies upon generating a large number of solutions of the congruence that can be tested.

Definition 12.37. *Given an integer n known to be composite (perhaps it has failed a primality test or perhaps it is a known public key in an RSA cryptosystem), let B denote the set $\{-1, p_1, \ldots, p_n\}$, where the p_i are distinct primes. We say x is a B-number for n if x^2 modulo n can be expressed as a product of powers of elements of B.*

Example 12.38. Let $n = 4633$ and choose $B = \{-1, 2, 3\}$. Then $67, 68, 69$ are B-numbers:

$$67^2 \equiv -144 \equiv -2^4 3^2 \pmod{4633},$$
$$68^2 \equiv -9 \equiv -3^2 \pmod{4633},$$

and

$$69^2 \equiv 128 \equiv 2^7 \quad (\text{mod } 4633).$$

Solve $s^2 \equiv t^2 \ (\text{mod } n)$:

$$(-77)^2 \equiv (67 \cdot 68)^2 \equiv 2^4 3^4 \equiv (2^2 3^2)^2 \equiv 36^2 \quad (\text{mod } 4633).$$

We check $\gcd(-77+36, n) = 41$ and $\gcd(-77-36, n) = 113$, giving nontrivial factors of n.

In general, given n, B, and several B-numbers $b_1, \ldots, b_k$, the problem is turned into one in linear algebra. Each b_i gives rise to a vector over the field $\mathbb{F}_2 = \mathbb{Z}/2\mathbb{Z}$ as follows. Assume that $B = \{-1, p_1, \ldots, p_{h-1}\}$; then each of the squares of the B-numbers has a unique factorization

$$(-1)^{e_0} p_1^{e_1} \cdots p_{h-1}^{e_{h-1}} \text{ with } e_i \in \mathbb{N},$$

and we associate to this the vector $(e_0, \ldots, e_{h-1})$ modulo 2. Doing this for each of the given B-numbers produces a $k \times h$ matrix over $\mathbb{F}_2$. We then look for linear dependence relations among the rows of the matrix.

Example 12.39. Let $n = 4633$ and $B = \{-1, 2, 3\}$. Then the calculation in Example 12.38 gives

$$\begin{bmatrix} 1 & 4 & 2 \\ 1 & 0 & 2 \\ 0 & 7 & 0 \end{bmatrix} \equiv \begin{bmatrix} 1 & 0 & 0 \\ 1 & 0 & 0 \\ 0 & 1 & 0 \end{bmatrix} \quad (\text{mod } 2)$$

and the first two rows give the desired relation.

Of course, a dependence relation will not necessarily yield a factorization, so the more B-numbers we can find, the greater will be our chances of success.

Example 12.40. Let $n = 1829$ and $B = \{-1, 2, 3, 5, 7, 11, 13\}$. Then

$$42, 43, 61, 74, 85, 86 \text{ are } B\text{-numbers.}$$

Factorizing them gives the vectors

$$42^2 \equiv -5 \cdot 13 \to (1, 0, 0, 1, 0, 0, 1),$$
$$43^2 \equiv 2^2 \cdot 5 \to (0, 2, 0, 1, 0, 0, 0),$$
$$61^2 \equiv 3^2 \cdot 7 \to (0, 2, 0, 0, 1, 0, 0),$$
$$74^2 \equiv -11 \to (1, 0, 0, 0, 0, 1, 0),$$
$$85^2 \equiv -7 \cdot 13 \to (1, 0, 0, 0, 1, 0, 1),$$
$$86^2 \equiv 2^4 \cdot 5 \to (0, 4, 0, 1, 0, 0, 0),$$

from which we find the following reduced matrix:

$$\begin{bmatrix} 1\,0\,0\,1\,0\,0\,1 \\ 0\,0\,0\,1\,0\,0\,0 \\ 0\,0\,0\,0\,1\,0\,0 \\ 1\,0\,0\,0\,0\,1\,0 \\ 1\,0\,0\,0\,1\,0\,1 \\ 0\,0\,0\,1\,0\,0\,0 \end{bmatrix}.$$

Notice that $r_2 = r_6$, so

$$(43 \cdot 86)^2 \equiv 2^6 5^2 \equiv (2^3 5)^2 \equiv 40^2.$$

Unfortunately, $43 \cdot 86 \equiv 40$, so we have found $40^2 \equiv 40^2$. Thus we find a solution of the congruence but no factor. Try another dependence relation:

$$r_1 + r_2 + r_3 + r_5$$

is the zero row. This gives a solution of the congruence,

$$(1459)^2 \equiv (b_1 b_2 b_3 b_5)^2 \equiv (2 \cdot 3 \cdot 5 \cdot 7 \cdot 13)^2 \equiv (901)^2.$$

We find

$$\gcd(1459 + 901, 1827) = 59,$$

giving a nontrivial factor of n.

Example 12.41. Let $n = 2201$, and compute a few candidates for B-numbers modulo n:

$$47^2 \equiv 8 = 2^3,$$
$$48^2 \equiv 103 \quad \text{has no small factors,}$$
$$49^3 \equiv 200 = 2^3 \cdot 5^2,$$
$$50^2 \equiv 299 = 13 \cdot 23,$$
$$51^2 \equiv 400 = 2^4 \cdot 5^2,$$
$$52^2 \equiv 503 \quad \text{has no small factors.}$$

So choose $B = \{-1, 2, 5\}$ and select as B-numbers $47, 49$, and 51. The factorizations give the matrix

$$\begin{bmatrix} 0\,1\,0 \\ 0\,3\,2 \\ 0\,4\,2 \end{bmatrix} \equiv \begin{bmatrix} 0\,1\,0 \\ 0\,1\,0 \\ 0\,0\,0 \end{bmatrix} \pmod 2,$$

which has the relation $r_1 = r_2$. This gives

$$(47 \cdot 49)^2 \equiv 1600 = 40^2;$$

we find that $47 \cdot 49 \equiv 102$ modulo n, and

$$\gcd(102 + 40, n) = 71$$

identifies the factor 71 of n.

One of the things that makes this method so effective is the ability of computers to do linear algebra – row reduction in particular – over $\mathbb{F}_2$ very rapidly.

12.8.3 Elliptic Curves

The third and final factorizing method we mention is the easiest to describe but by far the hardest to analyze. It was invented by Lenstra and is implemented in many computer packages. Suppose you are given an integer n that is known not to be prime. The idea is to choose an elliptic curve E defined over $\mathbb{Z}$, together with a rational nontorsion point P, and then try to compute the sequence of points $P, 2P, 3P, \ldots$ modulo n. This sounds simple, yet it is based on some hard analysis, and it does work for integers in a certain range. The idea behind the method is that adding points on an elliptic curve using the addition formulas always entails some division. Doing division modulo n can only be achieved by inverting an integer modulo n, and this is only possible if that integer is coprime to n. Thus *failure* to produce the next point in the sequence finds a factor of n.

In order to follow the discussion below, it may be helpful to review the explicit formulas for addition on an elliptic curve from Section 5.3 and recall that addition, multiplication, and division (via the Euclidean Algorithm) modulo n have polynomial complexity.

Example 12.42. Let $n = 21$, and choose the elliptic curve

$$E : y^2 + y = x^3 - x$$

together with the point $P = (0,0)$. The first few multiples of P modulo 21 can be computed without any problem:

$$2P = (1,0), 3P = (20,20), 4P = (2,18), 5P = (16,2), 6P = (6,14).$$

However, the computer will refuse to calculate $7P$ modulo 21. In order to find this value, it would need to invert 6 modulo 21; this is done using the Euclidean Algorithm, which rapidly detects the factor 3 of 21. Notice that if the initial point is $P = (0,0)$, then it is $\gcd(x(kP), n)$ that potentially reveals a factor of n.

Example 12.43. For a slightly more impressive example, let $n = 39701558597$. Working with the same elliptic curve and the same point, this time the computer will find

$$526P = (3341173047, 12476794460) \pmod{n},$$

but refuses to go any further. The reason is that

$$\gcd(3341173047, n) = 1049$$

yields a factor of n. The other factor, 37847053, is now easily found.

Lenstra's idea was that the flexibility of choosing curves E and points P, together with a suitable multiplier k, might make it possible to detect when computing kP modulo n becomes impossible – in other words, when you stumble onto a factor of n. There are much fuller accounts of this topic than we are able to give in the notes at the end of the chapter. An important aspect of the complexity of Lenstra's method is that the running time depends on the second largest prime factor of n, which may be much smaller than the square root of n. Rather than seek to analyze this method, we suggest the following as a worthwhile exercise.

Exercise 12.9. Using a computer package programmed with the arithmetic of elliptic curves, try to factorize some large integers of your choosing using Lenstra's method.

12.8.4 Elliptic Curve Factorizing in Practice

As an example of how Lenstra's method has been used in practice we include the following. A repeated theme in the book has been the phenomenon whereby an integer can be known as a composite with only a partial factorization (or none at all). On April 25th 1998, the complete factorization of the Mersenne number M_{589} was obtained. Since 589 is divisible by 19 and 31, two obvious factors present themselves; M_{19} and M_{31}. There are two others, the smallest of which is the 46 digit prime

$$2023706519999643990585239115064336980154410119.$$

The other prime factor has 227 decimal digits. In this instance, the second largest prime factor is quite a bit smaller than the square root of the number to be factored, which is an ideal situation for the application of Lenstra's method.

12.9 Complexity of Arithmetic in Finite Fields

For the results in this chapter, we used the complexity of the following operations over $\mathbb{F}_p = \mathbb{Z}/p\mathbb{Z}$:

$$C(\text{add } a \text{ to } b) = O(\log p);$$
$$C(\text{multiply } a \text{ by } b) = O((\log p)^2);$$
$$C(\text{invert } a \neq 0) = O((\log p)^3).$$

These estimates can be extended to cover arithmetic in finite fields. The elements of a finite field can be represented explicitly as polynomials over $\mathbb{F}_p$. The most important estimate we need is the complexity of performing the Euclidean Algorithm in $\mathbb{F}_p[x]$ for a fixed prime p. The proofs of the following results closely follow the earlier complexity arguments.

Theorem 12.44. *Suppose f and g are nonzero elements of $\mathbb{F}_p[x]$ whose degrees are bounded by n. Then we can find q, r in $\mathbb{F}_p[x]$ with $f = gq + r$ and $\deg r < \deg g$, with complexity $O\big(n^2(\log p)^3\big)$.*

Corollary 12.45. *Suppose $\mathbb{F}_q$ is a finite field, $q = p^r$, in which multiplication is determined by a monic irreducible polynomial of degree r (that is, $\mathbb{F}_q$ is presented as $\mathbb{F}_p[x]/f(x) \cdot \mathbb{F}_p[x]$ for a monic irreducible polynomial f of degree r). Then, for $a, b \neq 0$ in $\mathbb{F}_q$,*

(1) $\mathsf{C}(\text{add } a \text{ to } b) = O(r \log p) = O(\log q)$;
(2) $\mathsf{C}(\text{multiply } a \text{ by } b) = O\big(r^2(\log p)^2\big) = O\big((\log q)^2\big)$;
(3) $\mathsf{C}(\text{invert } b \neq 0) = O\big(r^3(\log p)^3\big) = O\big((\log q)^3\big)$;
(4) $\mathsf{C}(\text{find } b^n) = O\big(\log n(\log q)^2\big)$.

Exercise 12.10. Prove Theorem 12.44 and Corollary 12.45.

NOTES TO CHAPTER 12: For this chapter we leaned heavily on Koblitz's excellent account of computational number theory [90]. Consult the books of Bressoud and Wagon [21] and Bressoud [22] for more on this topic, as well as Cohen [32], Crandall and Pomerance [36], von zur Gathen and Gerhard [66], and the references therein. The lower bound (12.6) for Carmichael numbers is due to Alford, Granville, and Pomerance [2]. The RSA algorithm from Section 12.2 appeared in a paper of Rivest, Shamir and Adleman [129]. A few years before RSA was invented at MIT, Clifford Cocks in the UK invented a public-key cryptography scheme using similar ideas. His invention was classified until very recently. For details on the Agrawal–Kayal–Saxena algorithm, see their original paper [1], Bernstein's paper [12], or the announcement by Bornemann [17] and the references therein. Lenstra and Lenstra wrote an account of the Number Field Sieve in [99]. An excellent comparison of sieving methods, as well as some interesting history, can be found in a survey article by Pomerance [117]. The current state of the Cunningham Project may be found on Wagstaff's Web site [156].

References

1. M. Agrawal, N. Kayal, and N. Saxena, PRIMES is in P, www.cse.iitk.ac.in/users/manindra/primality.ps.
2. W. R. Alford, A. Granville, and C. Pomerance, There are infinitely many Carmichael numbers, *Ann. Math. (2)* **139**, no. 3 (1994), 703–722.
3. R. Apéry, Irrationalité de $\zeta(2)$ et $\zeta(3)$, *Astérisque* **61** (1979), 11–13.
4. T. M. Apostol, *Introduction to Analytic Number Theory*, Undergraduate Texts in Mathematics, Springer-Verlag, New York, 1976.
5. T. M. Apostol, *Modular Functions and Dirichlet Series in Number Theory*, second ed., Graduate Texts in Mathematics, vol. 41, Springer-Verlag, New York, 1990.
6. E. Artin, *The Gamma Function*, Translated by Michael Butler, Athena Series: Selected Topics in Mathematics, Holt, Rinehart and Winston, New York, 1964.
7. A. Baker, Linear forms in the logarithms of algebraic numbers. I, II, III, *Mathematika* **13** (1966), 204-216; **14** (1967), 102-107; **14** (1967), 220–228.
8. R. C. Baker, G. Harman, and J. Pintz, The difference between consecutive primes. II, *Proc. London Math. Soc. (3)* **83**, no. 3 (2001), 532–562.
9. K. Ball and T. Rivoal, Irrationalité d'une infinité de valeurs de la fonction zêta aux entiers impairs, *Invent. Math.* **146**, no. 1 (2001), 193–207.
10. E. J. Barbeau, *Pell's Equation*, Problem Books in Mathematics, Springer-Verlag, New York, 2003.
11. V. Bergelson, Ergodic Ramsey theory—an update, in M. Pollicott and K. Schmidt (eds.), *Ergodic theory of $\mathbb{Z}^d$ actions (Warwick, 1993-1994)*, London Math. Soc. Lecture Note Ser., vol. 228, Cambridge Univ. Press, Cambridge, 1996, pp. 1–61.
12. D. J. Bernstein, Proving primality after Agrawal–Kayal–Saxena, http://cr.yp.to/papers.html#aks.
13. F. Beukers, J. A. C. Kolk, and E. Calabi, Sums of generalized harmonic series and volumes, *Nieuw Arch. Wisk. (4)* **11**, no. 3 (1993), 217–224.
14. G. Billing and K. Mahler, On exceptional points on cubic curves, *J. London Math. Soc.* **15** (1940), 32–43.
15. Y. Bilu, G. Hanrot, and P. M. Voutier, Existence of primitive divisors of Lucas and Lehmer numbers, *J. Reine Angew. Math.* **539** (2001), 75–122, With an appendix by M. Mignotte.

16. B. J. Birch and H. P. F. Swinnerton-Dyer, Notes on elliptic curves. I, II, *J. Reine Angew. Math.* **212** (1963), 7–25; **218** (1965), 79–108.

17. F. Bornemann, PRIMES is in P: A Breakthrough for "Everyman", *Notices Amer. Math. Soc.* **50**, no. 5 (2003), 545–552.

18. P. Borwein, *Computational Excursions in Analysis and Number Theory*, CMS Books in Mathematics/Ouvrages de Mathématiques de la SMC, vol. 10, Springer-Verlag, New York, 2002.

19. B. Brent, An expansion of e^x off roots of one, *Fibonacci Quart.* **12** (1974), 208.

20. B. Brent, Functional equations with prime roots from arithmetic expressions for $\mathcal{G}_\alpha$, *Fibonacci Quart.* **12** (1974), 199–207.

21. D. Bressoud and S. Wagon, *A Course in Computational Number Theory*, Key College Publishing, Emeryville, CA, 2000.

22. D. M. Bressoud, *Factorization and Primality Testing*, Undergraduate Texts in Mathematics, Springer-Verlag, New York, 1989.

23. C. Breuil, B. Conrad, F. Diamond, and R. Taylor, On the modularity of elliptic curves over $\mathbb{Q}$: wild 3-adic exercises, *J. Amer. Math. Soc.* **14**, no. 4 (2001), 843–939.

24. V. Brun, La série $1/5 + 1/7 + 1/11 + 1/13 + 1/17 + 1/19 + 1/29 + 1/31 + 1/41 + 1/43 + 1/59 + 1/61 + \cdots$ où les dénominateurs sont nombres premiers jumeaux est convergente ou finie, *Bull. Sci. Math.* **43** (1919), 100–104, 124–128.

25. C. Caldwell, *The Prime Pages*, www.utm.edu/research/primes/.

26. J. W. S. Cassels, Mordell's finite basis theorem revisited, *Math. Proc. Cambridge Philos. Soc.* **100**, no. 1 (1986), 31–41.

27. J. W. S. Cassels, *Lectures on Elliptic Curves*, London Mathematical Society Student Texts, vol. 24, Cambridge University Press, Cambridge, 1991.

28. J. S. Chahal, *Topics in Number Theory*, The University Series in Mathematics, Plenum Press, New York, 1988.

29. K. Chandrasekharan, *Elliptic Functions*, Grundlehren der Mathematischen Wissenschaften [Fundamental Principles of Mathematical Sciences], vol. 281, Springer-Verlag, Berlin, 1985.

30. D. V. Chudnovsky and G. V. Chudnovsky, Sequences of numbers generated by addition in formal groups and new primality and factorization tests, *Adv. Appl. Math.* **7**, no. 4 (1986), 385–434.

31. P. A. Clement, Congruences for sets of primes, *Amer. Math. Monthly* **56** (1949), 23–25.

32. H. Cohen, *Advanced Topics in Computational Number Theory*, Graduate Texts in Mathematics, vol. 193, Springer-Verlag, New York, 2000.

33. J. B. Conrey, The Riemann Hypothesis, *Notices Amer. Math. Soc.* **50**, no. 3 (2003), 341–353.

34. G. Cornell and J. H. Silverman (eds.), *Arithmetic Geometry*, Papers from the Conference held at the University of Connecticut, Storrs, Connecticut, July 30–August 10, 1984, Springer-Verlag, New York, 1986.

35. G. Cornell, J. H. Silverman, and G. Stevens (eds.), *Modular Forms and Fermat's Last Theorem*, Papers from the Instructional Conference on Number Theory and Arithmetic Geometry held at Boston University, Boston, MA, August 9–18, 1995, Springer-Verlag, New York, 1997.

36. R. Crandall and C. Pomerance, *Prime Numbers: A Computational Perspective*, Springer-Verlag, New York, 2001.

37. J. Cremona, Elliptic curve data, www.maths.nott.ac.uk/personal/jec/.

38. J. T. Cross, In the Gaussian integers, $\alpha^4 + \beta^4 \neq \gamma^4$, *Math. Mag.* **66**, no. 2 (1993), 105–108.

39. H. Darmon, *Rational Points on Modular Elliptic Curves*, CBMS Regional Conference Series in Mathematics, vol. 101, Published for the Conference Board of the Mathematical Sciences, Washington, DC, 2004.

40. H. Davenport, *Multiplicative Number Theory*, Revised and with a preface by Hugh L. Montgomery, third ed., Graduate Texts in Mathematics, vol. 74, Springer-Verlag, New York, 2000.

41. P. Deligne, Preuve des conjectures de Tate et de Shafarevitch (d'après G. Faltings), *Astérisque* (1985), nos. 121-122, 25–41, (Seminar Bourbaki, Vol. 1983/84).

42. L. E. Dickson, *History of the Theory of Numbers. Vol. I: Divisibility and Primality*, Chelsea Publishing Co., New York, 1966.

43. L. E. Dickson, *History of the Theory of Numbers. Vol. II: Diophantine Analysis*, Chelsea Publishing Co., New York, 1966.

44. L. E. Dickson, *History of the Theory of Numbers. Vol. III: Quadratic and Higher Forms*, with a Chapter on the Class Number by G. H. Cresse, Chelsea Publishing Co., New York, 1966.

45. R. E. Dressler, A stronger Bertrand's postulate with an application to partitions, *Proc. Amer. Math. Soc.* **33** (1972), 226–228.

46. U. Dudley, History of a formula for primes, *Amer. Math. Monthly* **76** (1969), 23–28.

47. H. M. Edwards, *Riemann's Zeta Function*, Dover Publications Inc., Mineola, NY, 2001, reprint of the 1974 original [Academic Press, New York].

48. M. Einsiedler, G. R. Everest, and T. Ward, Primes in elliptic divisibility sequences, *London Math. Soc. J. Comput. Math.* **4** (2001), 1–15.

49. N. D. Elkies, On $A^4 + B^4 + C^4 = D^4$, *Math. Comp.* **51**, no. 184 (1988), 825–835.

50. N. D. Elkies, On the sums $\sum_{k=-\infty}^{\infty} (4k + 1)^{-n}$, *Amer. Math. Monthly* **110**, no. 7 (2003), 561–573.

51. P. Erdős, Beweis eines Satzes von Tschebyschef, *Acta Litt. Sci. Szeged* **5** (1932), 194–198.

52. P. Erdős, On a new method in elementary number theory which leads to an elementary proof of the prime number theorem, *Proc. Nat. Acad. Sci. U.S.A.* **35** (1949), 374–384.

53. Euclid, *The Thirteen Books of Euclid's Elements Translated from the Text of Heiberg. Vol. I: Introduction and Books I, II. Vol. II: Books III–IX. Vol. III: Books X–XIII and Appendix*, Translated with Introduction and Commentary by Thomas L. Heath, 2nd ed, Dover Publications Inc., New York, 1956.

54. G. R. Everest and H. King, Prime powers in elliptic divisibility sequences, *Comp. Math.* to appear, 2005.

55. G. R. Everest, G. McLaren, and T. Ward, Divisibility in elliptic divisibility sequences, arXiv:math.NT/0409540, 2004.

56. G. R. Everest, V. Miller, and N. Stephens, Primes generated by elliptic curves, *Proc. Amer. Math. Soc.* **132**, no. 1 (2004), 955–963.

57. G. R. Everest, P. Rogers, and T. Ward, A higher rank Mersenne problem, *Springer Lecture Notes in Computer Science* **2369** (2002), 95–107.

58. G. R. Everest, A. J. van der Poorten, I. Shparlinski, and T. Ward, *Recurrence Sequences*, Mathematical Surveys and Monographs, vol. 104, American Mathematical Society, Providence, RI, 2003.

59. G. R. Everest and T. Ward, *Heights of Polynomials and Entropy in Algebraic Dynamics*, Springer-Verlag, London, 1999.
60. J.-H. Evertse, On sums of S-units and linear recurrences, *Compositio Math.* **53**, no. 2 (1984), 225–244.
61. G. Faltings, Endlichkeitssätze für abelsche Varietäten über Zahlkörpern, *Invent. Math.* **73**, no. 3 (1983), 349–366.
62. G. Faltings, Erratum: "Finiteness theorems for abelian varieties over number fields", *Invent. Math.* **75**, no. 2 (1984), 381.
63. H. Furstenberg, On the infinitude of primes, *Amer. Math. Monthly* **62** (1955), 353.
64. H. Furstenberg, *Recurrence in Ergodic Theory and Combinatorial Number Theory*, M. B. Porter Lectures, Princeton University Press, Princeton, NJ, 1981.
65. J. M. Gandhi, Formulae for the nth prime, in J. H. Jordan and W. A. Webb (eds.), *Proceedings of the Washington State University Conference on Number Theory (Washington State Univ., Pullman, Wash., 1971)*, Dept. Math., Washington State Univ., Pullman, Wash., 1971, pp. 96–106.
66. J. von zur Gathen and J. Gerhard, *Modern Computer Algebra*, second ed., Cambridge University Press, Cambridge, 2003.
67. C. F. Gauss, *Disquisitiones Arithmeticae*, Translated and with a preface by Arthur A. Clarke, Revised by William C. Waterhouse, Cornelius Greither and A. W. Grootendorst and with a preface by William C. Waterhouse, Springer-Verlag, New York, 1986.
68. S. S. Gelbart and S. D. Miller, Riemann's zeta function and beyond, *Bull. Amer. Math. Soc. (N.S.)* **41**, no. 1 (2004), 59–112.
69. D. Goldfeld, Sur les produits partiels Eulériens attachés aux courbes elliptiques, *C. R. Acad. Sci. Paris Sér. I Math.* **294**, no. 14 (1982), 471–474.
70. D. Goldfeld, The elementary proof of the prime number theorem: an historical perspective, in D. Chudnovsky, G. Chudnovsky and M. B. Nathanson (eds.), *Number theory (New York, 2003)*, Springer-Verlag, New York, 2004, pp. 179–192.
71. S. W. Golomb, A direct interpretation of Gandhi's formula, *Amer. Math. Monthly* **81** (1974), 752–754.
72. W. T. Gowers, A new proof of Szemerédi's theorem, *Geom. Funct. Anal.* **11**, no. 3 (2001), 465–588.
73. B. Green and T. Tao, The primes contain arbitrarily long arithmetic progressions, arXiv:math.NT/0404188, 2004.
74. G. H. Hardy, *Divergent Series*, The Clarendon Press, Oxford University Press, Oxford, 1949.
75. G. H. Hardy and E. M. Wright, *An Introduction to the Theory of Numbers*, fifth ed., The Clarendon Press, Oxford University Press, New York, 1979.
76. H. Hasse, *Number Theory*, Translated from the third German edition and with a preface by Horst Günter Zimmer, Springer-Verlag, Berlin, 1980.
77. E. Hecke, *Lectures on the Theory of Algebraic Numbers*, Translated from the German by George U. Brauer, Jay R. Goldman and R. Kotzen, Graduate Texts in Mathematics, vol. 77, Springer-Verlag, New York, 1981.
78. K. Heegner, Diophantische Analysis und Modulfunktionen, *Math. Z.* **56** (1952), 227–253.
79. D. Husemoller, *Elliptic Curves*, With an Appendix by Ruth Lawrence, Graduate Texts in Mathematics, vol. 111, Springer-Verlag, New York, 1987.

80. A. E. Ingham, On the difference between consecutive primes, *Quart. J. Math. Oxford Ser. (2)* **8** (1937), 255–266.

81. G. J. O. Jameson, *The Prime Number Theorem*, London Mathematical Society Student Texts, vol. 53, Cambridge University Press, Cambridge, 2003.

82. W. Jänichen, Über die Verallgemeinerung einer Gaußschen Formel aus der Theorie der höheren Kongruenzen, *Sitzungsber. Berl. Math. Ges.* **20** (1921), 23–29.

83. G. J. Janusz, *Algebraic Number Fields*, second ed., Graduate Studies in Mathematics, vol. 7, American Mathematical Society, Providence, RI, 1996.

84. G. A. Jones and J. M. Jones, *Elementary Number Theory*, Springer Undergraduate Mathematics Series, Springer-Verlag London Ltd., London, 1998.

85. J. P. Jones, D. Sato, H. Wada, and D. Wiens, Diophantine representation of the set of prime numbers, *Amer. Math. Monthly* **83**, no. 6 (1976), 449–464.

86. D. Joyce, Euclid's Elements, aleph0.clarku.edu/~djoyce/.

87. Y. Katznelson, *An Introduction to Harmonic Analysis*, corrected ed., Dover Publications Inc., New York, 1976.

88. W. Keller, Fermat Factoring Status, www.prothsearch.net/fermat.html.

89. N. Koblitz, *Introduction to Elliptic Curves and Modular Forms*, Graduate Texts in Mathematics, vol. 97, Springer-Verlag, New York, 1984.

90. N. Koblitz, *A Course in Number Theory and Cryptography*, Graduate Texts in Mathematics, vol. 114, Springer-Verlag, New York, 1987.

91. L. J. Lander and T. R. Parkin, Counterexample to Euler's conjecture on sums of like powers, *Bull. Amer. Math. Soc.* **72** (1966), 1079.

92. L. J. Lander and T. R. Parkin, A counterexample to Euler's sum of powers conjecture, *Math. Comp.* **21** (1967), 101–103.

93. L. J. Lander, T. R. Parkin, and J. L. Selfridge, A survey of equal sums of like powers, *Math. Comp.* **21** (1967), 446–459.

94. S. Lang, *Elliptic Curves: Diophantine Analysis*, Grundlehren der Mathematischen Wissenschaften [Fundamental Principles of Mathematical Sciences], vol. 231, Springer-Verlag, Berlin, 1978.

95. S. Lang, *Elliptic Functions*, second ed., Graduate Texts in Mathematics, vol. 112, Springer-Verlag, New York, 1987, (with an appendix by J. Tate).

96. S. Lang, *Algebraic Number Theory*, second ed., Graduate Texts in Mathematics, vol. 110, Springer-Verlag, New York, 1994.

97. D. H. Lehmer, Factorization of certain cyclotomic functions, *Ann. Math.* **34** (1933), 461–479.

98. F. Lemmermeyer, *Reciprocity Laws From Euler to Eisenstein*, Springer Monographs in Mathematics, Springer-Verlag, Berlin, 2000.

99. A. K. Lenstra and H. W. Lenstra, Jr. (eds.), *The Development of the Number Field Sieve*, Lecture Notes in Mathematics, vol. 1554, Springer-Verlag, Berlin, 1993.

100. W. J. LeVeque, *Fundamentals of Number Theory*, Addison-Wesley Publishing Co., Reading, Mass.-London-Amsterdam, 1977.

101. E. Lutz, Sur l'equation $y^2 = x^3 - Ax - B$ dans les corps p–adic, *J. Reine Angew. Math.* **177** (1937), 237–247.

102. K. Mahler, On the Chinese remainder theorem, *Math. Nachr.* **18** (1958), 120–122.

103. K. Mahler, An application of Jensen's formula to polynomials, *Mathematika* **7** (1960), 98–100.

104. K. Mahler, On some inequalities for polynomials in several variables, *J. London Math. Soc.* **37** (1962), 341–344.

105. B. Mazur, Modular curves and the Eisenstein ideal, *Inst. Hautes Études Sci. Publ. Math.* (1977), no. 47, 33–186.

106. T. Metsänkylä, Catalan's conjecture: another old Diophantine problem solved, *Bull. Amer. Math. Soc. (N.S.)* **41**, no. 1 (2004), 43–57.

107. W. H. Mills, A prime-representing function, *Bull. Amer. Math. Soc.* **53** (1947), 604.

108. J. S. Milne, *Abelian varieties*, in G. Cornell and J. H. Silverman (eds.), *Arithmetic Geometry*, Papers from the Conference held at the University of Connecticut, Storrs, Connecticut, July 30–August 10, 1984, Springer-Verlag, New York, 1986, pp. 103–150.

109. P. Monsky, Simplifying the proof of Dirichlet's theorem, *Amer. Math. Monthly* **100**, no. 9 (1993), 861–862.

110. L. J. Mordell, On the rational solutions of the indeterminate equations of the third and fourth degrees, *Proc. Cambridge Phil. Soc.* **21** (1922), 179–192.

111. P. Moss, *Algebraic Realizability Problems*, Ph.D. thesis, University of East Anglia, 2003.

112. T. Nagell, Solution de quelque problèmes dans la théorie arithmétique des cubiques planes du premier genre, *Wid. Akad. Skrifter Oslo I* **1** (1935).

113. J. J. O'Connor and E. F. Robertson, History of mathematics archive, www-gap.dcs.st-and.ac.uk/~history/.

114. A. M. Odlyzko and H. J. J. te Riele, Disproof of the Mertens conjecture, *J. Reine Angew. Math.* **357** (1985), 138–160.

115. S. J. Patterson, *An Introduction to the Theory of the Riemann Zeta-Function*, Cambridge Studies in Advanced Mathematics, vol. 14, Cambridge University Press, Cambridge, 1988.

116. H. Poincaré, Sur les propriétés arithmétiques des courbes algébriques, *J. de Liouville* **7** (1901), 166–233.

117. C. Pomerance, A tale of two sieves, *Notices Amer. Math. Soc.* **43**, no. 12 (1996), 1473–1485.

118. A. J. van der Poorten, A proof that Euler missed...Apéry's proof of the irrationality of $\zeta(3)$: An informal report, *Math. Intelligencer* **1**, no. 4 (1978/79), 195–203.

119. A. J. van der Poorten, *Notes on Fermat's Last Theorem*, Canadian Mathematical Society Series of Monographs and Advanced Texts, John Wiley & Sons Inc., New York, 1996.

120. A. J. van der Poorten and H. P. Schlickewei, Additive relations in fields, *J. Austral. Math. Soc. Ser. A* **51**, no. 1 (1991), 154–170.

121. D. Ramakrishnan and R. J. Valenza, *Fourier Analysis on Number Fields*, Graduate Texts in Mathematics, vol. 186, Springer-Verlag, New York, 1999.

122. M. Reid, *Undergraduate Algebraic Geometry*, London Mathematical Society Student Texts, vol. 12, Cambridge University Press, Cambridge, 1988.

123. P. Ribenboim, "1093", *Math. Intelligencer* **5**, no. 2 (1983), 28–34.

124. P. Ribenboim, *Catalan's Conjecture: Are 8 and 9 the Only Consecutive Powers?*, Academic Press Inc., Boston, MA, 1994.

125. P. Ribenboim, *The New Book of Prime Number Records*, Springer-Verlag, New York, 1996.

126. P. Ribenboim, *Fermat's Last Theorem for Amateurs*, Springer-Verlag, New York, 1999.

127. H.-E. Richert, Über Zerfällungen in ungleiche Primzahlen, *Math. Z.* **52** (1949), 342–343.

128. B. Riemann, Über die Anzahl der Primzahlen unter einer gegebenen Grösse, *Monatsberichte der Berliner Akademie* (November 1859), 671–680, www.maths.tcd.ie/pub/HistMath/People/Riemann/.

129. R. L. Rivest, A. Shamir, and L. Adleman, A method for obtaining digital signatures and public-key cryptosystems, *Comm. ACM* **21**, no. 2 (1978), 120–126.

130. T. Rivoal, Irrationalité d'au moins un des neuf nombres $\zeta(5)$, $\zeta(7), \ldots, \zeta(21)$, *Acta Arith.* **103**, no. 2 (2002), 157–167.

131. P. Rogers, *Topics in Elliptic Divisibility Sequences*, M.Phil. thesis, University of East Anglia, 2002.

132. M. Roitman, On Zsigmondy primes, *Proc. Amer. Math. Soc.* **125**, no. 7 (1997), 1913–1919.

133. A. Schinzel, Primitive divisors of the expression $A^n - B^n$ in algebraic number fields, *J. Reine Angew. Math.* **268/269** (1974), 27–33.

134. H. P. Schlickewei, S-unit equations over number fields, *Invent. Math.* **102**, no. 1 (1990), 95–107.

135. K. Schmidt and T. Ward, Mixing automorphisms of compact groups and a theorem of Schlickewei, *Invent. Math.* **111**, no. 1 (1993), 69–76.

136. A. Selberg, An elementary proof of the prime-number theorem, *Ann. Math. (2)* **50** (1949), 305–313.

137. J.-P. Serre, *A Course in Arithmetic*, Graduate Texts in Mathematics, vol. 7, Springer-Verlag, New York, 1973.

138. V. Shoup, Searching for primitive roots in finite fields, *Math. Comp.* **58**, no. 197 (1992), 369–380.

139. J. H. Silverman, *The Arithmetic of Elliptic Curves*, Graduate Texts in Mathematics, vol. 106, Springer-Verlag, New York, 1986.

140. J. H. Silverman, Wieferich's criterion and the *abc*-conjecture, *J. Number Theory* **30**, no. 2 (1988), 226–237.

141. J. H. Silverman, Taxicabs and sums of two cubes, *Amer. Math. Monthly* **100**, no. 4 (1993), 331–340.

142. J. H. Silverman, *Advanced Topics in the Arithmetic of Elliptic Curves*, Graduate Texts in Mathematics, vol. 151, Springer-Verlag, New York, 1994.

143. J. H. Silverman and J. Tate, *Rational Points on Elliptic Curves*, Undergraduate Texts in Mathematics, Springer-Verlag, New York, 1992.

144. N. J. A. Sloane, An on-line version of the encyclopedia of integer sequences, *Electron. J. Combin.* **1** (1994), Feature 1, approx. 5 pp., www.research.att.com/~njas/sequences/.

145. M. Spivak, *Calculus*, 2 ed., Publish or Perish, Berkeley, CA, 1980.

146. H. M. Stark, A complete determination of the complex quadratic fields of class-number one, *Michigan Math. J.* **14** (1967), 1–27.

147. I. Stewart and D. Tall, *Algebraic Number Theory and Fermat's Last Theorem*, third ed., A K Peters Ltd., Natick, MA, 2002.

148. Z. Šuník, An ideal functional equation with a ring, *Mathematics Magazine* (October 2004), 310–313.

149. L. Szpiro, La conjecture de Mordell (d'après G. Faltings), *Astérisque* (1985), no. 121-122, 83–103, (Seminar Bourbaki, Vol. 1983/84).

150. J. T. Tate, Fourier analysis in number fields, and Hecke's zeta-functions, in J. W. S. Cassels and A. Fröhlich (eds.), *Algebraic Number Theory*, Proceedings of an Instructional Conference, Brighton, 1965, Thompson, Washington, DC, 1967, pp. 305–347.

151. P. L. Tchebychef, *Oeuvres. Tomes I, II*, Chelsea Publishing Co., New York, 1962.

152. P. L. Tchebychef, Mémoire sur les nombres premiers, *J. de Math. Pures. Appl.* **17** (1852), 366–390.

153. E. C. Titchmarsh, *The theory of the Riemann Zeta-Function*, edited and with a preface by D. R. Heath-Brown, second ed., The Clarendon Press, Oxford University Press, New York, 1986.

154. E. Trost, *Primzahlen*, Verlag Birkhäuser, Basel-Stuttgart, 1953.

155. J. B. Tunnell, A classical Diophantine problem and modular forms of weight 3/2, *Invent. Math.* **72**, no. 2 (1983), 323–334.

156. S. Wagstaff, The Cunningham Project, www.cerias.purdue.edu/homes/ssw/.

157. A. Weil, *Number theory: An Approach Through History, From Hammurapi to Legendre*, Birkhäuser Boston Inc., Boston, MA, 1984.

158. A. Weil, *Basic Number Theory*, Classics in Mathematics, Springer-Verlag, Berlin, 1995, Reprint of the second (1973) edition.

159. A. Weil, *Elliptic Functions According to Eisenstein and Kronecker*, Classics in Mathematics, Springer-Verlag, Berlin, 1999.

160. E. T. Whittaker and G. N. Watson, *A Course of Modern Analysis*, Cambridge Mathematical Library, Cambridge University Press, Cambridge, 1996.

161. A. Wiles, The Birch and Swinnerton-Dyer conjecture (Clay Mathematics Institute), www.claymath.org/prizeproblems/birchsd.pdf.

162. A. Wiles, Modular elliptic curves and Fermat's last theorem, *Ann. Math. (2)* **141**, no. 3 (1995), 443–551.

163. D. W. Wilson, The fifth taxicab number is 48988659276962496, *J. Integer Seq.* **2** (1999), Article 99.1.9, 1 HTML document.

164. H. Wußing, Implizite gruppentheoretische Denkformen in den *Disquisitiones Arithmeticae* von Carl Friedrich Gauß, in *Proceedings of the 2nd Gauss Symposium. Conference A: Mathematics and Theoretical Physics (Munich, 1993) (Berlin)*, Sympos. Gaussiana, de Gruyter, 1995, pp. 179–185.

165. M. Yabuta, A simple proof of Carmichael's theorem on primitive divisors, *Fibonacci Quart.* **39**, no. 5 (2001), 439–443.

166. D. Zagier, The Birch-Swinnerton-Dyer conjecture from a naive point of view, in G. van der Geer, F. Oort, and J. H. M. Steenbrink (eds.), *Arithmetic Algebraic Geometry (Texel, 1989)*, Progr. Math., vol. 89, Birkhäuser Boston, Boston, MA, 1991, pp. 377–389.

167. G. M. Ziegler, The great prime number record races, *Notices Amer. Math. Soc.* **51**, no. 4 (2004), 414–416.

168. K. Zsigmondy, Zur Theorie der Potenzreste, *Monatsh. Math.* **3** (1892), 265–284.

169. J. A. Zuehlke, Fermat's last theorem for Gaussian integer exponents, *Amer. Math. Monthly* **106**, no. 1 (1999), 49.

Index

Graduate Texts in Mathematics

(continued from page ii)